Artificial Intelligence for Biomass-based Biofuel Production

This book covers all current technological, industrial status, and future prospects of biofuel production with the concept of artificial intelligence including environmental and socioeconomic impact assessment of biofuel production from lignocellulosic biomass. It discusses the status and future scope of artificial intelligence for the advancement of biofuel research. It summarizes machine learning models in addressing the issues of biofuel supply chains, case studies, scientific challenges, and future directions.

Features:

- Covers the use of machine learning within the context of the processing of advanced biofuel feedstocks for biofuel production.
- Includes larger alcohols, ethers, levulinates, GTL fuels, and furans production using machine learning approach.
- Discusses how machine learning and biomass-based biofuel production can be integrated.
- Reviews sustainability and cost analysis of artificial intelligence–based biofuel production.
- Explores prediction of the potentiality of lignocellulosic biomass for biofuel applications.

This book is aimed at researchers and graduate students in energy and fuels, chemical engineering, and machine learning.

Artificial Intelligence for Biomass-based Biofuel Production

Current Status and Prospects

Edited by
Arindam Kuila and Anshuman Shastri

CRC Press
Taylor & Francis Group
Boca Raton London New York

CRC Press is an imprint of the
Taylor & Francis Group, an **informa** business

Designed cover image: © Shutterstock Images

First edition published 2026
by CRC Press
2385 NW Executive Center Drive, Suite 320, Boca Raton FL 33431

and by CRC Press
4 Park Square, Milton Park, Abingdon, Oxon, OX14 4RN

CRC Press is an imprint of Taylor & Francis Group, LLC

ISBN: 9781032916354 (hbk)
ISBN: 9781032945255 (pbk)
ISBN: 9781003571223 (ebk)

DOI: 10.1201/9781003571223

Typeset in Times
by Deanta Global Publishing Services, Chennai, India

Contents

Foreword

The production of biofuel from lignocellulose could become profitable if the costs associated with the production of biomass are very low. Each geographic region is distinguished by a great variety of flora. These can be found in many literature reports stating that the success of the biofuel production process is influenced by the biomass composition: proportions of cellulose, hemicellulose, and lignin, and the type of fiber arrangement, and thus also the plant species and the degree of its maturity. There are numerous examples of conversion of lignocellulosic biomass into biofuel. However, comparison of most of the data has proven to be difficult due to various amounts of substrates used for the process. Therefore, the use of literature data to design the biofuel production process at a greater scale is very complicated and problematic. Taking all these factors into account, it becomes very difficult to test all plant species that could potentially form a good source to produce biofuel.

A mathematical model based on experimental data can be a helpful tool in determining the suitability of a plant species for processing into biofuel. The production of biofuel is one of the biotechnological processes, the complexity of which is high and difficult to present via ready-made algorithms. Therefore, in such situations, it is perfect to use machine learning (ML), including, e.g., artificial neural networks (ANN) and random forests (RF), as a prediction tool for future biomass samples. This book consists of 14 chapters, which summarize a thorough understanding of the application of different ML models to regulate and monitor the production of biofuel from waste biomass through prediction, optimization, and real-time monitoring. The in-depth analysis of the most recent advancements in ML-assisted biofuel production methods has been provided. Moreover, the ML models in addressing the issues of biofuel supply chains, case studies, scientific challenges, and future direction in ML applications have also been summarized.

The chapters in this book are contributed by active researchers and teachers. The editors are senior researchers having worked and guided biofuel development and artificial intelligence–based work for several years and have done an excellent work in presenting in an exhaustive and highly simplified way the state of the art in these fields. I congratulate the editors and all contributors.

I recommend this book to students, researchers, and scientists from diverse fields due to the simplicity with which the material is presented. I also commend the publishers, CRC Press, USA, for choosing to publish a book on this important topic.

Prof. Ina Aditya Shastri,
Vice Chancellor, Banasthali Vidyapith (Deemed University), Rajasthan 304022, India

Preface

In the domain of renewable energy, biofuel production from lignocellulosic waste biomass represents a paragon of sustainable and potentially cost-effective fuel production. However, navigating the labyrinthine process of converting diverse biomass types into biofuel poses a colossal challenge, chiefly due to the inherent variability in biomass composition. Different plant species, each subject to regional variations and maturational stages, contain distinct proportions of cellulose, hemicellulose, and lignin, all of which have consequential effects on the biofuel yield. The vast heterogeneity in these biomass characteristics renders conventional computational approaches ill-suited for modeling the biomass-to-biofuel conversion process with high precision. It is against this backdrop that the application of machine learning (ML) models, notably artificial neural networks (ANNs), and random forests (RFs), emerges as a revolutionary technological advent. By employing machine learning, it becomes feasible to model complex, non-linear relationships that traditional mathematical equations cannot capture. This innovative approach enables a higher degree of predictive accuracy, fosters real-time process monitoring, and facilitates the optimization of biofuel yields, thereby allowing stakeholders to make data-driven, scientifically substantiated decisions.

This book aims to elucidate the ways in which these sophisticated ML algorithms can decipher the intricate interdependencies between various biomass constituents and the efficiency of the biofuel conversion process. This book covers lignocellulosic biomass, its composition, use of ML for biomass-based biofuel (bioethanol and biodiesel) production, life cycle assessment, cost analysis of artificial intelligence-based biofuel production, current status, and future scope of artificial intelligence for biomass-based biofuel production. This book has been written by renowned scientists who are working in this area for last several years. This book will be useful for researchers, engineers, and scientists in the domains of renewable energy, environmental biotechnology and engineering.

Dr. Arindam Kuila
Dr. Anshuman Shastri

About the Editors

Arindam Kuila is currently working as Assistant Professor at the Department of Bioscience & Biotechnology, Banasthali Vidyapith, Rajasthan, India. Previously, he worked as a research associate at Hindustan Petroleum Green R&D Centre, Bangalore, India. He completed his PhD from Agricultural & Food Engineering Department, the Indian Institute of Technology Kharagpur, India, in 2013 in the area of lignocellulosic biofuel production. He was awarded Petrotech Research Fellowship in 2008. He has completed one Indo-Brazil collaborative project funded by DBT, India. Presently, he is handling another project on scale-up study of lignocellulosic bioethanol production funded by DST. He has guided eight PhD students, currently five PhD students are working under his guidance. He has published 16 edited books, 21 book chapters, 42 papers in high-impact peer-reviewed journals, and filled 8 patents (3 patents granted). He is editorial board member and acted as a guest editor for several high-impact journals. He is a member of the Biotech Research Society of India and the Association of Microbiologists of India.

Anshuman Shastri is currently working as Director at the Centre for Artificial Intelligence, Banasthali Vidyapith, Rajasthan, India. He has vast experience in the area of artificial intelligence. He completed his PhD from Kent University, United Kingdom. Along with five additional nominations for major honors, he was awarded the International Young Scientist Award in 2020 and the International Young Researcher Award in 2021, all of which were accomplished before the age of 28. In addition, his profile is filled with accolades from extracurricular and academic pursuits, and he has won over 30 more medals, prizes, and recognitions for his outstanding work in quizzes, academics, cricket, and violin. He has published several high-impact peer-reviewed papers in the area of artificial intelligence.

Contributors

Gboyega Ebenezer Adebami
Department of Biological Sciences, College of Basic and Applied Sciences
Mountain Top University
Ogun State, Nigeria

Bukola Christianah Adebayo-Tayo
Department of Microbiology, Faculty of Sciences
University of Ibadan
Oyo State, Nigeria

Yvan Jesus Olortiga Asencios
Instituto do Mar
Universidade Federal de São Paulo (UNIFESP)
Santos, São Paulo, Brazil

Benjamin Ayodipupo Babalola
Biochemistry Division, Department of Chemistry
Purdue University
West Lafayette, Indiana, United States

Shubham Sunil Behele
Department of Biotechnology, School of Bioengineering, College of Engineering and Technology, Faculty of Engineering and Technology
SRM Institute of Science and Technology
Tamil Nadu, India

Krishika Lakhmichand Budhani
Department of Biotechnology, School of Bioengineering, College of Engineering and Technology, Faculty of Engineering and Technology
SRM Institute of Science and Technology
Tamil Nadu, India

Moumita Bishai
Department of Botany
Gurudas College
Kolkata, West Bengal, India

Arpan Das
Department of Microbiology
Maulana Azad College
Kolkata, West Bengal, India

Aditi Dey
School of Bioscience and Research
Sharda University
Greater Noida, Uttar Pradesh, India

Anthony Andrey Ramalho Diniz
Instituto do Mar
Universidade Federal de São Paulo (UNIFESP)
Santos, São Paulo, Brazil

Silas Olukunle Fagbenro
Department of Biological Sciences, College of Basic and Applied Sciences
Mountain Top University
Ogun State, Nigeria

Shefali Gulliya
Department of Biosciences and Biotechnology
Banasthali Vidyapith
Rajasthan, India

Samuel Jacob
Department of Biotechnology, School of Bioengineering, College of Engineering and Technology, Faculty of Engineering and Technology
SRM Institute of Science and Technology
Tamil Nadu, India

Elvis Fosso Kankeu
School of Chemical and Minerals Engineering, Faculty of Engineering
North-West University
Potchefstroom, South Africa

Ramalingam Kayalvizhi
Department of Biotechnology, School of Bioengineering, College of Engineering and Technology, Faculty of Engineering and Technology
SRM Institute of Science and Technology
Tamil Nadu, India

Mohd Haad Khan
Department of Biotechnology, School of Bioengineering, College of Engineering and Technology, Faculty of Engineering and Technology
SRM Institute of Science and Technology
Tamil Nadu, India

Andrea Komesu
Instituto do Mar
Universidade Federal de São Paulo (UNIFESP)
Santos, São Paulo, Brazil

Monika Kumari
Department of Biotechnology, School of Bioengineering, College of Engineering and Technology, Faculty of Engineering and Technology
SRM Institute of Science and Technology
Tamil Nadu, India

Ritu Kumari
Department of Botany, School of Bioengineering and Biosciences
Lovely Professional University
Phagwara, Punjab, India

Arindam Kuila
Department of Bioscience and Biotechnology
Banasthali Vidyapith
Rajasthan, India

Malhar Santosh Kapadnekar
Department of Biotechnology, School of Bioengineering, College of Engineering and Technology, Faculty of Engineering and Technology
SRM Institute of Science and Technology
Tamil Nadu, India

Luiza Helena da Silva Martins
Instituto de Saúde e Produção Animal
Universidade Federal Rural da Amazônia (UFRA)
Belém, Pará, Brazil

Thiago José Machado
TRIL LAB, IT Center
Universidade Federal da Paraíba
Paraíba, Brazil

Bilal Ahmad Mir
Department of Botany, School of Bioengineering and Biosciences
Lovely Professional University
Phagwara, Punjab, India

Khushboo Negi
Department of Biotechnology, Faculty of Engineering & Technology
Rama University
Kanpur, Uttar Pradesh, India

Soumya Pandit
School of Bioscience and Research
Sharda University
Greater Noida, Uttar Pradesh, India

Parul Parihar
Department of Biosciences and Biotechnology
Banasthali Vidyapith
Rajasthan, India

Eduardo Dellosso Penteado
Instituto do Mar
Universidade Federal de São Paulo (UNIFESP)
Santos, São Paulo, Brazil

Sheo Mohan Prasad
Ranajan Plant Physiology and Biochemistry Laboratory, Department of Botany
University of Allahabad
Prayagraj, Uttar Pradesh, India

Aman Deep Raju
Ranajan Plant Physiology and Biochemistry Laboratory, Department of Botany
University of Allahabad
Prayagraj, Uttar Pradesh, India

Johnatt Allan Rocha de Oliveira
Faculdade de Nutrição
Universidade Federal do Pará (UFPA)
Belém, Pará, Brazil

Shivani Sahu
Department of Bioscience and Biotechnology
Banasthali Vidyapith
Rajasthan, India

Moisés Dantas dos Santos
TRIL LAB, IT Center
Universidade Federal da Paraíba
Paraíba, Brazil

Falguni Sharma
Department of Automation
Banasthali Vidyapith
Rajasthan, India

Anshuman Shastri
Department of Automation
Banasthali Vidyapith
Rajasthan India

Adwika Singh
Department of Automation
Banasthali Vidyapith
Rajasthan, India

Garima Singh
School of Bioscience and Research
Sharda University,
Greater Noida, Uttar Pradesh, India

Rachana Singh
Ranajan Plant Physiology and Biochemistry Laboratory, Department of Botany
University of Allahabad
Prayagraj, Uttar Pradesh, India

Samakshi Verma
Department of Biotechnology, Faculty of Engineering & Technology
Rama University
Kanpur, Uttar Pradesh, India

1 An Overview of Lignocellulosic Biomass

Johnatt Allan Rocha de Oliveira, Eduardo Dellosso Penteado, Yvan Jesus Olortiga Asencios, Anthony Andrey Ramalho Diniz, Luiza Helena da Silva Martins, and Andrea Komesu

1.1 INTRODUCTION

Lignocellulosic biomass represents one of Earth's most abundant sources of renewable organic matter. It consists of plants and plant residues mainly composed of three structural polymers: cellulose, hemicellulose, and lignin. Due to its abundance and renewability, this type of biomass has attracted increasing interest in developing biofuels, bioproducts, and other forms of renewable energy (Blasi et al. 2023).

Lignocellulosic biomass can be defined as any plant material composed mainly of cellulose, hemicellulose, and lignin. It is, therefore, the fibrous part of plants, including trees, grasses, and agricultural crop residues. From a chemical point of view, lignocellulosic biomass is composed of a complex matrix of structural polymers, with these polymers playing specific roles in supporting the plant structure and its resistance to the external environment (Zoghlami and Paës 2019).

This type of biomass is considered one of the primary sources of raw material for biorefineries, as it is a renewable resource that can be processed to produce a wide range of products, including biofuels, bioplastics, and bioactive compounds. In addition, the use of lignocellulosic biomass is a sustainable alternative to the use of fossil fuels, as its conversion can generate fewer greenhouse gas emissions (Vasile and Baican 2023).

The chemical composition of lignocellulosic biomass is complex and varies according to the type of plant and its growth conditions. However, the main components are common to all plants and include cellulose, hemicellulose, and lignin, with some variations in proportions and structures (Tarasov et al. 2018). The cellulose is the most abundant polymer in lignocellulosic biomass, accounting for 30%–50% of the total composition. It is a linear polymer composed of D-glucose units linked

DOI: 10.1201/9781003571223-1

by β(1→4) bonds, forming long chains organized into highly ordered microfibrils. This structure gives cellulose high mechanical strength and a strong interaction with water. In the industrial context, cellulose can be broken down into simple sugars, such as glucose, which can then be fermented to produce biofuels, such as ethanol (Rongpipi et al. 2019).

Hemicellulose is a heterogeneous polymer representing between 15% and 30% of lignocellulosic biomass. Unlike cellulose, composed exclusively of glucose, hemicellulose is formed by different monosaccharides, including xylose, arabinose, mannose, galactose, and glucose. In addition, hemicellulose chains are branched and have a lower molecular weight than cellulose (Huang et al. 2021). This less-ordered structure makes hemicellulose more easily hydrolyzed by enzymes or chemical treatments, facilitating its conversion into fermentable sugars for the production of biofuels.

Lignin is the third principal component of lignocellulosic biomass, accounting for between 15% and 30% of its composition. It is a complex three-dimensional macromolecule composed of phenolic units, such as coniferyl, sinapyl, and p-coumaric alcohols (Vanholme et al. 2010). Lignin plays a fundamental role in the structure of the plant, providing rigidity, impermeability, and resistance to attacks by microorganisms. Although lignin is often seen as an obstacle in the conversion of biomass into biofuels due to its recalcitrance, it also has the potential to be transformed into high-value-added products, such as composite materials and aromatic chemicals (Alzagameem et al. 2018).

This chapter introduces lignocellulosic biomass, its sources, and its significance in sustainable bioenergy. It examines the structural characteristics of biomass, focusing on cellulose's role in plant rigidity, the complexity of hemicellulose, and lignin's protective properties. This chapter also explores pretreatment methods, enzymatic hydrolysis, and conversion technologies like fermentation and pyrolysis, emphasizing their applications in biofuel production. Lastly, it discusses the environmental impact of these processes and outlines strategies for sustainable biomass sourcing.

1.2 SOURCES OF LIGNOCELLULOSIC BIOMASS

Lignocellulosic biomass sources can be broadly divided into three main categories: agricultural residues, forestry residues, and dedicated energy crops. These sources vary in composition and availability, but they all have great potential for conversion into bioenergy and biochemical products (Oliveira et al. 2022, Saini et al. 2015). Agricultural residues are one of the most abundant sources of lignocellulosic biomass. These residues include by-products generated during crop harvesting, such as straw, husks, and stalks. Two common examples are wheat straw and rice straw (Silva et al. 2024, Oliveira et al, 2014).

Wheat straw is the residue left after the harvesting of wheat grain. It is composed primarily of cellulose, hemicellulose, and lignin and is a promising source for bioethanol production (Oliveira et al. 2015, Tufail et al. 2021). Studies have shown that, with proper treatment, wheat straw can be hydrolyzed to release fermentable sugars, which can then be converted into bioethanol or other bioproducts (Rusănescu et al. 2024).

Rice straw, similar to wheat straw, is an abundant - of rice production, especially in Asian countries. It is composed of approximately 40% cellulose, 30% hemicellulose, and 15% lignin (Sharma et al. 2023). Although rice straw is often burned in fields, resulting in carbon emissions, it has excellent potential for use in the production of bioethanol and biofertilizers.

Forest residues include wood chips, sawdust, and other by-products of the timber industry. These residues are critical in regions with extensive forest areas, such as Brazil, Canada, and the United States. Wood chips are a common by-product of the wood processing industry. They are generated during the felling and planting of timber and consist mainly of cellulose, hemicellulose, and lignin, with a composition that varies according to the type of wood (Braghiroli et al. 2020).

These chips can be converted into bioethanol or used for energy production through direct combustion or conversion to biogas. Sawdust is another by-product of the wood processing industry (Das et al. 2023). Although often discarded or used as low-quality fuel, sawdust has excellent potential for bioenergy production, primarily through pyrolysis or gasification. Studies have shown that sawdust can be efficiently converted into bio-oil, a liquid fuel used in boilers, or refined into biofuels (Bušić et al. 2018). Dedicated energy crops are plants grown specifically for biomass production. Unlike agricultural and forestry residues, these plants are chosen for their high biomass productivity and ability to grow on marginal or poor-quality land. Two common examples are switchgrass (*panicum virgatum*) and miscanthus (*miscanthus giganteus*).

Switchgrass is a perennial grass native to North America. It has been extensively studied as an energy crop due to its high productivity and adaptability to various climatic and soil conditions. It can grow on marginal soils where other food crops would not thrive and produce large amounts of biomass per hectare. Its composition includes approximately 40%–50% cellulose, 25%–30% hemicellulose, and 10%–15% lignin, making it an excellent feedstock for second-generation bioethanol production (Bai et al. 2022). Miscanthus is another perennial grass investigated as a promising energy crop. Native to Asia, this plant can grow up to 4 m in height and has a high biomass productivity. It is also efficient in using nutrients and water, making it suitable for cultivation on low-fertility lands (Weger et al. 2021).

1.3 STRUCTURAL CHARACTERISTICS OF LIGNOCELLULOSIC BIOMASS

Lignocellulosic biomass is a renewable resource generated by many different sectors, such as forestry, agriculture, and industry (Mujtaba et al. 2023). It forms the structural basis of the plant cell wall and is one of the most common biopolymers in nature, commonly known as natural or cellulosic fibers (Ilić et al. 2023; Chakhtouna et al. 2024). It comprises three main building blocks (cellulose, hemicellulose, and lignin), but it is also constituted by pectin, and protein, as well as extractives (lipids, tannins, resins, terpenes, terpenoids, steroids, flavonoids, and phenolic compounds) (Mujtaba et al. 2023; Awasthi et al. 2023; Okolie et al. 2021; Chakhtouna et al. 2024). Lignocellulosic feedstock contains organic and inorganic (mineral) matter, being heterogeneous materials, and the interaction between its components can considerably affect their conversion to biomaterials, biochemicals, and biofuels (Okolie et al. 2021).

1.3.1 Cellulose Structure and Function

Cellulose [$(C_6H_{10}O_5)_n$] is the most abundant natural polymer on Earth, often regarded also as the most abundant renewable organic resource available on the planet, and widely distributed in terrestrial plants, bacteria, and marine algae (Okolie et al. 2021; Deng et al. 2023; Khodayari et al. 2024; Delmer et al. 2024; Chakhtouna et al. 2024). It is the primary component of lignocellulose (40%–60%), defined as a long unbranched polymer of glucose, comprising glucose units linked by β-1,4-glycosidic bonds joined together by van der Waals forces, glycosidic linkages, and hydrogen bonds (Okolie et al. 2021; Deng et al. 2023; Mujtaba et al. 2023; Etale et al. 2023; Crosgrove 2024).

Its degree of polymerization is in the range of 1510–5500, and it is a fibrous substance with a strong hydrogen bonding holding the molecules together, which helps to maintain the structure of plant cell walls (Okolie et al. 2021). Depending on the nature of the raw feedstock, the degree of polymerization of cellulose varies, being considered relatively high (Chakhtouna et al. 2024). In its native form, cellulose generally exhibits a degree of crystallinity varying between the range of 40% and 70%, and in plant walls, most cellulosic fibers are present in the form of native cellulose (Chakhtouna et al. 2024). In land plants, a multiprotein cellulose synthase complex synthesizes cellulose, resulting in a long fibril of ~3 nm, which is commonly called the cellulose microfibril (Crosgrove et al. 2024). Every cell wall in the body of plants has the cellulose microfibril as an integral structural component, what is a key determinant of biological, physical, chemical, and structural properties of cell walls (Crosgrove et al. 2024). The cellulose microfibrils aggregate into larger macrofibrils that lend the cell wall its incredible strength, which rivals the tensile strength of steel (Baez and Bacete 2023).

Cellulose overcovered by lignin and hemicellulose moieties comprise the rigid cell wall structure of the plants (Dharmaraja et al. 2023). Regulating deposition rate and orientation of the cell wall material makes possible producing cell wall anisotropic properties that plants can use during their development of defense actions (Baez and Bacete 2023). Cellulose plays a critical role in plant morphogenesis, development, and growth (Baez and Bacete 2023)

1.3.2 Hemicellulose Complexity Diverse Polysaccharides

According to Delmer et al. (2024), providing a precise definition of hemicellulose is difficult. However, Okolie et al. (2021) define it as another biopolymer in lignocellulosic biomass, comprising polysaccharide mixtures, such as pentose and hexose sugars, and sugar acids, with its content highly dependent on the sources of the biomass. Hemicellulose [$(C_5H_{10}O_5)_m$] is the second polysaccharide after cellulose, and it can be categorized as xylans, mannans, and galactans based on the monosaccharide that predominates in their structure (Ilić et al. 2023; Dharmaraja et al. 2023). It comprises various branched polysaccharides of diverse sugar molecules, what includes xyloglucans, xylans, and glucomannans, with the inhomogeneous nature being the peculiarity of hemicellulose polysaccharides (Østby and Várnai 2023).

The hemicellulose structure has a backbone comprised by heteropolysaccharide polymer chains of C5 and C6 monomeric sugar units with short lateral branches (Dharmaraja et al. 2023). β-1,4-glycosidic bonds link the sugar units, which all through gives more rigidity to the outer surface of the cellulose and also provide it a more amorphous matrix (Dharmaraja et al. 2023). In the cell wall, hemicellulose is a branched polymer composed by different sugar units that binds well to cellulose (Etale et al. 2023). Hemicelluloses coat cellulose microfibrils, may also physically block the access to them, and also interconnect the cellulose microfibril network (Østby and Várnai 2023).

Common to woody and non-woody plants, hemicelluloses are a rather heterogeneous class of polysaccharides that contribute to rigidity and cell wall strength through interactions with cellulose and lignin (Balk et al. 2023). They are typically seen as the internal glue that hold together the cellulose microfibrils and lignin aggregates, that is, the structural elements of plant cell walls (Balk et al. 2023). Shape, architecture, assembly and properties of the primary cell wall are influenced by hemicellulose polysaccharides, and they also affect the deposition and bundling of cellulose fibrils (Oelmüller et al. 2023).

1.3.3 Lignin's Unique Properties

The lignin $[(C_{10}H_{11}O_{3.5})]$ is the third more rich organic heteropolymeric compound after cellulose, being liable for the rigidity in structural nature of plant cell walls and its high impermeability to water (hydrophobicity) (Dharmaraja et al. 2023; Ilić et al. 2023). It is a biopolymer that binds to cellulose and hemicelluloses with covalent bonds, playing also an important role in oxidative stress and microbial assault resistance in plant cell walls (Ilić et al. 2023). Plants have their strength and rigidity provided by the cross-linking of lignin molecules with cellulose and hemicellulose polymers (Okolie et al. 2021).

Lignin is formed from p-coumaryl (H unit) alcohol, coniferyl (G unit) alcohol, and sinapyl (S unit) alcohol, which are known as monolignols (alcoholic monomers), characterizing the lignin as a complex aromatic polymer (Utami et al. 2024; Riseh et al. 2024). Each of those units has the composition defined according to the cell types, species tissues, and developmental stages (Utami et al. 2024). Beneficial qualities such as antibacterial, adhesiveness, UV protection, and mechanical strength are provided to its derivatives by the functional moieties found in lignin's structure (Utami et al. 2024).

Lignin makes plant cell walls more woody and rigid, providing plants with structural support, and playing several crucial roles in plant biology, but specially in its defense against pathogens (Riseh et al. 2024). When plants detect the presence of pathogens (bacteria, fungi, or viruses), they activate genes responsible for lignin biosynthesis, through a series of biochemical reactions (Riseh et al. 2024). This defense mechanism increases the production and deposition of lignin in the affected tissues, what is known as lignification (Riseh et al. 2024). It helps to reinforce the cell walls, isolating the infected area, slowing or preventing the pathogen to spread in the plant, reducing its ability to cause disease (Riseh et al. 2024).

1.4 PRETREATMENT METHODS FOR LIGNOCELLULOSIC BIOMASS

As discussed in previous sections of this chapter, lignocellulosic biomass exhibits considerable recalcitrance, which limits its use in biotechnological processes. Since cellulose polysaccharides must be saccharified through acid or enzymatic hydrolysis, pretreatments are required to reduce this recalcitrance and fermentable sugars to produce biofuels or fine chemical products (Martins et al. 2015).

Several pretreatment techniques exist, including chemical, physical, mechanical, physicochemical, and biological methods (Solarte-Toro et al. 2019). This topic mainly covers mechanical and chemical processes.

1.4.1 Mechanical Pretreatment

Mechanical pretreatment of biomass is a physical procedure used to decrease particle size and enhance the material's surface area (to facilitate further processes like biochemical or thermochemical conversion) (Phojaroen et al. 2022).

Compaction, disintegration, grinding, and crushing are the most widely used methods. These methods are essential for breaking down complex structures, such as the cell wall of plants (composed of cellulose, hemicellulose, and lignin) (Vaezi et al. 2013), which improves the accessibility of microorganisms or enzymes responsible for hydrolysis and fermentation in biofuel production processes or other biotechnological products (Amamou et al. 2018).

This path is encouraging since it offers a more environmentally friendly solution. It can be incorporated into a biorefinery process because it does not require water or other chemicals that might harm the environment (Amamou et al. 2018).

Mechanical pretreatment helps deconstruct lignocellulosic biomasses' physical properties, increasing the surface area, reducing the cellulose crystallinity, and reducing the degree of polymerization (Sitotaw et al. 2021). In this topic, the approach will revolve around two main techniques: ball milling and grinding.

1.4.1.1 Ball Milling

One of the well-known physical pretreatments is ball milling, which reduces the size of biomass particles by breaking the cell wall and increasing the exposed surface area. The principle of mechanical action involves the application of impact and shear pressures, which produce repeated collisions between the balls and the biomass inside a rotating drum, which causes the particle size of the material to decrease (Sitotaw et al. 2021).

The high-speed balls collide with the biomass, fragmenting its fibers and breaking structural structures, such as lignin and the cellulose matrix. When rolling and sliding, they exert frictional forces that wear and tear the fibers, creating surfaces that are more accessible to enzymes without a hydrolysis process (Sitotaw et al. 2021).

The types of equipment involved are as follows.

- **Planetary ball mill:** This piece of equipment has drums that rotate around their axis and a central axis. This mill type generates high-impact energy and is efficient for biomass pretreatment on laboratory and industrial scales. It was used in the work of Yuan et al. (2015). Zakaria et al. (2014) investigated the use of the ball milling technique for the pretreatment of oil palm biomass to increase the efficiency of enzymatic hydrolysis. The results of the study showed that changing the biomass structure and particle size increased the accessibility of enzymes, leading to a more significant percentage of cellulose conversion into fermentable sugars (87%) and yields of plant bunches, glucose, and xylose (70.0% and 82.3%, respectively), this shows that the approach has potential for streamlining the procedures involved in producing biofuels.
- **Vibrating ball mills:** They use high-frequency vibration to move the balls and material inside the drum, causing a more intense and continuous grinding. This type is used for finer and more uniform fragmentation. Qu et al. (2017) investigated the fractionation and pretreatment of lignocellulosic biomass using ball milling and aqueous hydroxide solutions. This research demonstrated that, while the base helped in the solubilization of hemicellulose and the removal of lignin, mechanical grinding helped to improve the physical destructuring of the biomass. This combined approach resulted in greater accessibility of cellulose for enzymatic hydrolysis, increasing the efficiency of biomass conversion into fermentable sugars with yields of 98.48% achieved, with potential for application in biorefinery processes.

1.4.1.2 Grinding

Grinding is one of the techniques frequently used as physical pre-treatment to reduce particle size. This process involves the mechanical breakdown of lignocellulosic biomass into smaller particles.

There are several procedures for performing this breakdown, including the multi-cracker system, hammer mill, and agitated mill, which use several mechanical prin ciples to break down the material. The stirred mill uses high-speed agitated grinding media, applying impact and friction forces to grind the biomass finely, efficiently achieving tiny and homogenous particles (Tabil et al. 2011). The hammer mill uses rotating hammers that repeatedly collide with the biomass, breaking it into smaller pieces by direct impact, being an effective technique for coarser and larger-scale grinding (Barakat et al. 2014). The multi-cracker system uses discs or rollers that compress and crack the biomass by shearing and compression, offering an efficient reduction in particles of various sizes with lower energy consumption than impact methods (Thomas et al. 2012).

Physical pretreatment of biomass, like gridding and ball milling, has benefits such as increasing the material's surface area, making it easier for enzymes or chemical agents to access the material in later processes, and reducing the need for chemicals, which reduces harmful waste and environmental effects. The process's scalability and simplicity with readily available mechanical equipment are further benefits (Che Kamarludin et al. 2014; Khan et al. 2022).

However, this type of pretreatment has disadvantages, such as high energy consumption, especially for resistant biomass. In many cases, it is low-efficient on its own, not altering the chemical composition of the biomass, which makes the use of complementary treatments necessary. In addition, the high energy cost can limit its viability on larger industrial scales (Che Kamarludin et al. 2014; Sitotaw et al. 2021; Khan et al. 2022).

1.4.2 Chemical Pretreatment

Chemical pretreatment of lignocellulosic biomass aims to modify or remove structural components, such as lignin and hemicellulose, making the cellulose more accessible for subsequent processes, such as enzymatic hydrolysis. The most common methods are acids, alkalis, and organic solvents (Norrahim et al. 2021).

Treatment with acids, such as sulfuric acid, breaks down hemicellulose into simple sugars, while alkaline treatment, with sodium hydroxide, is effective in removing lignin and reducing the crystallinity of cellulose (Martins et al. 2015; Solarte-Toro et al. 2019).

Other methods include the use of organic solvents or oxidizing compounds. These treatments facilitate biomass degradation, promoting greater efficiency in producing biofuels and renewable chemicals (Norrahim et al. 2021). This topic will focus on acid and alkali pretreatments. Among the existing chemical pretreatments, acid has been extensively studied. This treatment can solubilize most hemicellulose in biomass but could be more effective in solubilizing lignin (Oriez et al. 2019).

Acid pretreatment of biomass involves the use of acids, usually sulfuric acid (diluted or concentrated), to hydrolyze hemicellulose and break down part of the lignocellulosic structure, releasing fermentable sugars (Oriez et al. 2019). The mechanism of action of acid pretreatment occurs due to hydronium ions that are capable of breaking intermolecular and intramolecular bonds between lignocellulosic biomass, composed of cellulose, hemicellulose, and lignin. This process can increase the porosity of the material, improving the capacity of the cellulose fraction to bind to cellulase enzymes, facilitating hydrolysis in later stages for biotechnological processes (Guo et al. 2012). The central mechanism is the breaking of glycosidic bonds in polysaccharides, mainly in hemicellulose, and the partial degradation of cellulose, resulting in the solubilization of sugars such as xylose and arabinose (Oriez et al. 2019). The advantages of acid pretreatment include high efficiency in solubilizing hemicellulose and creating favorable conditions for subsequent enzymatic hydrolysis. Disadvantages include the formation of inhibitory by-products such as furfural and acetic acid, which are corrosive and require resistant materials in equipment. Adequate treatment of acid waste is also needed to avoid environmental impacts (Saha et al. 2005; Martins et al. 2015; Johannes & Xuan, 2024). Chemical pretreatment with alkalis, such as NaOH (sodium hydroxide) and KOH (potassium hydroxide), is widely used to destructure lignocellulosic biomass, mainly by removing or modifying lignin (Martins et al. 2015). According to Amezcua-Allieri et al. (2017), the mechanism of action consists of cleaving ester bonds and saponifying lignin components, which solubilizes them and reduces the crystallinity of cellulose, increasing its accessibility for further procedures such as enzymatic hydrolysis. Alkali pretreatment of biomass has several benefits, including a straightforward method and excellent efficiency in removing lignin, which makes cellulose easier for enzymes to access. However, disadvantages include the need for large volumes of water for the washing process, the formation of by-products that can be inhibitors in later stages, and the cost of proper management of alkaline waste, in addition to potential environmental impacts if the disposal is not treated correctly (Martins et al. 2015; Johannes & Xuan, 2024). As we have shown, every pretreatment has benefits and drawbacks. Combining the techniques is one approach to increasing their efficacy. Therefore, we have also included methods such as steam explosion and hydrothermal (the most popular), which are considered physical–chemical pretreatments, to break down lignocellulosic biomass further (Nitsos et al. 2013; Auxenfans et al. 2017). In the steam explosion, the biomass is treated with high-pressure steam and then suddenly depressurized, causing the fibers to explode and promoting partial hydrolysis of the hemicellulose, in addition to altering the lignin (Auxenfans et al. 2017). During the hydrothermal treatment, the biomass is submerged in hot water under high pressure, which dissolves the hemicellulose and softens the biomass's structure by acting as a weak acid. Both techniques make the cellulose more accessible, which makes procedures like enzymatic hydrolysis easier (Nitsos et al. 2013).

1.5 ENZYMATIC HYDROLYSIS OF LIGNOCELLULOSIC BIOMASS

Lignocellulose can be converted into simple sugars either by employing enzymes or through chemical hydrolysis using acids. Acid hydrolysis of cellulose requires relatively high temperatures, which can lead to the breakdown of glucose into hydroxymethylfurfural and other by-products, ultimately reducing the overall glucose yield from the process (Ballesteros 2010). Enzymatic hydrolysis is increasingly favored due to its lower energy requirements, milder and non-corrosive operating conditions, and reduced production of fermentation inhibitors (Saini et. al. 2022, Maitan-Alfenas et al. 2015). However, several challenges remain in this process, including limited enzyme production, low hydrolysis efficiency, and high production costs, which hinder the economic viability of converting lignocellulose into biofuels (Wang and Lü, 2021). Biomass biodegradation is a complex process that involves multiple enzymes working together. The hydrolytic enzymes that play a role in the hydrolysis of lignocellulose consist of cellulase, hemicellulase, ligninase, and various auxiliary enzymes (Guo et al. 2022b).

1.5.1 Enzymes Involved in Hydrolysis

In natural environments, lignocellulose is decomposed by a diverse array of hydrolytic and oxidative enzymes generated by different fungi and bacteria. These microorganisms collaborate synergistically to break down cellulose, hemicellulose, and lignin (Pérez et al. 2002).

The major enzymes in all biomass conversion cocktails today are cellulases and hemicellulases (Stickel et al. 2014). Cellulases are a group of enzymes that catalyze the hydrolysis of cellulose, a major component of lignocellulosic biomass. Cellulose consists of long chains of glucose units linked by β-1,4-glycosidic bonds. Cellulases can be classified into three main types.

Endoglucanases: These enzymes cleave internal β-1,4-glycosidic bonds within the cellulose chain, creating free chain ends.

Exoglucanases: Also known as cellobiohydrolases, these enzymes remove cellobiose (a disaccharide made of two glucose units) from the ends of the cellulose chains.

β-glucosidases: These enzymes hydrolyze cellobiose and other short cello-oligomers into glucose, which can be utilized by microorganisms in fermentation processes.

The complete degradation of cellulose requires the collaborative action of endoglucanases (EC 3.2.1.4), cellobiohydrolases (EC 3.2.1.176), exoglucanases (EC 3.2.1.74), and β-glucosidases (EC 3.2.1.21) (Maitan-Alfenas et al. 2015). Several microorganisms are employed for the industrial production of cellulases, but filamentous fungi like *Trichoderma reesei*, *Acremonium* spp., *Aspergillus niger*, *Penicillium oxalicum*, and *Thermoascus aurantiacus* are among the most effective cellulase producers (Saini et al. 2022; Hemansi et al. 2018; Saini et al. 2015).

Cellulases are crucial for breaking down cellulose into glucose monomers, which can then be fermented to produce biofuels or other value-added products. This process enhances the overall efficiency of biomass conversion, making it more accessible for further degradation by other microorganisms.

Hemicellulases are a diverse group of enzymes that target hemicellulose, another significant component of lignocellulosic biomass. Hemicellulose is a complex polysaccharide made up of various sugars, including xylose, arabinose, mannose, and galactose, linked by different types of glycosidic bonds. Hemicellulases can be categorized into several types.

Xylanases: These enzymes specifically break down xylan, a major component of hemicellulose, into xylose and other monosaccharides.

Arabinases: These enzymes hydrolyze arabinose-containing hemicelluloses, breaking them down into simpler sugars.

Mannases: These enzymes degrade mannans, another type of hemicellulose, releasing mannose.

The complete enzymatic hydrolysis of xylan, which is the primary polymer found in hemicelluloses, necessitates the action of endo-β-1,4-xylanase (EC 3.2.1.8), which randomly cleaves the internal bonds of xylan to produce xylo-oligosaccharides. It also requires β-xylosidase (EC 3.2.1.37), which hydrolyzes the non-reducing ends of xylose chains to release xylose. Additionally, several accessory enzymes are involved, including α-L-arabinofuranosidase (EC 3.2.1.55), α-glucuronidase (EC 3.2.1.139), α-galactosidase (EC 3.2.1.22), acetylxylan esterase (EC 3.1.1.72), and ferulic acid esterase (EC 3.1.1.73) (Maitan-Alfenas et al. 2015).

Optimizing enzyme activity is crucial for maximizing the efficiency of enzymatic processes. Several strategies can be employed to enhance enzyme efficiency, such as the addition of bio-additives, immobilization techniques, and enzyme engineering.

Lignin, an undesirable component, can block cellulase from interacting with the substrate and cause enzymes to be adsorbed in a non-productive manner, reducing the efficiency of enzymatic hydrolysis. (Xu et al. 2021). Xu et al. (2021) explored the application of sophorolipid and whey protein to bind with lignin and decrease non-productive enzyme adsorption. The study found a positive relationship between the effectiveness of these additives and substrate concentration. At a solid content of 20% (w/v), sophorolipid and whey protein boosted glucose yield by 17.8% and 11.9%, respectively.

The high production cost of enzymes, which makes up around 30% of the total expenses in a biorefinery, can limit the economic feasibility of the process. Furthermore, because cellulase enzymes are soluble in the hydrolysis system, recovering and reusing them is challenging, which drives up operational costs (Xu et al. 2023). Immobilizing enzymes on solid supports can improve their stability and facilitate their recovery and reuse, while also ensuring proper enzyme orientation and reducing the risk of denaturation. Rashid et al. (2022) investigated the immobilization of cellulase on magnetic nickel nanostructures. Surprisingly, the immobilized cellulase showed an increase in activity within the first hour and retained 84% of its activity after 3 hours. Even after 10 usage cycles, the enzyme maintained 84% of its original activity. Additionally, the immobilized cellulase could be easily separated from the solution using a magnetic field.

Enzyme engineering significantly enhances the efficiency of enzymatic hydrolysis by modifying enzymes to improve their stability, specificity, and catalytic efficiency. Numerous efforts have been made to utilize genetic engineering for the production of a diverse array of cellulases at a reduced cost while achieving high biomass saccharification. Various microorganisms produce different cellulases, and some are being genetically modified to develop more effective cellulases. Dadwal et al. (2020) provide an overview of several successful genetic engineering approaches aimed at enhancing cellulase kinetics and cellulolytic efficiency.

Adopting these strategies can greatly boost enzyme efficiency, thereby enhancing the overall yield and productivity of enzymatic processes. Ongoing research and development in enzyme optimization methods are essential for progressing biotechnological applications in diverse industries.

1.5.2 Factors Affecting Hydrolysis Efficiency

Hydrolysis efficiency is influenced by various factors that affect enzymatic activity and substrate availability. Temperature plays a significant role in the rate of enzymatic reactions, as each enzyme has an optimal temperature range for maximum activity. For cellulases, optimal conditions are typically found at temperatures between 40°C and 50°C. Xylanase generally exhibits similar optimal assay conditions (Maitan-Alfenas et al. 2015). Enzyme immobilization can enhance thermal stability, with immobilized cellulase often displaying a higher optimum temperature compared to free cellulase. This difference is likely due to changes in the enzyme's physical and chemical properties induced by the immobilization process (Xu et al. 2023). Additionally, enhancing cellulase thermal stability can be achieved through screening for naturally occurring thermostable proteins, using molecular biology techniques to modify natural proteins for improved thermo-stability, and employing additives to protect enzymes from high temperatures (Ballesteros 2010).

The pH of the reaction environment is crucial for enzyme activity, as each enzyme has an optimal pH range. For cellulases, the optimal pH range is 4 to 5 (Maitan-Alfenas et al. 2015), and deviations from this can alter the enzyme's structure and reduce its activity. Maintaining the appropriate pH is essential for maximizing hydrolysis efficiency.

Another factor that affects hydrolysis efficiency is the characteristics of the substrate. Factors such as particle size, crystallinity, and the presence of lignin and hemicellulose can influence the accessibility of cellulase enzymes to cellulose.

Pretreatment that disrupts the lattice structure of lignocellulose can significantly improve the exposure of cellulose, enhancing enzyme accessibility. However, certain by-products generated from pretreatment methods, such as furans and phenolic compounds, can inhibit enzymatic hydrolysis and increase overall processing costs. For example, furfural and weak acids are commonly produced during ionic liquid pretreatment. Therefore, any strategy to enhance substrate accessibility must also consider the potential for generating inhibitors that can hinder enzymatic processes (Guo et al. 2023). Cañadas et al. (2023) explored the detoxification of pre-treated biomass using a combination of emerging green solvents and environmentally friendly extraction technologies. Their findings indicated that detoxification using green solvents could be an effective method to remove phenolic inhibitors, which could then be repurposed as antioxidants, while enhancing sugar production from the treated biomass.

By analyzing and refining these factors, the efficiency of hydrolysis processes can be improved, resulting in higher yields in applications like biofuel production and biorefinery operations.

1.6 CONVERSION TECHNOLOGIES FOR LIGNOCELLULOSIC BIOMASS

1.6.1 Biochemical Conversion

The new generations of refineries are known as biorefineries, composed of a set of processes that consist of the transformation of biomass materials into highly valuable products such as fuels and chemicals. Lignocellulosic biomass (such as agricultural residues, wood, and energy crops, among others) is widely available on our planet, is renewable, and eco-friendly, and can be transformed into fuels (such as 2G ethanol, hydrogen, Syngas, Fischer–Tropsch fuels), or even platform molecules such as glucose, and furfural, among other chemicals by chemical and/or biochemical processes (Brethauer & Studer, 2015). More specifically in biochemical conversion, enzymatic hydrolysis transforms lignocellulosic biomass into fermentable sugars, which can then be transformed (by additional process) into desired final products such as fuels and fine chemicals. The biochemical conversion of lignocellulosic biomass can produce a sugar platform biorefinery. The principal unit operations involved in this process are biomass pretreatment, enzymatic hydrolysis (to release fermentable sugars), and fermentation of the sugars (by suitable microorganisms) to the target molecules. Despite significant advances, the high cost of enzymes is still a hindrance to large-scale applications of biochemical conversion of lignocellulosic biomass. Thus, a process that combines (or integrates) enzyme production, enzymatic hydrolysis, and fermentation in a single step, would overcome this obstacle. Maybe in the future, genetically engineered superior microorganisms could be created to integrate all these processes. Furthermore, more research is needed to obtain yields and rates that are commercially attractive (Brethauer & Studer, 2015).

1.6.1.1 Fermentation Processes

Microorganisms can metabolize only simple molecules, for example, monosugars. Most of the research on the fermentation of cellulose (one of the main products obtained from lignocellulosic materials) is focused on ethanol production. Fermentation of cellulose is achieved following two steps: (1) hydrolysis of polysaccharides to monosaccharides and (2) conversion of monosaccharides to chemicals. Both steps use enzymes. High efficiency can be achieved through the combination of the previously mentioned steps (in a single reactor), using different enzymes such as cellulases and microbial species. The hydrolysis of lignocellulosic biomass converts cellulose into glucose, hemicellulose into pentoses (xylose and arabinose) and hexoses (glucose, galactose, and mannose) (Margaritis & Kilonzo, 2005).

In the simultaneous hydrolysis/fermentation of cellulose process, solvent production eliminates the problem of end-product inhibition, because glucose does not accumulate in this system and is converted into ethanol (or other chemical) immediately following hydrolysis. Moreover, monosugars obtained from the hydrolyzation of lignocellulosic biomass can be processed by dark fermentation to produce hydrogen (Kucharska et al. 2020).

1.6.1.2 Microbial Strains

Fermentation technology of glucose to obtain ethanol is well known and well established. The main drawback is the fermentation of pentose fractions, present in the hydrolysis of lignocellulosic materials. For example, the hydrolysis products obtained from wood contain high quantities of pentose, mainly xylose. Pentoses are not easily processed by *Saccharornyces cerevisiae*. Yeast species belonging to the genera *Pachysolen*, *Pichia*, and *Candida*, have the potential for pentose fermentation. Other candidates are some filamentous fungi such as *Fusarium oxysporum*, *Paecilomyces*, and *Mucor* (Margaritis & Kilonzo 2005, Basu 2010). Xylose can be fermented in two ways: direct conversion to ethanol or other chemicals or isomerizing to xylulose by xylose isomerase and then fermenting to ethanol. Growth rates and yields are influenced by strain variation and environmental conditions. The fermentation of xylose with yeast and bacteria releases low ethanol yields and not high xylose conversion rates. Molds improve ethanol yields from xylose, but the conversion rates are very slow. All these factors are not attractive economically (Margaritis & Kilonzo, 2005).

There are microorganisms capable of catalyzing the conversion of lignocellulosic biomass into precious products, for example into ethanol; in this case through lignin hydrolysis, cellulose hydrolysis, and fermentation of sugars. Microorganisms can come from pure culture or/mixed cultures, genetically modified microorganisms, or naturally existing microorganisms.

Many studies have been focused on the fermentation of lignocellulosic biomass to produce ethanol. Although many microbial species can perform the processes required for the conversion of lignocellulosic biomass to ethanol (biomass pretreatment, enzymatic hydrolysis, and fermentation of the sugars), only one species has been unquestionably reported, to hydrolyze lignin (under anaerobic conditions). For the fermentation process, pure culture is usually employed: the yeast *Saccharomyces cerevisiae* (Kucharska et al. 2020).

Although pure cultures are well established, well known, and consequently very used, the use of pure cultures has some disadvantages, for example, *S. cerevisiae* cannot metabolize xylose (the principal component in hemicellulose), and this limits the range of feedstock that can be used. Thus, requiring an extra process for the hydrolysis of cellulose; pure cultures also demand sterilization procedures, therefore increasing costs. In general, the rates of microbial processes are lower than the rates of physico-chemical processes (Dionisi et al. 2015).

1.6.2 Thermochemical Conversion

In thermochemical conversion, the lignocellulosic biomass is transformed into gases, which are after processed to synthesize desired chemicals or used directly.

1.6.2.1 Pyrolysis

Pyrolysis consists of heating biomass at high temperatures, in the absence of oxygen to produce gas, charcoal, and liquids. The liquid product also known as bio-oil, after hydrotreating can be transformed into green liquid fuels such as green diesel or green gasoline. In most cases, char is the target product (Basu, 2010). There are three types of pyrolysis: mild pyrolysis (Torrefaction), slow pyrolysis, and fast pyrolysis. Mild pyrolysis occurs at moderate temperatures, and the principal objective is to produce charcoal (usually used to produce energy). The fast pyrolysis produces principally liquid fuel (bio-oil); the slow pyrolysis produces some gas and solid char (Basu, 2010).

1.6.2.2 Gasification

The gasification process converts biomass into useful gaseous fuels or chemical feedstock. Gasification takes place in high temperatures and an oxygen-deficient environment; contrary to combustion where the process takes place in an oxidizing environment releasing as heat the total energy packed in the biomass; gasification releases Syngas as the principal product, among methane, char, and tar. In most cases, the gaseous products are the target (Basu, 2010).

The products obtained by biomass gasification consist principally of syngas, which can be used for hydrogen production, and/or Fischer–Tropsch synthesis, among other highly valued products (Olortiga Asencios & M. Assaf, 2023). Gasifiers are classified as follows: fixed bed, fluidized bed, and entrained-flow gasifier; the fluidized bed is the most used. Gasification can be carried out at temperatures ranging from 600°C to 1000°C, in the presence of oxidant gases such as H_2O (steam), O_2 (from air), CO_2, or a combination of them. The no-stoichiometric reaction is represented by Equation (1.1) (Kumar et al. 2009).

$$CH_xO_y(s)+O_2(g)+H_2O(v)=CH_4(g)+CO(g)$$
$$+CO_2(g)+H_2(g)+H_2O(v)+Char(s)+tar(l)+soot(s) \quad (1.1)$$

The physical state of each reactant in Equation (1.1) (solid, liquid, gas, or vapor in the case of water), is in parentheses. CH_xO_y, represents the principal elements of biomass. In the gasification process, the source of O_2 is usually air (21% O_2), H_2O in the form of steam. Tar is a liquid-viscose product mainly formed by monocyclic and polyclinic aromatic hydrocarbon substances. Other by-products can contain metal oxides and ash.

$$\text{Partial Oxidation reaction: } C + O_2 \rightarrow 2CO \quad (1.2)$$

$$\text{Total Oxidation reaction: } C + O_2 \rightarrow CO_2 \quad (1.3)$$

$$\text{Water gas shift reaction: } CO + H_2O = CO_2 + H_2 \quad (1.4)$$

$$\text{Steam reforming reaction: } CH_4 + H_2O \rightarrow CO + 3H_2 \quad (1.5)$$

$$\text{Water gas reaction } C+H_2O \rightarrow CO+H_2 \quad (1.6)$$

$$\text{Hydrogasification reaction: } C + 2H_2 \rightarrow CH_4 \quad (1.7)$$

$$\text{Boudouard reaction: } C+ CO_2 \rightarrow 2CO \quad (1.8)$$

Equations (1.2)–(1.8) described the common reactions that take place during the gasification process. Chen et al. (2015) and Xiao et al. (2020) studied the degradation kinetic of lignocellulosic biomass by thermogravimetric analyses (TGA), in a nitrogen environment. The thermal decomposition of the profile can be monitored by the weight loss of the biomass, which gives TGA profiles characterized by three stages: the first stage below 125°C, owing to dehydration of the biomass; the second stage ranging from 125°C to 500°C, known as active pyrolysis, related to the loss of the hemicellulose, cellulose, and part of lignin; and finally, the third stage above 500°C, known as being passive pyrolysis related to the loss of residual lignin. The elemental composition of biomass, the type of oxidizing, and the rate of heat transfer of the gasifier influence the kinetic of the degradations, and consequently, the temperature ranges mentioned previously.

1.7 APPLICATIONS AND ENVIRONMENTAL IMPACT

Nowadays, the humanity is currently grappling with the challenge of preserving and protecting the environment on Earth once the issues, such as air pollution, climate change, and shortages of food and energy are becoming increasingly urgent (Machineni, 2020; Radionova et al. 2022; Saravanan et al. 2022). To deal with the concerns over fossil fuel depletion, greenhouse gas emissions, and environmental degradation mount, a significant step has already been taken which is replace non-renewable fossil fuels to sustainable reusable and recoverable resources. In this sense, biofuels (e.g., biodiesel, bio-oil, bioethanol, biobutanol, biogas, hydrogen, syngas, and jetfuel) and biochemicals (e.g., propylene, ethylene, benzene, 5-hydroxymethylfurfural, levulinic acid, succinic acid, maleic acid, fumaric acid, phenols, and other aromatic compounds) derived from lignocellulose biomass (LCB) stand out as a promising solution to reduce our reliance on non-renewable resources due to the chemical attributes of its constituents and being most abundant organic materials widely distributed and readily accessible across the globe (Okolie et al. 2021a; Saravanan et al. 2022).

Lignocellulose biomass burning has powered human civilization for thousands of years (Raud et al. 2019). While it still serves as a primary energy source for many in developing and emerging economies, its traditional use is unsustainable. Inefficient combustion not only leads to energy loss but also releases harmful emissions with serious health consequences. However, today, modern technologies can convert organic matter into solid, liquid, and gaseous fuels, providing a more efficient and cleaner energy alternative to fossil fuels (Machineni, 2020; Okolie et al. 2021a; Radionova et al. 2022; Saravanan et al. 2022).

There are two primary conversion routes for producing biofuels and biochemicals from lignocellulosic biomass: biological and thermochemical pathways (Okolie et al. 2021a; Radionova et al. 2022; Saravanan et al. 2022). Through biological process, including anaerobic digestion and fermentation, microorganisms and enzymes break down the complex lignocellulose biomass, cellulose and hemicellulose, into sugar which can then be fermented to produce bioalcohols (bioethanol and biobutanol), biomethane biohydrogen, and other renewable fuels and chemicals (Okolie et al. 2021a; Radionova et al. 2022). Recent advancements in enzymatic degradation techniques have greatly improved the conversion of lignin, increasing lignin degradation efficiency by up to 40% over traditional methods, leading to greater biofuels yields (Ali et al. 2024). Furthermore, the advancement of microbial consortia has accelerated lignin degradation, resulting in substantial reductions in both processing time and costs (Becker and Wittmann 2019). In contrast, the thermochemical pathway, like pyrolysis, gasification, liquefaction, and carbonization, uses heat and chemical reactions to decompose the lignocellulosic biomass into sustainable fuels and chemicals (Okolie et al. 2021; Radionova et al. 2022).

Bio-oil, renewable and sustainable energy and chemical source, is a complex mixture of organic compounds and it is derived from lignocellulosic biomass through thermochemical processes such as pyrolysis and hydrothermal liquefaction (Okolie et al. 2021a; Liu and Yu, 2022). During these processes, cellulose and hemicellulose contribute to the formation of aldehydes, alcohols, acids, esters, ketones, oxygenated aromatics, and furans in bio-oils, while phenols, guaiacols, and syringols primarily result from lignin decomposition (Okolie et al. 2021a; Liu and Yu, 2022). Notably, the yield and composition of bio-oil are influenced by the operating conditions (temperature, vapor residence time, reaction duration, pressure, feed concentration, and reactor design) (Okolie et al. 2021a; Okolie et al. 2021b; Liu and Yu, 2022). One of the most promising uses of bio-oil is as a renewable fuel for power generation and heating, however an upgrading stream, like hydrodeoxygenation, catalytic cracking, and emulsification; is necessary to be used as a direct replacement for petroleum-based fuels due to its high-water content, acidity (low pH), high oxygen content, high viscosity and density, and thermal instability (Okolie et al. 2021a; Liu and Yu 2022).

Bio-oil can be refined into biochemicals used in the production of adhesives, resins, and polymers and it can also serve as a feedstock for producing hydrogen (Okolie et al. 2021a). Bio-oils generated from the liquefaction and pyrolysis of lignocellulosic biomass can be catalytically reformed to produce hydrogen, however carbon buildup on the catalyst and the variable composition of bio-oils pose significant obstacles to large-scale operation (Okolie et al. 2021a).

The production of biofuels through bio-oil upgrading is an environmentally friendly process, once the life cycle assessment findings reveal that the biofuel offers a 54.5% reduction in GHG emissions when compared to equivalent fossil fuels (Liu and Yu 2022). In regard the costs involved in the bio-oil and the final price of the produced biofuels, several techno-economic analysis studies have showed that the cost of biofuel produced from upgrading bio-oil obtained from lignocellulosic biomass are comparable to those of fossil fuels (Okolie et al. 2021; Okolie et al. 2021b; Liu and Yu, 2022).

Other biofuel produced in the thermochemical route, mainly conventional gasification or hydrothermal gasification, is the and syngas (consists of mostly H_2, CO, CO_2, and CH_4 gases) (Okolie et al. 2021a; Okolie et al. 2021b). Liquid biofuel, like bioethanol and biojet fuel, can be obtained upgranding syngas through Fischer–Tropsch process and biological conversion technology that uses anaerobic bacteria (Okolie et al. 2021a; Okolie et al. 2021b). In regardless with commercial application, there are a few projects producing syngas from lingnocellulosic biomass once the capital investment is high (Okolie et al. 2021a).

Bioethanol is one of the most widely studied biofuels derived from lignocellulosic biomass, because utilizes non-food biomass, including agricultural residues (e.g., corn stover, wheat straw), forestry residues, and dedicated energy crops such as switchgrass and miscanthus (Okolie et al. 2021a; Okolie et al. 2021b; Liu and Yu 2022). This reduces competition with food production and lowers the overall carbon footprint of biofuel production because the use of bioethanol as a transportation fuel can help lower greenhouse gas emissions compared to fossil fuels. The carbon dioxide emitted during the combustion of bioethanol is roughly equal to the carbon absorbed by plants during their growth, creating a more balanced carbon cycle (Okolie et al. 2021b). Another advantage of bioethanol from lignocellulosic biomass is the low toxicity, volatility and clean combustion properties (Okolie et al. 2021a).

The production of bioethanol from lignocellulosic feedstock involves several key steps: pretreatment, enzymatic hydrolysis (saccharification), fermentation by bacteria and yeasts to ethanol, and distillation (Okolie et al. 2021a; Okolie et al. 2021b; Liu and Yu 2022). Others processes used to produced bioethanol from lignocellulose biomas are convert the syngas (a mixture of H_2 and CO_2) to ethanol using Fischer–Tropsch catalysis or fermentation by *Clostridium sp* (Okolie et al. 2021a; Okolie et al. 2021b). Even these two pathways demonstrated comparable ethanol yield and energy efficiency, a higher fuel consumption and GHG emission was observed in biological via in the short term, but is projected to decrease over the long term; while the thermochemical routes have lower blue footprints (Mu et al. 2010). In another life cycle assessment comparing bioethanol from lignocellulosic biomass with fossil gasoline, the biofuel performed better in terms of fossil fuels depletion, global warming potential, and human toxicity potential, but worst in terms of ozone layer depletion, acidification, and eutrophication potential (Liu et al. 2020). In regardless economic point of view, a techno-economic analysis showed that minimum ethanol selling price for both methods very similar, 0.318 US$/L for biochemical and 0.329 US$/L for thermochemical route (Hakeem et al. 2023). However, these prices can change according the feedstock costs, pretreatment methods, and ethanol purification techniques (Hakeem et al. 2023).

Bioethanol from lignocellulose can be pioneer of bioethylene through the dehydration route, which already has an established commercial technology as previous described with ethylene yield of 56 g/kg of oats hulls (Skiba et al. 2017; Okolie et al. 2021a).

Butanol has 30% higher energy content than ethanol (29.2 MJ/L and 21.2 MJ/L, respectively) and can be used as a direct replacement for gasoline due to high octane number, low toxicity, less hazardous and less hygroscopy (Nanda et al. 2017; Okolie et al. 2021a; Banerjee et al. 2024). Acetone-butanol-ethanol (ABE) fermentation process uses lignocellulosic biomass to produce biobutanol, however the low yield and toxicity on microorganisms are ones of the drawbacks as a result of inefficient transport and utilization of fermentable sugars, the accumulation of toxic intermediates or excessive production of foreign proteins causing cellular stresses (Okolie et al. 2021a; Okolie et al. 2021b; Guo et al. 2022a). To address these limitations, various strategies have focused on modifying butanol-producing organisms, including mutagenesis, recombinant DNA technology, and proteomics (Okolie et al. 2021a; Okolie et al. 2021b; Guo et al. 2022). In relation with economics costs, the minimum butanol price of selling is higher than obtained with petrochemical process (U$ 2.77 and U$ 1.52, respectively) because relatively lower fuel yield and complex co-product portfolio which requires more intensive separation operations which are energy-intensive process (Hakeem et al. 2023; Dudek et al. 2024).

Biogas or biomethane have been extensively studied to be obtained from anaerobic digestion or biomethanation process of lignocellulosic biomass (Okolie et al. 2021a; Okolie et al. 2021b). This biofuel can serve of substitute of conventional gaseous fuels such as natural gas for combined heat and power generation due to low density, higher calorific value and greater thermal efficiency (Okolie et al. 2021; Okolie et al. 2021b). A significant reduction in greenhouse gas emission, over 60%, can be achieved if it is applied in heavy motor vehicles compared to compressed natural gas (Okolie et al. 2021a; Okolie et al. 2021b). However, the emission of NOx, hydrocarbon, and CO are marginally elevated due to the high concentration of N_2 in biogas and low warm-up temperatures of the engines with biogas causing the incomplete combustion (Okolie et al. 2021a; Okolie et al. 2021b).

Biogas production from lignocellulosic biomass occurs through anaerobic digestion, a biological process in which microorganisms break down organic material in the absence of oxygen. The biogas yield from lignocellulosic biomass depends on various factors, including the type of feedstock, the efficiency of the pretreatment method, and the operating conditions of the anaerobic digester (Okolie et al. 2021a; Boro et al. 2022). In general, lignocellulosic materials have lower biogas yields compared to more easily degradable substrates like food waste or animal manure, due to their high lignin content which is recalcitrant. However, after proper pretreatment (physical, biological—e.g., brown-rot and white-rot fungi- or chemical additives—NaOH and CaO), lignocellulosic biomass can produce significant amounts of biogas, with methane concentrations typically ranging from 50% to 70% (Okolie et al. 2021a; Boro et al. 2022). Other way to enhance anaerobic digestion of lignocellulosic biomass includes co-digestion with other substrate to improve the C/N ratio and the biodegradation (Okolie et al. 2021; Boro et al. 2022).

Biohydrogen production from lignocellulosic biomass is quite emerging and a variety of process have been studied such as dark, photo, and solid-state fermentation and microbial electrolysis cell (MEC) (Bhatia et al. 2021, Okolie et al. 2021a). Among all, dark fermentation stands out once it generates hydrogen at higher rates and lower costs (Bhatia et al. 2021). However, the process's yield and production rate are very low when compared to thermochemical processes (Rani et al. 2024). To address this drawback several efforts have being done to enhance the performance of biohydrogen production, including biomass pretreatment technologies, factors affecting the process efficiency, nanoparticles inclusion, bioaugmentation, and metabolic engineering techniques (Bhatia et al. 2021; Rani et al. 2024).

Biojet fuel derived from lignocellulosic biomass represents a promising solution for reducing greenhouse gas emissions in the aviation sector once the feedstock cost is inexpensive due to abundant available (Okolie et al. 2021a) thermochemical and biochemical processes require many stages to obtain the biojet fuel increasing the capital and operational costs impeding the development and commercial production process (Okolie et al. 2021; Penteado et al. 2024). One key method used to obtain is pyrolysis, where the biomass is heated in the absence of oxygen to produce bio-oil, which is then upgraded into jet fuel through catalytic processes (Penteado et al. 2024). Additionally, gasification can produce syngas, which is subsequently converted into biojet fuel via Fischer–Tropsch synthesis (Penteado et al. 2024).

In conclusion, the production of biofuels and biochemicals from lignocellulose biomass offers a promising pathway toward a more sustainable and resilient future. By reducing greenhouse gas emissions, utilizing waste materials, enhancing energy security, and creating economic opportunities, biofuels and biochemicals from plant-based materials have the potential to significantly contribute to global efforts to combat climate change and reduce our dependence on fossil fuels. However, overcoming the technical and economic challenges associated with biocompounds production will require continued research, innovation, and policy support. With the right investments and advancements, biofuels and biochemicals can become an integral part of a sustainable bioeconomy, helping to address some of the most pressing environmental and energy challenges of our time.

1.8 CONCLUSION

Lignocellulosic biomass is a crucial renewable resource with great potential for the sustainable production of biofuels, bioproducts, and chemicals. Composed mainly of cellulose, hemicellulose, and lignin, it offers a viable alternative to fossil fuels, helping to reduce greenhouse gas emissions and mitigate climate change. However, efficient conversion of this biomass faces technical challenges due to its complex structure, requiring advanced pretreatments such as milling, steam explosion, and chemical treatments. The efficiency of enzymatic hydrolysis also depends on factors such as temperature, pH, and by-product inhibition. Biochemical and thermochemical technologies have made it possible to obtain biofuels such as ethanol and biogas, as well as value-added products such as bio-oil and aromatic compounds from lignocellulosic biomass. However, economic challenges, especially the high costs of enzymatic processes and large-scale production, still need to be addressed. Lignocellulosic biomass, positioned as a strategic solution for the development of a sustainable bioeconomy, can be further empowered with the right public incentive policies and continuous investment in research. This technological route, with the support of such policies and investment, can solidify itself as an essential pillar for a greener and more resilient future.

REFERENCES

Ali, S., Rani, A., Dar, M.A., Qaisrani, M.M., Noman, M., Yoganathan, K., Asad, M., Berhanu, A., Barwant, M., Zhu, D. (2024). Recent advances in characterization and valorization of lignin and its value-added products: challenges and future perspectives. *Biomass*, *4*, 947–977. https://doi.org/10.3390/biomass4030053

Alzagameem, A., Khaldi-Hansen, B. E., Büchner, D., Larkins, M., Kamm, B., Witzleben, S., & Schulze, M. (2018). Lignocellulosic biomass as source for lignin-based environmentally benign antioxidants. *Molecules, 23*(10), 2664. https://doi.org/10.3390/molecules23102664

Amamou, S., Sambusiti, C., Monlau, F., Dubreucq, E., & Barakat, A. (2018). Mechano-enzymatic deconstruction with a new enzymatic cocktail to enhance enzymatic hydrolysis and bioethanol fermentation of two macroalgae species. *Molecules*, *23*(1), 174. https://doi.org/10.3390/molecules23010174.

Amezcua-Allieri, M. A., Sánchez Durán, T., & Aburto, J. (2017). Study of chemical and enzymatic hydrolysis of cellulosic material to obtain fermentable sugars. *Journal of Chemistry, 2017*(1), 5680105. https://doi.org/10.1155/2017/5680105.

Auxenfans, T., Crônier, D., Chabbert, B., & Paës, G. (2017). Understanding the structural and chemical changes of plant biomass following steam explosion pretreatment. *Biotechnology for biofuels, 10*, 1–16. https://doi.org/10.1186/s13068-017-0718-z.

Awasthi, M. K., Sar, T., Gowd, S. C., Rajendran, K., Kumar, V., Sarsaiya, S., Li, Y., Sindhu, R., Binod, P., Zhang, Z., Pandey, A., Taherzadeh, M. J. (2023). A comprehensive review on thermochemical, and biochemical conversion methods of lignocellulosic biomass into valuable end product, *Fuel, 342*, 127790. https://doi.org/10.1016/j.fuel.2023.127790

Baez, L. A., & Bacete, L. (2023). Cell wall dynamics: novel tools and research questions, *Journal of Experimental Botany, 74*(21), 6448–6467. https://doi.org/10.1093/jxb/erad310

Bai, J., Luo, L., Li, A., Lai, X., Zhang, X., Yu, Y., Wang, H., Wu, N., & Zhang, L. (2022). Effects of biofuel crop switchgrass (Panicum virgatum) cultivation on soil carbon sequestration and greenhouse gas emissions: a review. *Life (Basel), 12*(12), 2105. https://doi.org/10.3390/life12122105

Balk, M., Sofia, P., Neffe, A. T., & Tirelli, N. (2023). Lignin, the lignification process, and advanced, lignin-based materials, *International Journal of Molecular Sciences, 24*(14), 11668. https://doi.org/10.3390/ijms241411668

Ballesteros, M. (2010). Enzymatic hydrolysis of lignocellulosic biomass. In: *Bioalcohol Production* (pp. 159–177). Woodhead Publishing.

Banerjee, S., Pandit, C., Gundupalli, M.P., Pandit, S., Rai, N., Lahiri, D., Chaubey, K. K., & Joshi, S. J. (2024). Life cycle assessment of revalorization of lignocellulose for the development of biorefineries *Environment, Development and Sustainability, 26*, 16387–16418. https://doi.org/10.1007/s10668-023-03360-4

Barakat, A., Chuetor, S., Monlau, F., Solhy, A., & Rouau, X. (2014). Eco-friendly dry chemo-mechanical pretreatments of lignocellulosic biomass: impact on energy and yield of the enzymatic hydrolysis. *Applied Energy, 113*, 97–105. https://doi.org/10.1016/j.apenergy.2013.07.015Get rights and content

Basu, P. (2010). Introduction. In: *Biomass Gasification Design Handbook*. https://doi.org/10.1016/b978-0-12-374988-8.00001-5

Becker, J., & Wittmann, C. (2019). A field of dreams: lignin valorization into chemicals, materials, fuels, and health-care products. *Biotechnology Advances*, 37(6), 107360. https://doi.org/10.1016/j.biotechadv.2019.02.016

Bhatia, S. K., Jagtag, S. S., Bedekar, A. A., Bhatia, R. K., Rajendran, K., Pugazhendhi, A., Rao, C. V., Atabani, A. E., Kumar, G., & Yang Y. -H. (2021). Renewable biohydrogen production from lignocellulosic biomass using fermentation and integration of systems with other energy generation technologies. *Science of the total Environment, 765*, 144429. https://doi.org/10.1016/j.scitotenv.2020.144429

Blasi, A., Verardi, A., Lopresto, C.G., Siciliano, S., & Sangiorgio, P. (2023). Lignocellulosic agricultural waste valorization to obtain valuable products: an overview. *Recycling, 8*(4), 61. https://doi.org/10.3390/recycling8040061

Boro, M., Verma, A. K., Chettri, D., Yata, V. K., & verma, A. K. (2022). Strategies involved in biofuel production from agro-based lignocellulose biomass. *Environmental Technology & Innovation*, 28, 102679. https://doi.org/10.1016/j.eti.2022.102679

Braghiroli, F. L., & Passarini, L. (2020). Valorization of biomass residues from forest operations and wood manufacturing presents a wide range of sustainable and innovative possibilities. *Current Forestry Reports, 6*, 172–183. https://doi.org/10.1007/s40725-020-00112-9

Brethauer, S., & Studer, M. H. (2015). Biochemical conversion processes of lignocellulosic biomass to fuels and chemicals: a review. *Chimia, 69*(10). https://doi.org/10.2533/chimia.2015.572

Bušić, A., Marđetko, N., Kundas, S., Morzak, G., Belskaya, H., Ivančić Šantek, M., Komes, D., Novak, S., & Šantek, B. (2018). Bioethanol production from renewable raw materials and its separation and purification: a review. *Food Technology and Biotechnology, 56*(3), 289–311. https://doi.org/10.17113/ftb.56.03.18.5546.

Cañadas, R., Martín-Sampedro, R., González-Miquel, M., González, E. J., Ballesteros, I., Eugenio, M. E., & Ibarra, D. (2023). Green solvents extraction-based detoxification to enhance the enzymatic hydrolysis of steam-exploded lignocellulosic biomass and recover bioactive compounds. *Journal of Environmental Management*, *344*, 118448.

Chakhtouna, H., Benzeld, H., Zarl, N., Qalss, A. K., & Bouhfid, R. (2024). Recent advances in eco-friendly composites derived from lignocellulosic biomass for wastewater treatment, *Biomass Conversion and Biorefinery*, 14, 12085–12111. https://doi.org/10.1007/s13399-022-03159-9

Che Kamarludin, S. N., Jainal, M. S., Azizan, A., Safaai, N. S. M., & Mohamad Daud, A. R. (2014). Mechanical pretreatment of lignocellulosic biomass for biofuel production. *Applied Mechanics and Materials*, *625*, 838–841.

Chen, Z., Hu, M., Zhu, X., Guo, D., Liu, S., Hu, Z., Xiao, B., Wang, J., & Laghari, M. (2015). Characteristics and kinetic study on pyrolysis of five lignocellulosic biomass via thermogravimetric analysis. *Bioresource Technology, 192*. https://doi.org/10.1016/j.biortech.2015.05.062

Crosgrove, D. J. (2024). Structure and growth of plant cell walls, *Nature Reviews Molecular Cell Biology*, *25*, 340–358. https://doi.org/10.1038/s41580-023-00691-y

Crosgrove, D. J., Dupree, P., Gomez, E. D., Haigler, C. H., Kubicki, J. D., Zimmer, J. (2024). How many glucan chains form plant cellulose microfibrils? A mini review, *Biomacromolecules*. https://doi.org/10.1021/acs.biomac.4c00995

Dadwal, A., Sharma, S., & Satyanarayana, T. (2020). Progress in ameliorating beneficial characteristics of microbial cellulases by genetic engineering approaches for cellulose saccharification. *Frontiers in Microbiology, 11*, Article 1387.

Das, A., Das, S., Das, N., Pandey, P., Ingti, B., Panchenko, V., Bolshev, V., Kovalev, A., & Pandey, P. (2023). Advancements and innovations in harnessing microbial processes for enhanced biogas production from waste materials. *Agriculture, 13*, 1689. https://doi.org/10.3390/agriculture13091689.

Delmer, D., Dixon, R. A., Keegstra, K., & Mohnen, D. (2024). The plant cell wall – dynamic, strong, and adaptable – is a natural shapeshifter, *The Plant Cell*, 36, 1257–1311. https://doi.org/10.1093/plcell/koad325

Deng, W., Feng, Y., Fu, J., Guo, H., Guo, Y., Han, B., Jiang, Z., Kong, L., Li, C., Liu, H., Nguyen, P. T. T., Ren, P., Wang, F., Wang, S., Wang, Y., Wang, Y., Wong, S. S., Yan, K., Yan, N., Yang, X., Zhang, Y., Zhang, Z., Zeng, X., & Zhou, H. (2023). Catalytic conversion of lignocellulosic biomass into chemicals and fuels, *Green Energy & Environment*, *8*, 10–114. https://doi.org/10.1016/j.gee.2022.07.003

Dharmaraja, J., Shobana, S., Arvindnarayan, S., Francis, R. R., Jeyakumar, R. B., Saratale, R. G., Ashokkumar, V., Bhatia, S. K., Kumar, V., & Kumar, G. (2023). Lignocellulosic biomass conversion via greener pretreatment methods towards biorefinery application, *Bioresource Technology*, *369*, 128328. https://doi.org/10.1016/j.biortech.2022.128328

Dionisi, D., Anderson, J. A., Aulenta, F., Mccue, A., & Paton, G. (2015). The potential of microbial processes for lignocellulosic biomass conversion to ethanol: a review. *Journal of Chemical Technology and Biotechnology, 90*(3). https://doi.org/10.1002/jctb.4544

Dudek, K., Mokarrari, K. R., Aghamohamadi-Bosjin, S., Valdez-Vazquez, I., & Sowlati, T. (2024). Techno-economic analysis and strategic optimization of biobutanol production from lignocellulosic biomass in Mexico. *Biomass Conversion and Biorefinery*. https://doi.org/10.1007/s13399-024-05393-9.

Etale, A., Onyianta, A. J., Turner, S. R., & Eichhorn, S. J. (2023). Cellulose: a review of water interactions, applications in composites, and water treatment, *Chemical Reviews*, *123*, 2016–2048. https://doi.org/10.1021/acs.chemrev.2c00477

Guo, Y., Liu, Y., Guan, M., Tang, H., Wang, Z., Lin, L., & Pang, H. (2022b). Production of butanol from lignocellulosic biomass: recent advances, challenges, and prospects. *RSC Advance, 12*(29), 18848–18863. https://doi.org/10.1039/d1ra09396g.

Guo, F., Fang, Z., Xu, C. C., & Smith Jr, R. L. (2012). Solid acid mediated hydrolysis of biomass for producing biofuels. *Progress in Energy and Combustion Science*, *38*(5), 672–690. https://doi.org/10.1016/j.pecs.2012.04.001.

Guo, H., He, T., & Lee, D. J. (2022a). Contemporary proteomic research on lignocellulosic enzymes and enzymolysis: a review. *Bioresource Technology*, *344*, 126263.

Guo, H., Zhao, Y., Chang, J. -S., & Lee, D. -J. (2023). Enzymes and enzymatic mechanisms in enzymatic degradation of lignocellulosic biomass: A mini-review. *Bioresource Technology*, *367*, 128252.

Hakeem, I. G., Sharma, A., Sharma, T., Sharma, A., Joshi, J. B., Shah, K., Ball, A. S., & Surapaneni, A. (2023). Techno-economic analysis of biochemical conversion of biomass to biofuels and platform chemicals. *Biofuels, Bioproducts & Biorefining, 17*, 718–750. https://doi.org/10.1002/bbb.2463.

Hemansi, Gupta, R., Kuhad, R. C., & Saini, J. K. (2018). Cost-effective production of a complete cellulase system by newly isolated *Aspergillus niger* RCKH-3 for efficient enzymatic saccharification: medium engineering by overall evaluation criteria approach (OEC). *Biochemical Engineering Journal*, *132*, 182–190.

Huang, L. Z., Ma, M. G., Ji, X. X., Choi, S. E., & Si, C. (2021). Recent developments and applications of hemicellulose from wheat straw: a review. *Frontiers in Bioengineering and Biotechnology*, 9, 690773. https://doi.org/10.3389/fbioe.2021.690773

Ilić, N., Milić, M., Beluhan, S., & Dimitrijević-Branković, S. (2023). Cellulases: from lignocellulosic biomass to improved production, *Energies*, *16*(8), 3598. https://doi.org/10.3390/en16083598

Johannes, L. P., & Xuan, T. D. (2024). Comparative analysis of acidic and alkaline pretreatment techniques for bioethanol production from perennial grasses. *Energies*, *17*(5), 1048. https://doi.org/10.3390/en17051048.

Khan, M. U., Usman, M., Ashraf, M. A., Dutta, N., Luo, G., & Zhang, S. (2022). A review of recent advancements in pretreatment techniques of lignocellulosic materials for biogas production: opportunities and limitations. *Chemical Engineering Journal Advances*, *10*, 100263. https://doi.org/10.1016/j.ceja.2022.100263.

Khodayari, A., Hirn, U., Spirk, S., Ogawa, Y., Seveno, D., & Thielemans, W. (2024). Advancing plant cell wall modelling: atomistic insights into cellulose, disordered cellulose, and hemicelluloses – a review, *Carbohydrate Polymers*, *343*, 122415. https://doi.org/10.1016/j.carbpol.2024.122415

Kucharska, K., Słupek, E., Cieśliński, H., & Kamiński, M. (2020). Advantageous conditions of saccharification of lignocellulosic biomass for biofuels generation via fermentation processes. *Chemical Papers, 74*(4). https://doi.org/10.1007/s11696-019-00960-1

Kumar, A., Jones, D. D., & Hanna, M. A. (2009). Thermochemical biomass gasification: a review of the current status of the technology. *Energies*, *2*(3). https://doi.org/10.3390/en20300556

Liu, F., Short, M. D., Alvarez-Gaitan, J. P., Guo, X., Duan, J., Saint, C., Chen, G., & Hou, L. (2020). Environmental life cycle assessment of lignocellulosic ethanol-blended fuels: a case study. *Journal of Cleaner Production, 245*, 118933. https://doi.org/10.1016/j.jclepro.2019.118933

Liu, W. J., & Yu, H. Q. (2022). Thermochemical conversion of lignocellulosic biomass into mass-producible fuels: emerging technology progress and environmental sustainability evaluation. *ACS Environmental Au*, *2*(2), 98–114. https://doi.org/10.1021/acsenvironau.1c00025

Machineni, L. (2020). Lignocellulosic biofuel production: review of alternatives. *Biomass Conversion and Biorefinery,* 10, 779–791. https://doi.org/10.1007/s13399-019-00445-x

Maitan-Alfenas, G. P., Visser, E. M., & Guimarães, V. M. (2015). Enzymatic hydrolysis of lignocellulosic biomass: converting food waste in valuable products. *Current Opinion in Food Science*, *1*, 44–49.

Margaritis, A., & Kilonzo, P. M. (2005). Production of ethanol using immobilised cell bioreactor systems. https://doi.org/10.1007/1-4020-3363-x_22

Martins, L. H. S., Rabelo, S. C., & da Costa, A. C. (2015). Effects of the pretreatment method on high solids enzymatic hydrolysis and ethanol fermentation of the cellulosic fraction of sugarcane bagasse. *Bioresource Technology*, *191*, 312–321. https://doi.org/10.1016/j.biortech.2015.05.024.

Mu D, Seager T, Rao PS, Zhao F. (2010). Comparative life cycle assessment of lignocellulosic ethanol production: biochemical versus thermochemical conversion. *Environmental Management*, 46, 565–578. https://doi.org/10.1007/s00267-010-9494-2

Mujtaba, M., Fraceto, L. F., Fazeli, M., Mukherjee, S., Savassa, S. M., Medeiros, G. A., Pereira, A. E. S., Mancini, S. D., Lipponen, J., & Vilaplana, F. (2023). Lignocellulosic biomass from agricultural waste to the circular economy: a review with focus on biofuels, biocomposites and bioplastics, *Journal of Cleaner Production*, 402, 136815. https://doi.org/10.1016/j.jclepro.2023.136815

Nanda, S., Golemi-Kotra, D., McDermott, J.C., Dalai, A. K., Gökalp, I., & Kozinski, J. A. (2017). Fermentative production of butanol: perspectives on synthetic biology. *New Biotechnology, 37*, 210–221. https://doi.org/10.1016/j.nbt.2017.02.006

Nitsos, C. K., Matis, K. A., & Triantafyllidis, K. S. (2013). Optimization of hydrothermal pretreatment of lignocellulosic biomass in the bioethanol production process. *ChemSusChem*, *6*(1), 110–122. https://doi.org/10.1002/cssc.201200546.

Norrrahim, M. N. F., Ilyas, R. A., Nurazzi, N. M., Rani, M. S. A., Atikah, M. S. N., & Shazleen, S. S. (2021). Chemical pretreatment of lignocellulosic biomass for the production of bioproducts: an overview. *Applied Science and Engineering Progress*, *14*(4), 588–605. https://doi.org/10.14416/j.asep.2021.07.004.

Oelmüller, R., Tseng, Y. -H., & Gandhi, A. (2023). Signals and their perception for remodeling, adjustment and repair of the plant cell wall, *International Journal of Molecular Sciences*, *24*(8), 7417. https://doi.org/10.3390/ijms24087417

Okolie, J. A., Nanda, S., Dalai, A. K., & Kozinski, J. A. (2021). Chemistry and specialty industrial applications of lignocellulosic biomass, *Waste and Biomass Valorization*, *12*, 2145–2169. https://doi.org/10.1007/s12649-020-01123-0

Okolie, J. A., Mukherjee, A., Nanda, S., Dalai, A. K., & Kozinski, J. A. (2021a). Next-generation biofuels and platform biochemicals from lignocellulosic biomass. *International Journal of Energy Research*, 45, 14145–14169. https://doi.org/10.1002/er.6697

Okolie, J. A., Nanda, S., Dalai, A. K., & Kozinki, J. A. (2021b). Chemistry and Specialty Industrial Applications of Lignocellulosic Biomass, *Waste Biomass Valorization, 12*, 2145–2169. https://doi.org/10.1007/s12649-020-01123-0

Oliveira, J. A. R., de Paula Menezes Barbosa, P., & Macêdo, G. A. (2022). High concentrate flavonoids extract from citrus pomace using enzymatic and deep eutectic solvents extraction. *Foods, 11*, 3205. https://doi.org/10.3390/foods11203205

Oliveira, J. A. R., Komesu, A., & Maciel Filho, R. (2014). Hydrothermal pretreatment for enhancing enzymatic hydrolysis of seeds of açaí (Euterpe oleracea) and sugar recovery. *Chemical Engineering Transactions, 37*, 787–792.

Oliveira, J. A. R., Martins, L. H. S., Komesu, A., & Maciel Filho, R. (2015). Evaluation of alkaline delignification (NaOH) of açaí seeds (Euterpe oleracea) treated with dilute H_2SO_4 and effect on enzymatic hydrolysis. *Chemical Engineering Transactions, 43*, 499–504.

Olortiga Asencios, Y. J., & M. Assaf, E. (2023). Green hydrogen Production from the oxidative reform of clean biogas over Ni/MgO-Nb2O5 catalysts. *Journal of Science & Sustainable Engineering, 1*(1). https://doi.org/10.34024/jsse.2023.v1.15904

Oriez, V., Peydecastaing, J., & Pontalier, P. Y. (2019). Lignocellulosic biomass fractionation by mineral acids and resulting extract purification processes: conditions, yields, and purities. *Molecules, 24*(23), 4273. https://doi.org/10.3390/molecules24234273.

Østby, H., Várnai, A. (2023). Hemicellulolytic enzymes in lignocellulose processing. *Essays in Biochemistry*, 67, 533–550. https://doi.org/10.1042/EBC20220154

Penteado, E. D., Diniz, A. A. R., Asencios, Y. J. O., da Silva Martins, L. H., de Oliveira, J. A. R., & Komesu, A. (2024). Production of biojet fuel. In: *Biojet Fuel: Current Technology and Future Prospect* (pp. 37–60). Springer, Singapore. https://doi.org/10.1007/978-981-99-8783-2_3

Pérez, J., Munoz-Dorado, J., De la Rubia, T. D. L. R., & Martinez, J. (2002). Biodegradation and biological treatments of cellulose, hemicellulose and lignin: an overview. *International Microbiology, 5*, 53–63.

Phojaroen, J., Jiradechakorn, T., Kirdponpattara, S., Sriariyanun, M., Junthip, J., & Chuetor, S. (2022). Performance evaluation of combined hydrothermal-mechanical pretreatment of lignocellulosic biomass for enzymatic enhancement. *Polymers, 14*(12), 2313. https://doi.org/10.3390/polym14122313.

Qu, T., Zhang, X., Gu, X., Han, L., Ji, G., Chen, X., & Xiao, W. (2017). Ball milling for biomass fractionation and pretreatment with aqueous hydroxide solutions. *ACS Sustainable Chemistry & Engineering*, *5*(9), 7733–7742. https://doi.org/10.1021/acssuschemeng.7b01186.

Radionova, M. V, Bozieva, A. M., Zharmukhamedov, S. K., Leong, Y. K., Lan J. C. W., Veziroglu, A., Veziroglu, T. N., Tomo, T., Chang, J.S., & Allakhverdiev, S. I. (2022). A comprehensive review on lignocellulosic biomass biorefinery for sustainable biofuel production. *International Journal of Hydrogen Energy, 47*(3), 1481–1498. https://doi.org/10.1016/j.ijhydene.2021.10.122

Rani, P., Yadav, D. K., Yadav, A., Bishnoi, N. R., Kumar, V., Ram, C., Pugazhendhi, A., & Kumar, S. S. (2024). Frontier in dark fermentative biohydrogen production from lignocellulosic biomass: challenges and future prospects. *Fuel, 366*, 121187. https://doi.org/10.1016/j.fuel.2024.131187

Rashid, S. S., Mustafa, A. H., Ab Rahim, M. H., & Gunes, B. (2022). Magnetic nickel nanostructure as cellulase immobilization surface for the hydrolysis of lignocellulosic biomass. *International Journal of Biological Macromolecules, 209*, 1048–1053.

Raud, M., Kikas, T, Sippula, O., & Shurpali, N. J. (2019). Potentials and challenges in lignocellulosic biofuel production technology. *Renewable and Sustainable Energy Reviews, 111*, 44–56. https://doi.org/10.1016/j.rser.2019.05.020

Riseh, R. S., Fathi, F., Lagzian, A., Vatankhah, M., Kennedy, J. F. (2024). Modifying lignin: a promising strategy for plant disease control. *International Journal of Biological Macromolecules, 271*, 132696. https://doi.org/10.1016/j.ijbiomac.2024.132696

Rongpipi, S., Ye, D., Gomez, E. D., & Gomez, E. W. (2019). Progress and opportunities in the characterization of cellulose—an important regulator of cell wall growth and mechanics. *Frontiers in Plant Science, 9*, 1894. https://doi.org/10.3389/fpls.2018.01894

Rusănescu, C. O., Ciobanu, M., Rusănescu, M., & Dinculoiu, R. L. (2024). Pretreatments applied to wheat straw to obtain bioethanol. *Applied Sciences, 14*, 1612. https://doi.org/10.3390/app14041612

Saha, B. C., Iten, L. B., Cotta, M. A., & Wu, Y. V. (2005). Dilute acid pretreatment, enzymatic saccharification and fermentation of wheat straw to ethanol. *Process Biochemistry, 40*(12), 3693-3700. https://doi.org/10.1016/j.procbio.2005.04.006

Saini, J. K., Saini, R., & Tewari, L. (2015). Lignocellulosic agriculture wastes as biomass feedstocks for second-generation bioethanol production: Concepts and recent developments. *3 Biotech, 5*(4), 337–353. https://doi.org/10.1007/s13205-014-0246-5

Saini, J. K., Kaur, A., & Mathur, A. (2022). Strategies to enhance enzymatic hydrolysis of lignocellulosic biomass for biorefinery applications: a review. *Bioresource Technology, 360*, 127517.

Saravanan, A., Senthil Kumar, P., Jeevanantham, S., Karishma, S., & Vo, D. V. N. (2022). Recent advances and sustainable development of biofuels production from lignocellulosic biomass. *Bioresource Technology, 344*, 126203. https://doi.org/10.1016/j.biortech.2021.126203

Sharma, N., Allardyce, B. J., Rajkhowa, R., & Agrawal, R. (2023). Rice straw-derived cellulose: a comparative study of various pre-treatment technologies and its conversion to nanofibres. *Scientific Reports, 13*(1), 16327. https://doi.org/10.1038/s41598-023-43535-7

Silva, T.A.L., Varão, L.H.R., Pasquini, D. (2024). Lignocellulosic Biomass. In: Thomas, S., Hosur, M., Pasquini, D., Jose Chirayil, C. (eds) *Handbook of Biomass*. Springer, Singapore. https://doi.org/10.1007/978-981-19-6772-6_5-1.

Sitotaw, Y. W., Habtu, N. G., Gebreyohannes, A. Y., Nunes, S. P., & Van Gerven, T. (2021). Ball milling as an important pretreatment technique in lignocellulose biorefineries: a review. *Biomass Conversion and Biorefinery, 13*(17), 15593–15616. https://doi.org/10.1007/s13399-021-02018-3.

Skiba, E. A., Baibakova, O. V., Budaeva, V. V., Pavlov, I. N., Vasilishin, M. S., Makarova, E. I., Sakovich, G. V., Ovchinnikova, E. V., Banzaraktsaeva, S. P., Vernikovskaya, N. V., & Chumachenko, V. A. (2017). Pilot technology of ethanol production from oat hulls for subsequent conversion to ethylene. *Chemical Engineering Journal, 329*, 178-186. https://doi.org/10.1016/j.cej.2017.05.182

Solarte-Toro, J. C., Romero-García, J. M., Martínez-Patiño, J. C., Ruiz-Ramos, E., Castro-Galiano, E., & Cardona-Alzate, C. A. (2019). Acid pretreatment of lignocellulosic biomass for energy vectors production: a review focused on operational conditions and techno-economic assessment for bioethanol production. *Renewable and Sustainable Energy Reviews, 107*, 587-601. https://doi.org/10.1016/j.rser.2019.02.02.

Stickel, J. J., Elander, R. T., Mcmillan, J. D., & Brunecky, R. (2014). Enzymatic hydrolysis of lignocellulosic biomass. In: Bisaria, V. S., & Konde, A. (eds) *Bioprocessing of Renewable Resources to Commodity Bioproducts*.

Tabil, L., Adapa, P., & Kashaninejad, M. (2011). Biomass feedstock pre-processing-part 1: pretreatment. *Biofuel's Engineering Process Technology, 18*, 411–439.

Tarasov, D., Leitch, M., & Fatehi, P. (2018). Lignin–carbohydrate complexes: properties, applications, analyses, and methods of extraction: a review. *Biotechnology for Biofuels, 11*, 269. https://doi.org/10.1186/s13068-018-1262-1

Thomas, M., Vrij, M., Zandstra, T., & Van der Poel, A. F. B. (2012). Grinding performance of wheat, maize and soybeans in a multicracker system. *Animal Feed Science and Technology, 175*(3-4), 182–192. https://doi.org/10.1016/j.anifeedsci.2012.05.002.

Tufail, T., Saeed, F., Afzaal, M., Ain, H. B. U., Gilani, S. A., Hussain, M., & Anjum, F. M. (2021). Wheat straw: A natural remedy against different maladies. *Food Science & Nutrition, 9*(4), 2335–2344. https://doi.org/10.1002/fsn3.2030

Utami, R., Tran, M. H., & Lee, E. Y. (2024). Mini-review on lignin-based self-healing polymer, *International Journal of Biological Macromolecules, 279*, 135295. https://doi.org/10.1016/j.ijbiomac.2024.135295.

Vaezi, M., Pandey, V., Kumar, A., & Bhattacharyya, S. (2013). Lignocellulosic biomass particle shape and size distribution analysis using digital image processing for pipeline hydro-transportation. *Biosystems Engineering, 114*(2), 97–112. https://doi.org/10.1016/j.biosystemseng.2012.11.007.

Vanholme, R., Demedts, B., Morreel, K., Ralph, J., & Boerjan, W. (2010). Lignin biosynthesis and structure. *Plant Physiology, 153(3)*, 895–905. https://doi.org/10.1104/pp.110.155119

Vasile, C., & Baican, M. (2023). Lignins as promising renewable biopolymers and bioactive compounds for high-performance materials. *Polymers (Basel), 15(15)*, 3177. https://doi.org/10.3390/polym15153177

Wang, T., & Lü, X. (2021). Overcome saccharification barrier: advances in hydrolysis technology. In *Advances in 2nd generation of bioethanol production* (pp. 137–159). Woodhead Publishing.

Weger, J., Knápek, J., Bubeník, J., Vávrová, K., & Strašil, Z. (2021). Can miscanthus fulfill its expectations as an energy biomass source in the current conditions of the Czech Republic?—Potentials and barriers. *Agriculture, 11*, 40. https://doi.org/10.3390/agriculture11010040

Xiao, R., Yang, W., Cong, X., Dong, K., Xu, J., Wang, D., & Yang, X. (2020). Thermogravimetric analysis and reaction kinetics of lignocellulosic biomass pyrolysis. *Energy, 201*. https://doi.org/10.1016/j.energy.2020.117537

Xu, C., Alam, M. A., Wang, Z., Chen, H., Zhang, J., Huang, S., ... & Xu, J. (2021). Mechanisms of bio-additives on boosting enzymatic hydrolysis of lignocellulosic biomass. *Bioresource Technology, 337*, 125341.

Xu, C., Tong, S., Sun, L., & Gu, X. (2023). Cellulase immobilization to enhance enzymatic hydrolysis of lignocellulosic biomass: an all-inclusive review. *Carbohydrate Polymers*, 121319.

Yuan, Z., Long, J., Wang, T., Shu, R., Zhang, Q., & Ma, L. (2015). Process intensification effect of ball milling on the hydrothermal pretreatment for corn straw enzymolysis. *Energy Conversion and Management, 101*, 481–488. https://doi.org/10.1016/j.enconman.2015.05.05.

Zakaria, M. R., Fujimoto, S., Hirata, S., & Hassan, M. A. (2014). Ball milling pretreatment of oil palm biomass for enhancing enzymatic hydrolysis. *Applied Biochemistry and Biotechnology, 173*, 1778–1789. https://doi.org/10.1007/s12010-014-0964-5.

Zoghlami, A., & Paës, G. (2019). Lignocellulosic biomass: Understanding recalcitrance and predicting hydrolysis. *Frontiers in Chemistry, 7*, 874. https://doi.org/10.3389/fchem.2019.00874.

2 Lignocellulosic Biomass-based Biofuel Production

Eduardo Dellosso Penteado, Yvan Jesus Olortiga Asencios, Anthony Andrey Ramalho Diniz, Luiza Helena da Silva Martins, Johnatt Allan Rocha de Oliveira, and Andrea Komesu

2.1 INTRODUCTION

The global population is expected to increase from 7.4 billion to 10 billion by 2055 (Ashokkumar et al. 2022; Heleniak T. 2021). During this time, energy demand is projected to grow at an average annual rate of 0.7% until 2030 (IEA 2023). In addition, conventional energy sources are diminishing rapidly, with estimates suggesting that the remaining reserves of crude oil, coal, and natural gas will be exhausted in approximately 50, 132, and 50.9 years, respectively (Uthandi et al. 2022; BP 2019). This highlights the urgent need to transition to sustainable energy alternatives. Many countries around the world have established goals for increasing the production of renewable fuels and integrating them with existing petroleum-based fuels. As a result, the demand for renewable energy is rapidly expanding. Projections suggest that the share of renewables in the energy mix will rise significantly, from 25% in 2017 to an estimated 85% by 2050 (Uthandi et al. 2022; IRENA 2018).

Biofuel production from lignocellulosic biomass is an increasingly viable option for promoting sustainable energy solutions. Lignocellulosic biomass typically refers to the plant material derived from terrestrial sources, characterized by its composition of cellulose, hemicellulose, and lignin (Lee et al. 2023). This type of biomass is among the most plentiful renewable energy resources available, with estimates suggesting an annual global output of approximately 200 billion tons (Shrivastava and Sharma 2022; Liang et al. 2020). Utilizing this abundant resource for biofuel production can significantly contribute to reducing reliance on fossil fuels while supporting efforts to combat climate change.

There are various conversion processes available that can change different types of biomass into a range of fuels and chemicals. This transformation typically involves multiple steps necessary to produce value-added products (Velvizhi et al. 2023). The

 DOI: 10.1201/9781003571223-2

significant expenses associated with converting lignocellulosic biomass into liquid fuel pose a challenge to the large-scale commercial production of biofuels derived from these materials (Jatoi et al. 2021). Developing efficient methods and technologies for biomass conversion not only helps reduce costs but also enhances the overall viability of lignocellulosic biofuels in the energy market. Researchers are continuously exploring innovative approaches to streamline these processes and improve their economic feasibility.

This chapter explores biofuel production from lignocellulosic biomass, focusing on its ability to generate gas, liquid, and solid fuels, as well as electricity through microbial technologies. It begins with an examination of gas fuels like syngas, biogas/biomethane, and biohydrogen. The discussion then moves to liquid biofuels such as bioethanol, biodiesel, and emerging alternatives like biojet fuel and biobutanol, which are vital for reducing greenhouse gas emissions. Solid biofuels, particularly biochar, are highlighted for their renewable energy potential and environmental benefits. Finally, this chapter addresses microbial fuel cells in converting organic materials into electricity.

By examining the various facets of lignocellulosic biomass-based biofuel production, this chapter aims to highlight its significance in meeting the world's energy demands while addressing environmental concerns. The exploration of these biofuels not only underscores their potential for reducing reliance on fossil fuels but also showcases the vital role of sustainable practices in achieving a resilient energy future.

2.2 GAS FUEL

Lignocellulosic biomass can be converted into gaseous fuels, which play a significant role in advancing renewable energy. These fuels are produced using both thermochemical and biochemical methods and have broad applications in areas such as power generation, heating, and transportation. The main gaseous biofuels generated from lignocellulosic biomass are syngas, biogas or biomethane, and biohydrogen.

2.2.1 Syngas

Synthesis-gas (or syngas) is a gas mixture of CO and H_2, with various H_2/CO ratios (= 1, 2, 3 ...). Syngas, when processed by the water–gas shift reaction (WGSR, reaction 2.1), represents a principal source of hydrogen, a promising energy vector that can be used, for example, in fuel cells; however, hydrogen can also be used by direct combustion to generate energy (Asencios et al. 2016; Ruckenstein & Hu 1999).

Syngas can also be used as feedstock in Fischer–Tropsch process (FT) for producing liquid fuels (also known as gas-to-liquid process, GTL), for methanol and dimethyl ether synthesis, among other highly valued products. FT is a very attractive route for obtaining clean fuels (free of S, N, etc.), which emit fewer harmful pollutants. The H_2/CO ratio of syngas can be regulated according to the different purposes required for a given process (Lv et al. 2007). Syngas can also be directly used as a fuel in power plants and power production (steam cycle, combustion in gas turbines or internal combustion engines, co-combustion, and others). Syngas can also be used to produce ethanol, methanol, and ammonia (Ciliberti et al. 2020).

$$\text{WGSR: } CO + H_2O \leftrightarrow CO_2 + H_2 \ \Delta H^0_{298K} = -41 kJ.mol^{-1} \tag{2.1}$$

$$\text{SMR:} CH_4 + H_2O \rightarrow CO + 3H_2 \ \Delta H^0_{298K} = 225.4 kJ.mol^{-1} \tag{2.2}$$

$$\text{DRM:} CH_4 + CO_2 \rightarrow 2CO + 2H_2 \ \Delta H^0_{298K} = 260.5 kJ.mol^{-1} \tag{2.3}$$

$$\text{POM:} CH_4 + 0.5O_2 \rightarrow CO + 2H_2 \ \Delta H^0_{298K} = -22.6 kJ.mol^{-1} \tag{2.4}$$

The current production of hydrogen in the world is mainly obtained by syngas produced through steam methane reforming (SMR, reaction 2.2). Industrial SMR uses methane from natural gas, producing syngas with an H_2:CO ratio of 3:1. Alternatively, dry reforming of methane (DRM, reaction 2.3) uses CO_2 producing H_2:CO of 1:1. The main disadvantage is that both routes are based on fossil fuels and require energy (they are endothermic reactions), DRM is a very attractive route because two warming gases are converted into highly valued product: syngas. Another well-studied route to produce syngas is the partial oxidation of methane (POM, reaction 2.4), however, the principal challenges in this reaction are the hot spot formed in the catalyst (leading to sintering and related problems), and the pure oxygen needed for this reaction, however, the exothermal nature of this reaction is a great advantage (Jung et al. 2021; Olortiga Asencios & M. Assaf 2023).

Currently, most of the hydrogen consumed worldwide comes from syngas produced from methane (natural gas) via steam reforming of methane (SRM), the industrial process uses a catalyst of nickel supported on alumina. In this scenario, more efficient processes and sustainable hydrogen production routes are needed to enable the introduction of hydrogen into the world's energy matrix and reduce greenhouse gases, which is vital to reducing global warming and complying with international environmental agreements by 2030, Paris Agreement; UN, UNFCCC (Cordonier Segger, 2016).

SRM, DRM, and POM (reactions 2.2–2.4), can be carried out using biogas and biomethane as sources of methane. Alternatively, syngas can be obtained by the gasification of lignocellulosic biomass. The gasification of lignocellulosic material to syngas takes place in a reactor (known as gasifier), at high temperatures ranging 800°C–1500°C. The reaction takes place in the presence of a low concentration of oxygen. Gasification can be carried out with oxygen; however, air, water steam, or their mixture are also employed for this process. Syngas obtained in this way contain, among H_2 and CO, H_2, and CO_2. The big challenge in this process is to clean and regulate the final quality of syngas. Figure 2.1 summarizes the information described in the previous paragraphs (Ciliberti et al. 2020).

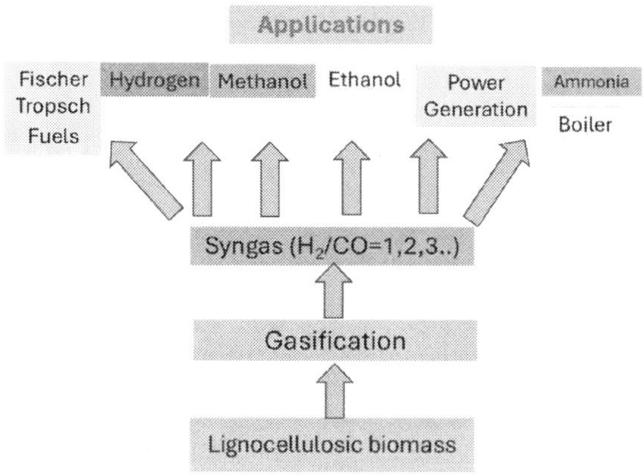

FIGURE 2.1 Syngas production and applications.

2.2.2 Biogas/Biomethane

The potential in self-sustainability has attracted attention for the possibility of producing biogas from lignocellulosic biomass via anaerobic digestion (Tamilselvan and Selwynraj 2024). According to Sun et al. (2021), energy crops and crop residues have a large potential to produce 37.2 billion cubic meters of biogas, what the European Biomass Association estimates that accounts for 47.7% of the total potential from agriculture and waste to produce biogas. Therefore, biogas has been viewed as a promising energy source that can be converted to methanol, used as a heat and power source, or even compressed and used as a car fuel (Karthikeyan et al. 2024). Lignocellulose can be considered the most abundant bioresource, with approximately 200 billion tons generated per year (Chettri et al. 2024). It comprises almost 63% of the total waste mass, and its utilization for energy and fuel production could not only advance the path for a sustainable closed circular bioeconomy but also reduce environmental stress, as the improper disposal of waste can lead to loss of aqueous systems, degradation of atmospheric quality, and disturbances in ecological balance (Singh et al. 2022). Biogas can be produced from a wide range of substrates, which influences its quality and yield. Those substrates can be classified according to their source, combination, and characteristics. They are important to define the processing and pretreatment methods applied to the biogas production (Chettri et al. 2024). Lignocellulosic biomass is primarily composed by cellulose, hemicellulose, and lignin (Miskin and Hinge 2024). According to Chettri et al. (2024), a general composition of lignocellulose is approximately 51% of cellulose and 31% of hemicellulose, and lignin holds these two other components together. The complexity of this structure poses a challenge to biogas production, as the lignin component acts as a physical barrier, what makes hard the access to the cellulose and hemicellulose by the microbial consortia, demanding a pretreatment to enhance the biomass digestibility (Miskin and Hinge 2024). Anaerobic digestion (AD) is a result of synergic actions taken by several bacteria species, in a chain of biological processes, under controlled environmental conditions, to degrade complex organic compounds in the absence of oxygen (Chettri et al. 2024; Dinca et al. 2024). It is a well-known method that makes possible converting organic materials into biogas and organic nutrients (Karthikeyan et al. 2024). It is an interesting processing choice, as it can produce biogas enriched with hydrogen or methane, which can be used as source of bioenenergy (Dutra et al. 2024).

The entire process of anaerobic digestion comprises four-phase distinct biochemical processes: hydrolysis, acidification, acetogenesis, and methanogenesis (Mirabi et al. 2024; Karthikeyan et al. 2024; Ma et al. 2021). The initial phase is hydrolysis, where the complex organic substances (proteins, carbon hydrates, and lipids) are broken down into soluble organic molecules, such as sugar, fatty acids, amino acids, and other related chemicals (Karthikeyan et al. 2024). Equation (2.5) shows the generic hydrolysis reaction (Uddin and Wright 2022).

$$(C_6H_{10}O_5)_n + n\ H_2O \rightarrow n\ C_6H_{12}O_6 + n\ H_2 \tag{2.5}$$

The second stage is acidification or acidogenesis, where the feedstock generated from the hydrolysis phase is broken down by acidogenic bacterial. In the reactor, an acidic environment is generated by fermentative microorganisms, which produce H_2, H_2S, CO_2, NH_3, organic acids, and also shorter volatile acids with some traces of other by-products (Olatunji et al. 2021). For the hydrolysis and acidogenesis processes, which are the methane fermentation processes, there are 20–30 species of microorganisms involved, which include *Butyrivibria*, *Chloroflexi Phyla*, *Clostridium*, *Firmicutes*, *Thermotogae*, *Thermotoga* genera, and others (Ma et al. 2021). Chemical reactions in the acidogenesis stage are illustrated by Equations (2.6)–(2.8) (Uddin and Wright 2022).

$$C_6H_{12}O_6 \leftrightarrow 2CH_3CH_2OH + 2CO_2 \tag{2.6}$$

$$C_6H_{12}O_6 + 2H_2 \leftrightarrow 2CH_3CH_2COOH + 2H_2O \tag{2.7}$$

$$C_6H_{12}O_6 \rightarrow 3CH_3COOH \tag{2.8}$$

The third stage of AD is acetogenesis, where acetogenic bacteria, called acetogens, transform the products from the previous phase and some of the long-chain fatty acids, produced from the hydrolysis phase into acetate, H_2, and CO_2, and also break down the feedstock to make possible the methane production by methanogens (Uddin and Wright 2022; Olatunji et al. 2021). Chemical reactions that occur during the acetogenesis stage are represented in Equations (2.9)–(2.11) (Uddin and Wright 2022).

$$CH_3CH_2COO^- + 3H_2O \leftrightarrow CH_3COO^- + H^+HCO_3^- + 3H_2 \tag{2.9}$$

$$C_6H_{12}O_6 + 2H_2O \leftrightarrow 2CH_3COOH + 2CO_2 + 4H_2 \tag{2.10}$$

$$CH_3CH_2OH + 2H_2O \leftrightarrow CH_3COO^- + 3H_2 + H^+ \tag{2.11}$$

Methanogenesis is the final stage of AD, where the methanogenic bacteria use all the intermediate products generated in the previous stages to produce methane (Uddin and Wright 2022). For this process, about 10–20 species are involved, what includes *Methanobacterium*, *Methanobrevibacter*, *Methanoculleus*, *Methanosarcina*, *Methanospirillum*, *Methanothermobacter*, among others (Ma et al. 2021). As in the presence of oxygen methanogenic bacteria cannot survive, this stage has to be strictly anaerobic, and its chemical reactions are shown in Equations (2.12)–(2.14) (Uddin and Wright 2022).

$$CH_3COOH \rightarrow CH_4 + CO_2 \tag{2.12}$$

$$CO_2 + 4H_2 \rightarrow CH_4 + 2H_2O \tag{2.13}$$

$$2CH_3CH_2OH + CO_2 \rightarrow CH_4 + 2CH_3COOH \tag{2.14}$$

To convert lignocellulosic biomass to biogas it is also needed to perform its pretreatment, which alters the structure of the biomass by decomposing the lignocellulosic matrix, eliminating lignin (Poddar et al. 2022). Pretreatment technologies for making lignocellulosic biomass amenable to AD include biological, physical, chemical, thermal, and combined pretreatment methods in addition to some emerging technologies (Poddar et al. 2022). There are many pretreatment strategies for valorization of lignocellulosic biomass, as it is essential to fractionate the biomass into its chemical constituents that can be further processed not only in biogas, but also in liquid biofuels, biofertilizer, bioplastic, lubricants, industrial enzymes, organic acids, among other by-products through a biorefinery approach, for example (Chaudhary et al., 2024).

2.2.3 Biohydrogen

Being the most abundant element on earth, hydrogen is colorless, odorless, and is considered a clean fuel because it releases water during its burning and emits no greenhouse gases (Okolie et al. 2021). Its energy content (141.9 kJ/g) is greater by almost 2.6 times than methane (55.5 kJ/g), three times than gasoline (47.5 kJ/g), and 3.2 times than diesel (44.8 kJ/g), having the highest gravimetric energy density among all the fuels (~33.3 kWh/kg) (Malik et al. 2023; Al-Mahgari et al. 2023). The advantages of its application include the fact that it can be used directly as fuel in power engines, fuel cells or also for the production of green chemicals and liquid hydrocarbons fuels through the Fischer–Tropsch process (Okolie et al. 2021).

Through specific biological, thermochemical, electrolytic, electrocatalytic, and photocatalytic conversion routes, it is possible producing hydrogen both from renewable and petrochemical resources (Okolie et al. 2021). From fossil fuels, hydrogen has been produced from natural gas, heavy oil, coal, and naphtha, using technologies like pyrolysis, steam reforming, and gasification techniques (Chandran and Mohan 2023). It is a molecule that does not exit readily in nature, so it must be extracted from hydrocarbons or some other hydrogen-containing molecules, as water is an example (Malik et al. 2023).

Producing hydrogen from water electrolysis can be considered a clean method of generation, but it requires a considerable amount of high-grade energy, what can economically compromise this route of production (Chandran and Mohan 2023). In a general way, hydrogen is produced via fossil fuels, which are non-renewable and emit large quantity of CO_2, leading to climate change, environmental degradation, and pollution (Roy and Pramanik 2024; Chandran and Mohan 2023). According to Jara-Cobos et al. (2023), methods to produce hydrogen are predominantly based on natural gas or other fossil fuel sources, which require abundant energy and emit a significant amount of CO_2 into the atmosphere.

Feedstocks can be used to generate hydrogen, which includes biomass and water (Malik et al. 2023). The most abundant feedstock found on the Earth is the lignocellulosic biomass, which is composed of rich aromatic and high carbohydrate polymer, and can be considered as any organic substance derived from the photosynthesis process, directly or indirectly (Chandran and Mohan 2023; Jara-Cobos et al. 2023). According to Lepage et al. (2021), there are several types of technology underway to produce hydrogen (H_2) from (lignocellulosic) biomass, which they categorize in biological conversions, thermochemical processes, and potentially electrochemically assisted production.

Biological processes that can be used to generate hydrogen from biomass include dark fermentation, photo-fermentation and bio-photolysis (Malik et al. 2023; Okolie et al. 2021). Jara-Cobos et al. (2023) consider dark fermentation the most promising technique to generate hydrogen through biomass conversion. According to the authors, anaerobic bacteria can carry out hydrogen production, being grown in a substrate rich in carbohydrates or a dark substrate. They also point out that this method makes possible obtaining biohydrogen, but also acetic, lactic, butyric, and propionic acids, as well as solvents such as methanol, ethanol, and acetone. Some microbes that can be used in dark fermentation are *Ethanoligenens*, *Clostridium*, and *Desulfovibrio* species (Gupta et al. 2024).

Photo-fermentation involves bacteria (photosynthetic microorganisms) that produce hydrogen from organic matter, using solar energy (Gupta et al. 2024; Jara-Cobos et al. 2023). Purple non-sulfur bacteria can use substrates, such as volatile fatty acids, simple sugars, and agricultural and industrial wastes in the presence of light, under anaerobic conditions, being capable of reducing H^+ ions to gaseous H_2, what also reduces the power obtained from the oxidation that happens to organic compounds (Gupta et al. 2024; Jara-Cobos et al. 2023). This process has inspired the development of several photobioreactors (PBRs) to improve the production of biohydrogen in continuous operation (Gupta et al. 2024). Some anaerobic bacteria have been studied to carry out photo-fermentation, such as genera *Rhodobacter*, *Rhodopseudomonas*, *Rhodobium*, and *Rhodospirillum*, which can be used to convert organic acids into carbon dioxide and hydrogen (Gupta et al. 2024).

Bio-photolysis is a process that directly or indirectly involves photosynthesis, which is used to convert solar into chemical energy (Ivanenko et al. 2024). Direct bio-photolysis is a combination of biological and chemical processes, where the microbial photosynthesis mechanism uses the water molecule to transfer solar energy into molecular hydrogen (Sivaramakrishnan et al. 2021). Green microalgae and cyanobacteria are capable of direct bio-photolysis, and this route is used by some efficient hydrogen producing algae strains, such as *Scenedesmus* spp., *Chlorella* spp., and *Chlorococcum* spp. (Ivanenko et al. 2024; Sivaramakrishnan et al. 2021).

Indirect bio-photolysis requires two stages to produce biohydrogen, separated in space and time (Ivanenko et al. 2024; Hitam and Jalil 2023). First, carbon dioxide is fixed by photosynthetic cells, with the formation of carbohydrates and oxygen (Ivanenko et al. 2024). In anaerobic conditions, sugars are converted into organic acids and carbon dioxide, and next, during further fermentation, those organic acids are decomposed to H_2 and CO_2, being those two stages described in Equations (2.15) and (2.16) (Ivanenko et al. 2024).

$$6H_2O + 6CO_2 \rightarrow (C_6H_{12}O_6)_n + 6O_2 \tag{2.15}$$

$$(C_6H_{12}O_6)_n + 12H_2O \rightarrow 12H_2 + 6CO_2 \tag{2.16}$$

According to Machineni et al. (2023), direct and indirect bio-photolysis systems generate pure hydrogen (H_2), while photo and dark fermentations systems generate a mixed biogas, which is primarily composed by carbon dioxide (CO_2) and hydrogen (H_2), and a lesser quantity of methane (CH_4), carbon monoxide (CO), hydrogen sulfide (H_2S) or ammonia (NH_3). According to Lepage et al. (2021), the most advanced technology for hydrogen production from biomass is thermochemical conversion. The same authors highlight that the three primary thermochemical routes are pyrolysis, gasification, and aqueous phase reforming. Gasification is a process where fossil or non-fossil fuels (gaseous, liquid, or solid) are decomposed into useful gases and chemicals (Rangel et al. 2023). Pyrolysis is part of gasification as it is one of the reactions occurring during gasification, but it is performed in the absence of oxidizing agents or with a limited supply, what does not allow biomass gasification to an appreciable extent (Rangel et al. 2023).

Gasification is a thermochemical route for hydrogen production, where the biomass having less energy density is changed in some gaseous fuel using simple chemical reactions, where these gaseous fuels consist mainly of hydrogen (Mishra et al. 2023). In this process, the main steps are drying of biomass, pyrolysis, and reduction, and is followed by combustion, being represented by the chemical reaction in Equation (2.17) (Mishra et al. 2023).

$$\begin{aligned}&\textit{Biomass materials} + \textit{Air} \rightarrow H_2 + CO_2 + CH_4\\&+ CO + N_2 + \textit{Tar} + \textit{Char} + H_2O + \textit{Hydrocarbons}\end{aligned} \tag{2.17}$$

Where tar is a blended hydrocarbon, insignificant in content, and char is a solid residue, black-colored, obtained after the carbonization process (Mishra et al. 2023; Kwapinski 2019). In pyrolysis process, solid biomass (carbonaceous or organic biomass) is heated in the absence of oxygen, aiming at getting gaseous compound, liquid fuels and solid char (Mishra et al. 2023). Depositions of the forest, sewage sludge, crop depositions and organic metropolitan waste are the main categories of lignocellulosic biomass appropriate for pyrolysis (Mishra et al. 2023). Pyrolysis can be classified in slow pyrolysis, fast pyrolysis, and flash pyrolysis, depending mainly on temperature of the reaction, heating rate, time for residence, and so on (Mishra et al. 2023).

The aqueous phase reforming (APR) makes possible generating hydrogen molecules from organic matter in aqueous solution, once that, under APR operating conditions, the substrate is first converted to syngas, where this gas product is mainly composed of CO_2 and H_2 (Justicia et al. 2024). Next, other side reactions, such as Fischer–Tropsch or alkane formation, can take place to generate both liquid and gas by-products (Justicia et al. 2024). Electrolysis is a promising method to obtain pure hydrogen from biomass (Li et al. 2023). Two different technologies can be used with this purpose, which are microbial electrolysis cells (MEC) and proton exchange membrane electrolysis cells (PEMEC) (Li et al. 2023; Umer et al. 2024; Li et al. 2024). MEC can produce hydrogen from various substrates, and it is composed of two electrodes, cathode, and anode, which can be installed in a single chamber or two separated ones (Sharma et al. 2023).

MEC can use a broad range of biomass substrates/feedstock, but biogenic microbes cannot directly process the biomass, requiring a pretreatment or fermentation to produce low-molecular-weight substances (Umer et al. 2024). Additionally, electrode materials are expensive and low hydrogen yield provide significant challenge to scaling up this technology (Umer et al. 2024). PEMEC devices include a polymeric membrane that makes possible transferring protons to generate a significant amount of hydrogen gas with high-purity (Singh et al. 2023; Umer et al. 2024). PEMEC has being considered a promising method to generate pure hydrogen, but to enhance its efficiency and reduce the required electric potential, researchers have been exploring various catalysts and alcoholic substrates (Umer et al. 2024).

2.3 LIQUID FUEL

Biofuel is the term used for liquid fuels or fuels in the form of gases generally used in transportation. Although they can be produced from raw materials from renewable sources, greater attention is paid to those made from biomass (Demirbas 2008). The use of these fuels, in general, is relevant to the environment, as they emit fewer gases that are responsible for increasing the greenhouse effect than fuels obtained from petroleum derivatives. Among the liquid fuels, this topic will mainly address bioethanol (obtained from the alcoholic fermentation of sugars), biodiesel (obtained from the transesterification of vegetable oils) (Demirbas 2008), biojet fuel, biobutanol, and bio-oil. The main contemporary sustainability challenge is integrating new renewable options for energy production and preserving the environment's quality (Rebequi et al. 2023).

2.3.1 Bioethanol

Green ethanol, or bioethanol (CH_3CH_2OH or EtOH), is defined as a liquid biofuel being produced from carbohydrate sources and fermentation processes of these raw materials, and depending on the raw material, this bioethanol can be classified as first, second, and third generation (Melendez et al. 2022). First-generation bioethanol comes from raw materials from groups 1 and 2. Group 1 includes raw materials that include glucose and sucrose monosaccharides, and group 2 contains polysaccharides such as starch. Second-generation bioethanol will have lignocellulosic biomass as raw materials (containing cellulose, hemicellulose, and lignin). Third-generation bioethanol already includes algae cellulose (Guimarães et al. 2023).

Bioethanol is considered the fuel of the future because it comes mainly from renewable sources. It has a higher octane number than gasoline, so when mixed with gasoline, it can reduce the emission of gases such as NO_x, CO_2, and hydrocarbons during combustion; such elements contribute to the increase in the greenhouse effect. In addition, bioethanol has optimal fuel characteristics for the combustion engine and energy production (Shah and Rehan 2014). Biomasses are the most promising raw materials for bioethanol production, as they do not compete with food production and reduce the environmental impacts caused by the accumulation of these materials on the planet. For example, we have lignocellulosic biomasses. However, some bottlenecks, such as the pretreatment and saccharification stages, still need improvements to consolidate this process Shah and Rehan (2014). The pretreatment of biomass will deconstruct its primary fibers (cellulose, hemicellulose, and lignin), which together form a recalcitrant material resistant to the action of enzymes during saccharification. Thus, the pretreatment stage is crucial for obtaining bioethanol from these raw materials. These pretreatments can be classified as physical, chemical, biological, or combined; the choice of this stage depends on the condition and objectives required.

To produce bioethanol through the alcoholic fermentation process, you first need a carbon source, such as sugars, which can be obtained from raw materials from groups 1 to 4. Alcoholic fermentation is a super important step in the bioethanol production process because if it is not well controlled, it can compromise the success of obtaining the final product (Scheffers 1987). According to Scheffers (1987), alcoholic fermentation will occur mainly through yeasts capable of metabolizing sugars such as glucose through respiratory or fermentative pathways and even a combination of these two. The simplified reaction is shown below (Equations 2.18–2.19):

$$C_6H_{12}O_6 + 6\ O_2 \rightarrow 6\ CO_2 + 6\ H_2O\ \text{(respiration)} \tag{2.18}$$

$$C_6H_{12}O_6 \rightarrow C_2H_5OH + 2\ CO_2\ \text{(alcoholic fermentation)} \tag{2.19}$$

The cytosol is the enzymatic system where fermentation occurs, while the ability to breathe is involved with the mitochondria. Thus, when there is an absence of O_2, only fermentation continues. Yeasts such as *Saccharomyces telluric* v.d. Walt are deficient in the process of breathing, being only capable of fermenting, even in the presence of oxygen. However, most yeasts are able to perform respiration (Scheffers 1987). The success of bioethanol fermentation depends on the choice of the microorganism capable of carrying out the process. *Saccharomyces cerevisiae* yeasts are the most widely used to perform this function. However, they limit production because they have a low tolerance to the alcohol concentration to which they are exposed and only metabolize mono and disaccharides (Melendez et al. 2022), as well as hexose sugars, not being effective for pentose sugars.

According to Nassir and Al-Sahlany (2021), most microorganisms can produce bioethanol. Still, the ability to tolerate moderate concentrations of ethanol in the substrate will depend on the type of microorganism used, and this is also true for the sugars that will be fermented. For example, Lactobacilli will be better able to ferment lactose, while yeasts such as *Saccharomyces cerevisiae* will ferment glucose better. Some will break down polysaccharides such as cellulose into simple sugars to produce sugars, while others can ferment pentoses and bioethanol. Some research has studied different microorganisms to improve the efficiency of bioethanol production; among them, we list some that will be explained in more detail below. According to Jwad and AI-Haidari (2017), studies on yeasts have intensified since bioethanol is produced mainly by the action of this class of microorganisms. *Saccharomyces cerevisiae* is the most consolidated so far, but studies carried out on the action of *Saccharomyces carlsbergensis* for ethanol production have been considered. Hu et al. (2005) studied the ethanol tolerance of two mutant microorganisms, *Schizosaccharomyces pombe* and *Saccharomyces cerevisiae* (called SPSC fusant for short) in ethanol concentrations of 18% (v/v) at a temperature of 30°C. These authors found that palmitic acid present in the composition of yeast plasma membranes may be related to good ethanol tolerance; those yeasts with higher concentrations of this fatty acid presented more excellent resistance to ethanol. Kamini and Gunasekaran (1987) produced bioethanol by studying a monoculture of Kluyveromyces fragilis NRRL 665 and a coculture of *Zymomonas mobilis* using lactose as substrate, starting from a substrate concentration of 200 g/L. The monoculture of *Kluyveromyces fragilis* produced 36 g/L of ethanol, while the coculture reached a concentration of 55.2 g/L of bioethanol. However, the coculture had an increase in residual sugar (galactose); it was still more efficient than using only the monoculture.

2.3.2 Biodiesel

Several studies have examined the use of natural vegetable oils as an alternative source of energy in recent decades (Kalayasiri et al. 1996), as they are promising raw materials for the production of biodiesel. Biodiesel is a natural alternative to petroleum diesel that can be obtained from renewable sources such as vegetable oils, animal fats, and fats used in food preparation (cooking). In chemistry, it is characterized as a monoalkyl ester of fatty acids from naturally occurring lipids (Ramos et al. 2003). It can be obtained, together with glycerin, through the oxidation reaction in the presence of ethanol or methanol. The triacylglycerols (or triglycerides) are combined, and an acid or base catalyst is needed to carry out the reaction (Ramos et al. 2003). The transesterification technique is often used to reduce viscosity in the production of biodiesel. This procedure reduces the viscosity of oil or fat using an acid or base catalyst in the presence of methanol or ethanol (Mathiyazhagan et al. 2011). Through the transesterification reaction, biodiesel can be produced from fats of vegetable and animal origin. The molar ratio of alcohol influences transesterification, the presence of water, and the content of Free Fatty acids, in addition to the reaction temperature, catalyst concentration, and stirring speed (Mathiyazhagan et al. 2011). We have already mentioned the main raw materials used in biodiesel production. However, its progress is limited due to the scarcity of raw materials, low economic gains, long life cycles, and the effect on the price of agricultural products, availability of fertile land, and food security (Zulqarnain et al. 2021).

Microbial lipids from oleaginous microorganisms are great substitutes for traditional raw materials in biodiesel manufacturing. However, it is essential to identify inexpensive sources of nutrients for these oleaginous microbes' growth to be sustainable and commercially viable resources (Zulqarnain et al. 2021).

Some interesting works on biodiesel production will be described in the following text.

Liu and Zhao (2007) developed a new biodiesel production method where direct methanolysis was promoted by using cellular biomass acid from oleaginous yeasts and filamentous fungi. The methanol dosage and the catalyst concentration, temperature, and reaction time were optimized. The authors obtained a 98% yield through this process. Lignocellulosic biomasses are promising alternatives for producing biodiesel because they are cheap and renewable. Thus, they can be used to produce low-cost microbial oil that can be used to produce biodiesel (Chintagunta et al. 2021).

Wang et al. (2021a) describe that there are two obstacles in the industrialization of oil production by oleaginous microorganisms from lignocellulosic biomass: (1) Develop adequate pretreatment methods to enhance the effectiveness and reduce the production of inhibitors, to produce more effective compounds to be used by microorganisms. (2) Select or modify microorganisms that reproduce rapidly and accumulate products of interest in the presence of inhibitors.

Microorganisms capable of synthesizing and accumulating lipids exceeding 20% of the cell's dry weight are commonly referred to as oleaginous microorganisms. Microbial lipids, called single-cell oils (SCOs), are oils produced from carbohydrates by oleaginous microorganisms under certain circumstances and stored in the cells. Triacylglycerols (TAGs) represent the most common form of SCOs (Wang et al. 2021b). Oleaginous microorganisms, including yeasts, fungi, algae, and bacteria, are abundant and varied. In particular, yeasts can metabolize a wider variety of carbon sources, which makes them the most suitable for industrial lipid production. As reported in the literature, only 4% of the 1600 yeast species have been recognized as oleaginous microorganisms. Thus, the search for new high-yield microorganisms is crucial for industrialization (Wang et al. 2021a).

2.3.3 Biojet Fuel

The growing demand for sustainable fuels and pressure to reduce greenhouse gas emissions have driven research and development into alternatives to traditional fossil fuels in aviation. Biojet, also known as biojet fuel, is one such alternative that has gained prominence. This renewable fuel is produced from biomass and is designed to be used directly in aircraft engines without major modifications to existing infrastructure.

The use of biojets in commercial aviation has gradually increased, with several collaborative initiatives being initiated by governments, airlines, and refineries. Currently, several biojet production routes have been approved by the American Society for Testing and Materials (ASTM) for use in blends with conventional jet fuel, such as the HEFA (Hydroprocessed Esters and Fatty Acids) process and the Fischer–Tropsch (FT) process. HEFA is the most widely used, accounting for over 90% of global biojet production (Peters et al. 2023). Commercial aircraft operating on a blend of biojet and fossil fuel have flown thousands of flights since 2011, when the first authorization for using biokerosene in commercial operations was granted. Airlines such as United Airlines, KLM, and Lufthansa have pioneered the use of biojet on commercial flights. United Airlines, for example, was the first US airline to operate scheduled flights using biojet, supplied by AltAir Fuels (Dyk and Saddler 2021). Despite these advances, biojet still faces significant challenges. The biggest of these is the cost of production, which is still substantially higher than that of conventional jet fuel. Studies show that biojets can be up to three times more expensive than fossil fuels, mainly due to the cost of raw materials and complex conversion processes. HEFA, for example, relies heavily on vegetable oils such as palm or soybean oil, which are commodities subject to price volatility and environmental concerns. This dependence on specific raw materials also puts biojet production in direct competition with food production, raising questions about long-term sustainability (Lim et al. 2023). Another challenge is limited production capacity. Although several biojet refineries operate, more than the production scale is needed to meet the growing global demand for sustainable aviation fuels. In 2019, global biojet production was estimated at just 50 million liters, compared to the 300 billion liters of fossil fuel consumed by global aviation. This mismatch in supply and demand creates a significant bottleneck for the widespread adoption of biojet. The regulatory landscape is adjusting to support the growth of biojet. One example is the International Civil Aviation Organization's (ICAO) Carbon Offsetting and Reduction Scheme for International Aviation (CORSIA) program, which came into effect in 2021. CORSIA sets targets to stabilize net CO_2 emissions from international aviation from 2020 onwards, encouraging sustainable fuels such as biojet. ICAO's long-term goal is to reduce emissions from the sector by 50% by 2050, compared to 2005. In addition, the European Union (EU) has established the "Green Deal," a policy that aims to make the continent carbon neutral by 2050. This strategy involves increasing the production and use of advanced biofuels, including biojet. The ReFuelEU Aviation Initiative, launched in 2021, aims to oblige airlines operating in the EU to use an increasing percentage of sustainable aviation fuels (Grimme 2023). In the United States, the Federal Aviation Administration (FAA) is investing in research and development through the CLEEN (Continuous Lower Energy, Emissions, and Noise) program, which supports technological innovations to reduce emissions in aviation, including the use of biojet. The program's goals are to increase domestic production of sustainable fuels and reduce the cost of production, making biojet more economically competitive. The prospects for biojet are closely linked to technological advancement, innovation in feedstocks, and the implementation of effective public policies. The coming years promise to be a period of accelerated growth in the production and use of biojets as the technology matures and the pressure to reduce emissions from the aviation sector increases. A key area of innovation in the biojet field is the development of new, more efficient production routes that are less dependent on traditional feedstocks. Technologies such as Alcohol-to-Jet (AtJ), which converts alcohols (such as ethanol or butanol) into jet fuel, have shown great potential. This technology allows using a wide range of feedstocks, including lignocellulosic residues and gaseous industrial wastes. LanzaTech, for example, is developing processes to convert waste gases from steel industries into biojet (ICAO 2024). Another promising innovation is the use of algae as a feedstock. Algae can grow on non-arable land and can be cultivated using brackish or wastewater, making them a sustainable source of biomass. In addition, algae have a high growth rate and can produce large quantities of lipids, which are converted into hydrocarbons for biojet. In the long term, new conversion technologies, such as more efficient catalysts and metabolic engineering processes, are expected to improve the efficiency and reduce the cost of biojet production. Research in synthetic biology is being conducted to optimize microorganisms capable of converting biomass into usable compounds for biojet production, accelerating the development of biotechnological solutions (Dahal et al. 2021). For biojet to become a viable alternative on a large scale, it will be necessary to expand the production infrastructure. Developing biofuel refineries with the capacity to produce biojet on a large scale is crucial. The recent expansion of the Neste refinery in Finland, which aims to increase the production of biofuels, including biojet, is an example of how the industry responds to the growing demand (Panoutsou et al. 2021). In addition, for biojets to reach the necessary scale, promoting sustainable biomass supply chains is essential. This includes developing integrated biomass cultivation systems that minimize environmental impact and ensure a constant feedstock supply. Implementing financial incentive policies, such as subsidies and tax credits, and introducing carbon pricing systems will make biojet competitive with fossil fuels. Carbon pricing, in particular, can play a crucial role in internalizing the environmental costs associated with GHG emissions, making biojet a more economically attractive option for Airlines (Chiaramonti et al. 2014). The success of these policies depends on effective coordination between governments, international bodies, and industry. Existing initiatives, such as CORSIA and the EU Green Deal, can serve as platforms for implementing stricter regulations, mandating the increased use of sustainable fuels. Biojet production involves several technological routes, each with its advantages and challenges. The main technological routes include HEFA Process (Hydroprocessed Esters and Fatty Acids). The HEFA process is one of the most established routes for biojet production. In this process, vegetable oils or animal fats are hydrotreated to remove oxygen and produce hydrocarbons suitable for use as jet fuel. HEFA technology is widely used due to its ability to produce fuels with properties very similar to conventional jet fuel. However, a disadvantage of this process is its reliance on vegetable oils and fats, which sometimes still compete with food-grade fuels (Vela-García et al. 2022). HEFA has been implemented in several industrial plants around the world. The AltAir Fuels refinery in the United States is a prominent example of this technology used to supply biojets to airlines, including United Airlines. The Fischer–Tropsch process, a thermochemical conversion technology, holds significant promise for biojet production from a variety of feedstocks, including forestry residues and agricultural waste. In this process, biomass is first converted into syngas (a mixture of carbon monoxide and hydrogen) through gasification. The syngas is then transformed into liquid hydrocarbons in a catalytic Fischer-Tropsch reaction (Gillingham and Stock, 2018). While the FT process holds significant potential for large-scale biojet production, it is not without its challenges. The process requires complex infrastructure and significant investment. Furthermore, conversion efficiency remains a hurdle compared to other production routes. Biojet offers a promising solution to decarbonize the aviation sector, but its large-scale adoption depends on technological advances, effective policies, and overcoming economic challenges. The field of biojet technology is ripe with potential and excitement, and continued innovation is key to unlocking this potential. With the right regulatory support and continued innovation, biojet is expected to play a growing role in global efforts to reduce carbon emissions and achieve sustainability in aviation. The coming decades will be critical to cementing biojet as an integral part of the energy transition, contributing to a lower carbon and more efficient future in the aviation sector (Why et al. 2019).

2.3.4 Biobutanol

Biobutanol (C_4H_9OH) is a colorless, highly flammable alcohol with four distinct isomers: n-butanol (1-butanol), 2-butanol (sec-butanol), 2-methyl-1-propanol (iso-butanol), and 2-methyl-2-propanol (tert-butanol). While these isomers share the same chemical formula, they differ in production methods, energy content, and combustion characteristics (Dharmaraja et al., 2020). Biobutanol holds significant potential as a biofuel due to its properties being similar to gasoline, with an air-to-fuel ratio of 12.2, an energy density of 29.2 MJ/L, and a research octane number of 96. Blends of butanol with gasoline, up to 80% v/v, are compatible with current internal combustion engines without major modifications (Kumar et al. 2024). This makes biobutanol a promising alternative for reducing dependence on fossil fuels while ensuring compatibility with existing fuel infrastructure. Currently, the majority of butanol production worldwide relies on petroleum-derived chemicals, which makes the industry vulnerable to fluctuations in crude oil prices. In 2021, the global market volume for n-butanol, the most commonly commercialized isomer, reached approximately 5.2 million metric tons. This market is projected to continue growing over the next decade (Serrano et al. 2024). However, the commercial production of biobutanol primarily relies on starch- or sugar-based crops, which is classified as first-generation biobutanol. Due to the high cost of feedstock, this type of biobutanol is not competitive with fossil fuels (Jin et al. 2022). The reliance on food crops raises concerns about food security and market fluctuations, further impacting its economic viability compared to traditional fuels. As a result, there is a growing interest in developing second-generation biobutanol from lignocellulosic biomass, which could provide a more sustainable and cost-effective alternative. Biobutanol production primarily utilizes the Acetone–Butanol–Ethanol (ABE) fermentation process, commonly known as the Weizmann process (Kallarakkal et al. 2021). This method involves various strains of *Clostridium* species, which are recognized for their effectiveness in fermenting carbohydrates into biobutanol, acetone, and ethanol. The process has a long-standing history and has been successfully scaled up for industrial applications, demonstrating its viability as a commercial method for biobutanol production (Vamsi Krishna et al. 2022). The process typically involves several stages: hydrolysis of lignocellulosic biomass to release fermentable sugars, followed by fermentation under anaerobic conditions to produce butanol.

Many agricultural residues have been explored and utilized for biobutanol production through fermentation processes (Karthick and Nanthagopal, 2021). Common feedstocks include corn stover, sugarcane bagasse, wheat straw, and rice husks, which are rich in lignocellulosic materials. These residues are advantageous as they are abundant and typically considered waste products, making them cost-effective alternatives to food-based feedstocks (Kumar et al. 2020; Zhang et al. 2021). As evidenced by the recent studies, the exploration of various lignocellulosic biomass and agricultural waste materials for biobutanol production demonstrates significant potential for sustainable biofuel development. Zetty-Arenas et al. (2021) investigated the co-fermentation of mixed sugars from sugarcane bagasse hemicellulosic hydrolysate (CHH) and molasses (SCM) for butanol production using *Clostridium saccharoperbutylacetonicum* DSM 14923. The study found that a CHH-to-SCM ratio of 25:75 led to optimal sugar consumption (97%) and the production of 7.8 g/L butanol and 9.8 g/L ABE after 30 hours of fermentation. Su et al. (2022) investigated the effectiveness of using orange peel waste as a cofactor in ABE fermentation, resulting in *Clostridium* sp. *BOH3* producing 17.7 g/L of butanol and 27.6 g/L of solvents from rice waste. This method significantly enhanced production efficiency and reduced feedstock costs by up to 97.7%, while also accelerating the fermentation process. The study by Muharja et al. (2023) investigated the effects of various treatments on extracting reducing sugars and biobutanol from cocoa pod husk, achieving a sugar concentration of 9.24 g/L and butanol and ABE concentrations of 20.4 g/L and 54.4 g/L, respectively. The incorporation of surfactants and L-cysteine significantly improved enzymatic hydrolysis and fermentation efficiency, showcasing effective strategies for valorizing cocoa pod husk into biofuels. Tekin et al. (2023) investigates the potential of industrial tea waste for biobutanol production using *Clostridium beijerinckii*. Key parameters such as initial sugar loading, fermentation time, nitrogen sources, and mineral salts were optimized, resulting in maximum ABE production of 9.73 g/L and biobutanol yield of 6.21 g/L with specific nutrient concentrations. This research is significant as it presents tea waste as a promising feedstock for biobutanol, showing improved productivity and yield compared to standard media. Luo et al. (2023) investigated the use of butyric acid as a catalyst for the pretreatment of corn stover, achieving significant xylan removal (83.17%) and a fermentable sugar yield of 97.16%. The study found that this pretreatment enhanced the enzymatic digestibility and crystalline index of the biomass. Additionally, the residual butyric acid was effectively reused as a co-substrate for biobutanol production, yielding 8.38 g/L of butanol and 12.63 g/L of solvents, demonstrating a promising strategy for co-producing fermentable sugars and biofuels from lignocellulosic biomass. Chen et al. (2024) demonstrated that combining different lignocellulosic biomass types, such as sugarcane bagasse, corncob, and pine with wheat bran, enhances nitrogen content in the hydrolysate, leading to increased biobutanol production. Integrating pervaporation with ABE fermentation further reduced product inhibition, achieving a total solvent concentration of 14.51 g/L. Collectively, these studies pave the way for more efficient and cost-effective strategies in biobutanol production, contributing to the advancement of renewable energy solutions. The production of biobutanol from lignocellulosic biomass faces various challenges that need addressing for improved sustainability and cost-effectiveness. Key obstacles include the inherent resistance of lignocellulosic materials, which complicates the effective release of fermentable sugars, necessitating advanced pretreatment methods. Additionally, optimizing fermentation processes to boost biobutanol yields while minimizing product inhibition is crucial. Economic viability is often hindered by high processing costs, especially when compared to traditional fossil fuels. Overall, advancements in biotechnology and operational efficiencies are essential for realizing the potential of lignocellulosic biomass as a renewable energy source.

2.3.5 Bio-oil

Bio-oil is a liquid product derived from the thermal decomposition of biomass, composed mainly of oxygenates such as acids, aldehydes, ketones, and phenols. Since the beginning of the 21st century, bio-oil has attracted significant attention as a renewable alternative to fossil fuels due to its potential for use in various energy and chemical applications. This topic explores the historical development of bio-oil, the current scenario of its production and use, the primary conversion technologies, prospects, and the challenges its widespread adoption faces (Cao et al. 2017). The history of bio-oil is intrinsically linked to the development of pyrolysis, one of the critical technologies for converting biomass into liquid fuels. The concept of pyrolysis dates back to experiments in the 19th century when scientists began to explore the thermal decomposition of organic materials. However, bio-oil production from biomass only gained industrial relevance in the 1980s, amid growing concerns about dependence on fossil fuels and the need to diversify energy sources (Lachos-Perez et al. 2023).

In the 1990s and 2000s, researchers began focusing on developing fast pyrolysis technologies capable of converting biomass into bio-oil efficiently and with high yields. During this period, several laboratories and industries, especially in North America and Europe, carried out pilot tests and laid the foundations for the commercial use of bio-oil in boilers and engines. The hope was that bio-oil, with its potential to replace petroleum derivatives partially, could significantly reduce our dependence on fossil fuels and mitigate environmental concerns (Su et al. 2022). The increased interest in bioenergy, driven by greenhouse gas emission reduction targets, has brought new opportunities for bio-oil in the context of the low-carbon economy. In addition, bio-oil production has been considered a promising alternative for valorizing agricultural and forestry residues, such as sugarcane straw, corn bagasse, and wood residues. The production and use of bio-oil still need to be improved compared to other renewable energy sources, such as biodiesel and ethanol. However, advances in research and development of conversion technologies and improvements in policies to support biofuels have increased the viability of bio-oil as part of the global energy matrix. Notably, countries such as the United States, Canada, Finland, and Brazil have been at the forefront of initiatives to increase bio-oil production and its integration into sectors such as power generation, transportation, and the chemical industry, shaping the global bio-oil landscape (Zerva et al. 2021).

In the United States, bio-oil is included in the portfolio of advanced biofuels under the Renewable Fuel Standard (RFS). This program encourages the production of biofuels that significantly reduce carbon emissions compared to fossil fuels. Likewise, the US Department of Energy (DOE) continually invests in research into optimizing pyrolysis processes and upgrading bio-oil to increase its applicability in the transportation sector and refineries. In Europe, initiatives such as the European Green Deal and the Renewable Energy Directive (RED II) also provide a favorable regulatory framework for producing and using bio-oil. The European Union has set ambitious targets to reduce its greenhouse gas emissions, and bio-oil, derived from agricultural and forestry residues, has the potential to play an essential role in the transition to a circular economy (Wang et al. 2022a). Bio-oil is obtained mainly through thermochemical processes, decomposing biomass into liquid, solid (biochar), and gaseous fractions. The most commonly used methods are pyrolysis, hydrothermal liquefaction, and gasification, followed by liquid synthesis. Each technology offers different advantages and challenges, depending on the type of biomass used, the operating conditions, and the desired applications (Kang et al. 2013). Fast pyrolysis is the most common process for bio-oil production, accounting for up to 70%–75% of bio-oil yield from dry biomass. The process involves rapidly heating the biomass (around 450°C–600°C) without oxygen, which causes its thermal decomposition into liquid, solid, and gaseous fractions. The main characteristic of fast pyrolysis is the extremely short residence time of a few seconds, which is required to maximize bio-oil production (Li et al. 2004). Bio-oil resulting from fast pyrolysis is a dark, viscous liquid composed of a complex mixture of oxygenated compounds. However, technical limitations, such as its chemical instability and high acidity, make it difficult to use directly as a fuel in conventional engines. Therefore, upgrading processes is necessary to increase its stability and fuel properties.

Bio-oil upgrading involves catalytic reactions, such as hydrodeoxygenation (HDO), which aims to reduce the oxygen content, increasing its stability and calorific value. Hydrodeoxygenation is a process similar to the hydrotreating of fossil fuels, which removes oxygen through reactions with hydrogen under high pressures and temperatures, resulting in more stable hydrocarbons (Tippayawong et al. 2008).

Hydrothermal liquefaction (HTL) is a thermochemical process that converts wet biomass into bio-oil under subcritical or supercritical water conditions (200°C–374°C and 5–25 MPa). Unlike pyrolysis, HTL can directly process biomass with high moisture content, eliminating the need to pre-dry the feedstock and reducing processing costs (Gollakota et al. 2018). Hydrothermal liquefaction's main advantage is producing a bio-oil with lower oxygen content and more excellent stability than bio-oil from pyrolysis. In addition, the process is flexible and can be applied to different types of biomass, including algae, agro-industrial residues, and forestry residues. However, the process requires high pressure and superheated water, which increases operational complexity and investment costs. Gasification is another thermochemical process used to convert biomass into bio-oil, although it is most often associated with biogas and electricity production. Gasification occurs at temperatures above 700°C under partial oxidation conditions, resulting in the production of synthesis gas (syngas), a mixture of carbon monoxide (CO) and hydrogen (H_2) (Osman et al. 2021). Syngas can be transformed into bio-oil using the Fischer–Tropsch (FT) process, which converts syngas into long-chain liquid hydrocarbons. The FT process is widely used in the petrochemical industry but can also be adapted to produce biofuels from biomass-derived syngas.

Bio-oil production via gasification and FT synthesis is advantageous regarding feedstock flexibility, as the process can utilize a variety of agricultural and forestry residues. However, the conversion process involves multiple steps and requires substantial infrastructure investments, which makes this technological route more suitable for large-scale applications. Bio-oil prospects are promising, especially in the global transition to a low-carbon economy. Technologies for bio-oil production are continuously evolving, emphasizing improving the efficiency of conversion processes, reducing production costs, and increasing the quality of the final product. The following explores some trends and innovations that may shape the future of bio-oil (Maytorena et al. 2024). One of the most promising research areas is the development of more efficient catalysts for bio-oil upgrading. Bio-oil's chemical stability and high oxygen content limit its direct use as a fuel, making hydrodeoxygenation (HDO) processes essential for its large-scale application.

Researchers are exploring new types of heterogeneous catalysts, such as metal oxides and precious metals, which can increase the selectivity of reactions and reduce the energy required for deoxygenation. In addition, using regenerative and long-life catalysts can help reduce the operational costs associated with bio-oil upgrading, making the process more economical (Cheng et al. 2016). An important trend is the integration of biorefineries that use bio-oil as a platform for the production of biofuels and value-added chemicals. Biorefinery is a concept inspired by the oil refinery model, where multiple products are obtained from a single feedstock. Bio-oil can be used as a precursor not only for fuels but also for chemical products such as solvents, resins, and phenolic compounds. Integrating biorefineries increases the economic viability of bio-oil production, allowing for product diversification and resource optimization. This approach aligns with the principles of the circular economy, in which waste is valued and reused in different production chains (Arias et al. 2023).

In addition to its traditional use as fuel in boilers and engines, bio-oil is being explored for new applications. Bio-oil can be used as a raw material for the production of bioplastics, lubricants, and chemical additives, expanding its market beyond the energy sector (Sánchez-Borrego et al 2021). Chemical companies are investigating bio-oil use in polymerization processes to create biodegradable plastics, while the automotive sector is exploring bio-oil use in lubricating oils and hydraulic fluids. These new markets could increase the demand for bio-oil significantly as sustainability and the circular economy gains importance in global industries (Okolie et al. 2023). Although bio-oil has excellent potential to contribute to the sustainable energy matrix, several challenges still need to be overcome before it becomes a viable alternative on a large scale. One of the most significant limitations of bio-oil is its chemical instability, caused by the high content of oxygenated compounds, such as acids and alcohols. These compounds make bio-oil corrosive and unstable during storage, which limits its application in internal combustion engines and turbines (Ameh et al. 2024).

2.4 SOLID FUEL

Solid biofuels derived from lignocellulosic biomass play a crucial role in sustainable energy production, especially in the context of heat generation, soil amendment, and carbon sequestration. One of the most prominent solid biofuels is biochar, a product of biomass pyrolysis.

2.4.1 Biochar

Biochar is a carbonaceous material that can be used to produce energy (a sustainable fuel); however, biochar can also be used in soil remediation and improvement, wastewater treatment, and carbon capture and sequestration. Biochar can be defined as charcoal, in general terms, a carbon-rich material produced principally by pyrolysis (in the absence or minimal presence of oxygen) of carbonaceous materials such as plants, including lignocellulosic materials. Figure 2.2 summarizes all this information (Bhattacharya et al. 2024).

According to the International Biochar Initiative (IBI), biochar is a solid substance generated through the pyrolysis of biomass in an oxygen-limited environment. (Laird et al. 2009). The term biochar is equivalent to charcoal when biochar plays the same role as charcoal when used for specific purposes (Sohi et al. 2010).

Another term, often found in the literature, is hydrochar, which is a carbonaceous material (similar to biochar), however, it is obtained by a different process. Hydrochar is made through a hydrothermal carbonization process (carbonization

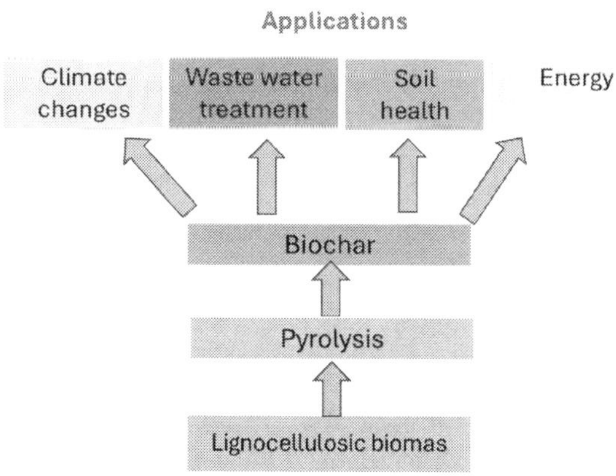

FIGURE 2.2 Biochar production application.

made in the presence of steam, at high pressures and temperatures), resulting in a slurry. The physicochemical properties of biochar and hydrochar are quite different (Zhang et al. 2018). Biochar properties such as porosity (surface area), cation exchange capacity, acidity/basicity, surface charge (point of zero charge, pH_{PZC}), and superficial functional groups depend on feedstock material, synthetic route as well as operating conditions employed during the preparation method. Most of the research reported on biochar focuses on the pyrolysis of biomass. However, in the current literature, further methods are reported such as gasification, hydrothermal carbonization, etc. (Maitlo et al. 2022). The pyrolysis of biomass (e.g., lignocellulosic materials), produces three main components, thus including a solid (biochar), liquid (bio-oil), and gas (syngas). The syngas (H_2/CO) produced by pyrolysis also includes CO_2, and other low molecular gases, after processing the syngas produced in this way can be used in gas engines. The bio-oil is composed of water, alcohols, aliphatic and aromatic hydrocarbons, phenolic compounds, and nitrogenous compounds such as pyridine, pyrazine, amines, among others. Bio-oil can also be used as fuel in boilers (Maitlo et al. 2022; Kant Bhatia et al. 2021).

Another way to produce biochar is by gasification of biomass. According to Kant Bhatia et al. (2021), the surface area of biochar produced by this process is smaller than the biochar produced by the pyrolysis process; and less important functional groups such as hydroxyl, carbonyl, and carboxyl groups are formed. Biochar has a wide range of applications, making it a valuable resource in various fields. In agriculture applications, biochar has found applications in soil health improvement and plant yield. The benefit of increasing yields without the addition of synthetic fertilizers, or other additives, is very important for sustainable agriculture. The porous nature of biochar favors water and water-soluble nutrient retention, and beneficial soil organisms. Biochar also increases the cation exchange capacity, surface area, and pH of the soil. In some cases, biochar is mixed with compost or other compostable materials (Das et al. 2023).

The use of fossil fuels is leading our planet to global warming, in this regard, biochar can help to overcome this problem, as biochar can reduce emissions of greenhouse gases from burning fossil fuels or even from biomass combustion, as biochar is a stable porous material that can stay in the soil for many years. Biochar acts as an adsorbent material, drawing carbon dioxide from the atmosphere, thus acting as a carbon sink on agricultural lands. Furthermore, biochar obtained from biomass releases carbon dioxide, but at a low rate when compared to biomass combustion (Ghanbarpour Mamaghani et al. 2023; Liu et al. 2022).

There are many ways for the application of biochar in renewable energy production, for example, biochar can be used as a catalyst in transesterification reactions to produce biodiesel (Chanathaworn et al. 2022; Zhao et al. 2023), as an electrode (anodic/cathodic) in a microbial fuel cell (Balajii & Niju, 2019), as catalytic support in syngas production (Wang et al. 2022b), in anaerobic digestion for energy production (biochar enhances electron transfer) (Sangrulkar et al. 2023), and directly used as fuel to produce energy (Maitlo et al. 2022). In general terms, the studies reported in the current literature suggest that the utilization of biochar in various processes improves efficiency and productivity; however, there are still many of its applications that need to be realized, also, the influence of the preparation method/starting material on the final properties of biochar is now very clear, thus suggesting that more studies are needed.

2.4.2 Electricity (Microbial Fuel Cell/Microbial Electrolysis Cell)

Bioelectricity production from lignocellulosic biomass involves converting organic plant material into electricity. Several key technologies like direct combustion, co-firing, gasification, anaerobic digestion, and pyrolysis, and emerging technologies, such as fuel cell, can be used to achieve, each with several advantages and disadvantages (Evans et al. 2010; Haq et al. 2021).

Direct combustion is the most straightforward and widely used method for generating electricity from biomass once it is the cheapest process (Haq et al. 2021). Lignocellulosic biomass is directly burning in a furnace or boiler to produce heat, which can then be used to produce steam which drives a turbine connected to a generator, producing electricity (Haq et al. 2021). The drawback of this process are the requirement of high-quality biomass feedstock and high ash content, which needs disposal and causes serious health problems (Patel et al. 2016). To decrease the environmental impacts of the directly combustion, the co-firing of lignocellulosic biomass with fossil fuels has been applied (Awasthi and Bhaskar, 2019). However, these are the most inefficient process (less than 15%) because the operational conditionals are uncontrolled (Evans et al. 2010; Wang et al. 2021a).

More efficient methods such as advanced gasification, pyrolysis, and anaerobic digestion have emerged as promising solutions to produce high-quality biomass-based fuels that can be used as an energy source for turbines (Liu et al. 2020; Wang et al. 2021b). In gasification, lignocellulosic biomass is heated in a low-oxygen environment, producing syngas (a mixture of carbon monoxide, hydrogen, and methane) (Haq et al. 2021). In pyrolysis, the lignocellulose biomass is thermally decomposed at high temperatures in the absence of oxygen, resulting in bio-oil, syngas, and biochar (Haq et al. 2021). In anaerobic digestion, microorganisms break down organic material in the absence of oxygen, producing biogas (mainly methane and carbon dioxide) (Wang et al. 2021a). The gas products and the bio-oil produced in can be burned in a gas turbine or internal combustion engine to generate electricity. These technologies, not only optimize the conversion of biomass into fuels but also improve energy yields, achieving more than 30% of efficiency, and reduce waste and emissions (Evans et al. 2010). Even being zero net of carbon process, the energy generation from lignocellulosic biomass can vary from many factors, such as the technologies used, for example, when gasification was used to produce energy, a reduction of 50% of the greenhouse gas emission was observed compared to direct burning of this feedstock (Shen et al. 2015; Balcioglu et al. 2023).

Even there are many lignocellulose biomasses plants to generate electricity in operation in the world, it has been globally observed over the past two decades that some units have either been canceled (planned or commissioned but never constructed and operated) or have failed (operated for several years but not reaching their intended lifespan) (Han and Chen 2024). The reasons for this are the high capital cost necessary to build several operation units in unique plant and huge amount of biomass is required due to low energy density of biomass which increase the operational costs (collection, transportation storage) (Liu et al., 2020; Wang et al., 2021a).

However, among the different methods for bioelectricity production from lignocellulose biomass, the microbial fuel cell (MFC), which convert chemical energy stored in the feedstock to electricity using microorganisms to catalyst the process, has been more attractive than them due to ambient operation conditions, high energy efficiency, and the capability to co-generate bioelectricity alongside bioremediation (Liu et al., 2020; Sani et al.; 2021; Wang et al., 2021a; Shrivastava and Sharma, 2022; Sharma et al., 2023). MFC are a bioelectrochemical devices where electrogenic bacteria can transfer electrons derived from the oxidation of organic feedstock to electrode (called the anode) and then travel through an external circuit to another electrode (the cathode), where they combine with protons and reducing agent, like oxygen to form water (Sani et al.; 2021; Wang et al., 2021b; Shrivastava and Sharma, 2022; Sharma et al., 2023). This electron flow through the external circuit generates electrical power. To keep the neutral charge, protons pass through a membrane that separates the anode chamber from the cathode chamber, typically called a proton exchange membrane (PEM) (Sani et al.; 2021; Wang et al., 2021a; Shrivastava and Sharma, 2022; Sharma et al., 2023a).

Lignocellulose biomass has directly fed the anode chamber in MFC as a source of nutrient and organic matter (Zhao et al, 2017; Wang et al., 2021a). However, as can be seen in Table 2.1, the power outputs are too little for economically feasibility because this material has a recalcitrant structure, that impede its direct breakdown by microorganisms (Sani et al., 2021, Wang et al., 2021b). A critical step to improve the performance lignocellulose biomass on MFC, up to 25%, is the pretreatment (see Table 2.1), where the complex structure of this material is broken in simple compounds, such as saccharides, degraded phenolic compounds, organic acids including acetic acid from hydrolysis of acetyl groups, and furan derivative (Yoshimura et al., 2018; Wang et al., 2021a; Shrivastava and Sharma, 2022; Sharma et al., 2023b). Additionally, it is worthily to mention that several parameters affect the power density obtained by MFC, like, temperature, microbial community, structure of electrode, proton exchange membrane/ion exchange membrane. Because of this the challenge of this technologies is the competition for the substrate between the exoelectrogenic and methanogenic and fermentative microorganisms (Sani et al., 2021).

Another use of lignocellulose biomass in MFC is to produce biocompounds in the cathode chamber that enhance the oxygen reduction reactions and the biodegradation of recalcitrant compounds (Shrivastava and Sharma, 2022). White rot fungi have been applied in this compartment of MFC to produce enzymes, like laccase that oxidize difficult compounds, such as phenolic substance that are harmful to microbial community (Pal and Sharma, 2020). In addition, fungi can release some substance that serve as mediator and catalyze the reduction reaction and enhancing the performance of the catholyte and improved the power density produced by MFC (Pal and Sharma, 2020; Shrivastava and Sharma, 2022, Ottoni et al., 2023). In some cases, the power outputs of MFC using the fungi can be doubled than the MFC control without

TABLE 2.1
Summary of Studies Reporting the Use of Untreated and Pretreated Lignocellulosic Biomass as Feedstock in MFC

Lignocellulose Biomass	Pretreatment	Maximum Power Density	References
Cattle manure	No	36.6 mW/m^2	Lee and Nirmalakhandan (2011)
Rice straw	No	190 mW/m^2	Gurung and Oh (2015)
Cassava peels	No	29 mW/m^3	Adekunle et al. (2016)
Rice bran	No	16.5 mW/m^2	Yoshimura et al. (2018)
Corn bran	No	23.75 mW/m^2	Mohd Zaini Makhtar and Tajarudin (2020)
Banana peel	No	12.65 mW/m^2	Mohd Zaini Makhtar and Tajarudin (2020)
Palm oil mill effluent	No	22.03 mW/m^2	Mohd Zaini Makhtar and Tajarudin (2020)
Sugarcane bagasse	No	8.78 W/m^3	
Corn cobs	No	6.73 W/m^3	
Bamboo	Yes	578 mW/m^2	Dai et al. (2021)
Corn straw	Yes	660 mW/m^2	Wang et al. (2017)
Orange peel	Yes	358.8 mW/m^2	Miran et al. (2016)

fungus in the cathode chamber (Ottoni et al., 2023). From this perspective, the cost of MFC can be reduce because the metal-costly catalysts can be replaced through these fungi making the continuous-flow systems usable in large scale (Shrivastava and Sharma, 2022, Ottoni et al., 2023).

More recently, lignocellulose biomass has been transformed into electrode material in MFC with excellent electrical conductivity through straightforward carbonization once it is rich in carbon atoms and it has a porous structure that improve the microorganisms' adhesion, the electrode reaction, and electron transfer (Wang et al., 2021a; Katam et al., 2023). In the literature, the electrode derived from lignocellulose biomass, such as palm fiber, water hyacinth and silver grass, have obtained up to four times higher energy than traditional carbonaceous electrodes, for example, carbon cloth (Allam et al., 2020; Rethinasabapathy et al., 2020; Song et al., 2024).

Looking ahead, integrating the electricity with other chemical compounds productions has the potential to revolutionize the MFC technology (Moradian et al., 2021; Shrivastava and Sharma, 2022). Several biofuels, such as bioethanol, biobutanol, and hydrogen, have been studied together with the bioelectricity generated from MFC (Offei et al., 2019; Moradian et al., 2020; Shrivastava and Sharma, 2023)

Therefore, electricity production from lignocellulosic biomass offers an eco-friendly way to reducing reliance on fossil fuels power demand. Although there are many technological pathways, the efficiency and scalability of these systems are key challenges being addressed through research and innovation.

2.5 FINAL REMARKS

Lignocellulosic biomass is an abundant, renewable, and carbon-neutral feedstock to produce biofuels and bioenergy to addressing global energy demands and reducing dependence on fossil fuels. Over the past few decades, significant strides have been made in developing technologies to convert lignocellulosic biomass into these compounds, specifically the biochemical and thermochemical routes. Even though, there are some biorefineries of lignocellulose biomass producing bioenergy and biofuels, there are several barriers that must overcome to become these processes competitive with fossil fuels.

One of the this is the recalcitrance of lignocellulose. The tightly bonded structure of cellulose, hemicellulose, and lignin makes it resistant to microbial and enzymatic degradation needing a pretreatment phase to break down the lignin matrix and make the cellulose and hemicellulose more accessible. Various pretreatment methods, including chemical, physical, and biological techniques, have been developed to address this challenge, but they often add complexity and cost to the process. Finding efficient, low-cost pretreatment methods remains a critical area of ongoing research, specially using nanomaterials. Another drawback of these pathways is the microbes used in fermentation and in anaerobic digestion, such as yeast and bacteria to produce biogas/biomethane, biohydrogen, bioethanol, biobutanol that show low yields. To solve this problem, genetically engineered strains are being developed to enhance their ability to ferment a broader range of sugars and tolerate the harsh conditions often associated with lignocellulosic feedstocks. Despite these advancements, the need for robust, cost-effective microbial systems that can efficiently convert lignocellulosic sugars into biofuels is still a major obstacle. In thermochemical conversion methods, like pyrolysis used to produce bio-oil, which can be refined into fuel, and gasification that generates syngas for fuel synthesis, the high energy inputs required for these processes and the need for further refinement of the resulting biofuels represent challenges that must be overcome to make this process economic viability. Other challenge in economic aspect is the costs of biomass collection, transportation, pretreatment, enzymatic hydrolysis, and downstream processing that are still higher than the production costs of petroleum-based fuels. Although government policies and subsidies can help bridge this gap in the short term, long-term solutions will require technological innovations that significantly reduce production costs.

Economies of scale may offer one path to improved cost efficiency. As production facilities grow larger and more integrated, the costs associated with biofuel production may decrease. Additionally, biorefineries that produce not only biofuels but also other valuable co-products, such as biochemicals, bio-plastics, and bio-fertilizers, could enhance the economic sustainability of the industry. Such integrated systems would allow for the full utilization of lignocellulosic biomass, minimizing waste and maximizing profitability. Beyond economic considerations, the environmental impacts of lignocellulosic biofuel production are crucial to its long-term success. Compared to first-generation biofuels, lignocellulosic biofuels generally have a lower environmental impact, particularly in terms of land use and water consumption. However, large-scale cultivation of dedicated energy crops could still have unintended environmental consequences, such as loss of biodiversity, soil degradation, and changes in land use patterns. Careful management practices, including sustainable agricultural practices and crop rotation, will be necessary to ensure that lignocellulosic biofuels are truly environmentally friendly.

The future of lignocellulosic biofuel production lies in continued innovation and collaboration between academia, industry, and government. Breakthroughs in biotechnology, materials science, and systems integration will be key to overcoming current limitations of biological and thermochemical technologies. Public policy can play a significant role in fostering the development of biofuel technologies by providing incentives for research, development, and deployment. At the same time, industry partnerships will be essential for scaling up production and reducing costs.

Therefore, lignocellulosic biomass-based biofuel production has the potential to revolutionize the energy landscape, providing a renewable and environmentally friendly alternative to fossil fuels. While significant challenges remain in terms of technological efficiency, cost, and sustainability, ongoing research and development efforts are bringing the field closer to realizing its full potential. With continued innovation and commitment, lignocellulosic biofuels could play a central role in the transition to a more sustainable energy future.

REFERENCES

Adekunle, A., Gariepy, Y., Lyew, D., & Raghavan, V. (2016). Energy recovery from cassava peels in a single-chamber microbial fuel cell. *Energy Sources, Part A: Recovery, Utilization, and Environmental Effects, 38*(17), 2495–2502. https://doi.org/10.1080/15567036.2015.1086909

Allam, F., Elnouby, M., El-Khatib, K. M., El-Badan, D. E., & Sabry, S. A. (2020). Water hyacinth (Eichhornia crassipes) biochar as an alternative cathode electrocatalyst in an air-cathode single chamber microbial fuel cell. *International Journal of Hydrogen Energy, 45*(10), 5911–5927. https://doi.org/10.1016/j.ijhydene.2019.09.164

Al-Mahgari, A. M., Al-Nimr, M. A., & Khashan, S. A. (2023). Pressurized green hydrogen from water electrolysis: compression before or after electrolysis? A comparison among different configurations. *Journal of Energy Storage, 73*, 109251. https://doi.org/10.1016/j.est.2023.109251

Ameh, V. I., Ayeleru, O. O., Nomngongo, P. N., & Ramatsa, I. M. (2024). Bio-oil production from waste plant seeds biomass as pyrolytic lignocellulosic feedstock and its improvement for energy potential: a review. *Waste Management Bulletin, 2*, 32–48. https://doi.org/10.1016/j.wmb.2024.03.002

Arias, A., Feijoo, G., & Moreira, M. T. (2023). Biorefineries as a driver for sustainability: Key aspects, actual development and future prospects. *Journal of Cleaner Production, 418*, 137925. https://doi.org/10.1016/j.jclepro.2023.137925

Asencios, Y. J. O., Marcos, F. C. F., Assaf, J. M., & Assaf, E. M. (2016). Oxidative-reforming of methane and partial oxidation of methane reactions over $NiO/PrO_2/ZrO_2$ catalysts: effect of nickel content. *Brazilian Journal of Chemical Engineering, 33*(3). https://doi.org/10.1590/0104-6632.20160333s20150056

Ashokkumar, V., Venkatkarthick, R., Jayashree, S., Chuetor, S., Dharmaraj, S., Kumar, G., ... & Ngamcharussrivichai, C. (2022). Recent advances in lignocellulosic biomass for biofuels and value-added bioproducts-A critical review. *Bioresource Technology, 344*, 126195.

Awasthi A., & Bhaskar T. (2019). Chapter 11 - Combustion of lignocellulosic biomass. In: *Biofuels: Alternative Feedstocks and Conversion Processes for the Production of Liquid and Gaseous Biofuels*. Editor: Pandey, A, Larroche, C., Dussap, C-G, Gnansounou, E., Khanal, S. K., Ricke, S. https://doi.org/10.1016/B978-0-12-816856-1.00011-7

Balajii, M., & Niju, S. (2019). Biochar-derived heterogeneous catalysts for biodiesel production. *Environmental Chemistry Letters, 17*(4). https://doi.org/10.1007/s10311-019-00885-x

Balcioglu, G., Jeswani, H. K., Azapagi, A. (2023). A sustainability assessment of utilising energy crops for heat and electricity generation in Turkey. *Sustainable Production and Consumption, 41*, 134–155. https://doi.org/10.1016/j.spc.2023.08.003

Bhattacharya, T., Khan, A., Ghosh, T., Kim, J. T., & Rhim, J. W. (2024). Advances and prospects for biochar utilization in food processing and packaging applications. *Sustainable Materials and Technologies, 39*. https://doi.org/10.1016/j.susmat.2024.e00831

BP. (2019). *Crude Oil, Coal, Natural Gas: A Report on BP Statistical Review of World Energy 2019.*

Cao, L., Zhang, C., Chen, H., Tsang, D. C. W., Luo, G., Zhang, S., & Chen, J. (2017)l. Hydrothermal liquefaction of agricultural and forestry wastes: state-of-the-art review and future prospects. *Bioresource Technology*. http://dx.doi.org/10.1016/j.biortech.2017.08.196.

Chanathaworn, J., Yatongchai, C., & Samanman, S. (2022). Leucaena–derived biochar for biodiesel production. *Molekul, 17*(1). https://doi.org/10.20884/1.jm.2022.17.1.5604

Chandran, E. M., & Mohan, E. (2023). Sustainable biohydrogen production from lignocellulosic biomass sources – metabolic pathways, production enhancement, and challenges. *Environmental Science and Pollution Research, 30*, 102129–102157. https://doi.org/10.1007/s11356-023-29617-z

Chaudhary, G., Chaudhary, N., Saini, S., Gupta, Y., Vivekanand, V., & Panghal, A. (2024). Assessment of pretreatment strategies for valorization of lignocellulosic biomass: path forwarding towards lignocellulosic biorefinery. *Waste and Biomass Valorization, 15*, 1–36. https://doi.org/10.1007/s12649-023-02219-z

Chen, X., Li, Y., Li, X., Shi, J., & Liu, L. (2024). Exploring the potential of multiple lignocellulosic biomass as a feedstock for biobutanol production. *Fuel, 357*, 129697.

Cheng, S., Wei, L., Zhao, X., & Julson, J. (2016). Application, deactivation, and regeneration of heterogeneous catalysts in bio-oil upgrading. *Catalysts, 6*(12), 195. https://doi.org/10.3390/catal6120195

Chettri, D., Verma, A. K., Ghosh, S., & Verma, A. K. (2024). Biogas from lignocellulosic feedstock: current status and challenges. *Cleaner Production and Sustainable Processes for Environmental Remediation, 31*, 1–26. https://doi.org/10.1007/s11356-023-29805-x

Chiaramonti, D., Prussi, M., Buffi, M., & Tacconi, D. (2014). Sustainable bio kerosene: process routes and industrial demonstration activities in aviation biofuels. *Applied Energy, 136*, 767–774. https://doi.org/10.1016/j.apenergy.2014.08.065

Chintagunta, A. D., Zuccaro, G., Kumar, M., Kumar, S. J., Garlapati, V. K., Postemsky, P. D., ... & Simal-Gandara, J. (2021). Biodiesel production from lignocellulosic biomass using oleaginous microbes: prospects for integrated biofuel production. *Frontiers in Microbiology, 12*, 658284. https://doi.org/10.3389/fmicb.2021.658284.

Ciliberti, C., Biundo, A., Albergo, R., Agrimi, G., Braccio, G., de Bari, I., & Pisano, I. (2020). Syngas derived from lignocellulosic biomass gasification as an alternative resource for innovative bioprocesses. *Processes, 8*(12). https://doi.org/10.3390/pr8121567

Cordonier Segger, M. C. (2016). Advancing the paris agreement on climate change for sustainable development. *Cambridge International Law Journal, 5*(2). https://doi.org/10.4337/cilj.2016.02.03

Dahal, K., Brynolf, S., Xisto, C., Hansson, J., Grahn, M., Grönstedt, T., & Lehtveer, M. (2021). Techno-economic review of alternative fuels and propulsion systems for the aviation sector. *Renewable and Sustainable Energy Reviews, 151*, 111564. https://doi.org/10.1016/j.rser.2021.111564

Dai, H. N., Nguyen, T. -A. D., Le, L. -P. M., Tran, M. V., Lan, T. -H., & Wang, C.-T. (2021). Power generation of Shewanella oneidensis MR-1 microbial fuel cells in bamboo fermentation effluent. *International Journal of Hydrogen Energy, 46*, 16612–16621. https://doi.org/10.1016/j.ijhydene.2020.09.264

Das, S. K., Ghosh, G. K., & Avasthe, R. (2023). Application of biochar in agriculture and environment, and its safety issues. *Biomass Conversion and Biorefinery, 13*(2). https://doi.org/10.1007/s13399-020-01013-4

Demirbas, A. (2008). The importance of bioethanol and biodiesel from biomass. *Energy Sources, Part B, 3*(2), 177–185. https://doi.org/10.1080/15567240600815117

Dharmaraja, J., Shobana, S., Arvindnarayan, S., Vadivel, M., Atabani, A. E., Pugazhendhi, A., & Kumar, G. (2020). Biobutanol from lignocellulosic biomass: bioprocess strategies. In *Lignocellulosic Biomass to Liquid Biofuels* (pp. 169–193). Academic Press.

Dinca, M. N., Ferdes, M., Paraschiv, G., & Zabava, B. S. (2024). Enzymatic pretreatment of lignocellulosic biomass for biogas production – review. *Engineering for Rural Development, 22*, 908–913. https://doi.org/10.22616/ERDev.2024.23.TF181

Dutra, J. C. F., Passos, M. F., Moretti, E. R., Nascimento, L. A. S., Silva, A. J., Silva, T. F., Aguiar, R. H., Rodrigues, L. S., Mockaitis, G. (2024). Methane production from lignocellulosic biomass using hydrothermal pretreatment. *Biomass Conversion and Biorefinery, 14*, 3699–3713. https://doi.org/10.1007/s13399-022-02604-z

Evans, A., Strezov, V., & Evans, T. J. (2010). Sustainability considerations for electricity generation from biomass. *Renewable and Sustainable Energy Reviews, 14*(5), 1419–1427. https://doi.org/10.1016/j.rser.2010.01.010

Ghanbarpour Mamaghani, Z., Hawboldt, K. A., & MacQuarrie, S. (2023). Adsorption of CO_2 using biochar - Review of the impact of gas mixtures and water on adsorption. *Journal of Environmental Chemical Engineering, 11*(3). https://doi.org/10.1016/j.jece.2023.109643

Gillingham, K., & Stock, J. H. (2018). The cost of reducing greenhouse gas emissions. *Journal of Economic Perspectives, 32*, 53–72.

Gollakota, A. R. K., Kishore, N., & Gu, S. (2018). A review on hydrothermal liquefaction of biomass. *Renewable and Sustainable Energy Reviews, 81*(Part 1), 1378–1392. https://doi.org/10.1016/j.rser.2017.05.178.

Grimme, W. (2023). The introduction of sustainable aviation fuels—A discussion of challenges, options and alternatives. *Aerospace, 10*, 218. https://doi.org/10.3390/aerospace10030218

Guimaraes, C. E. C., Neto, F. S., de Castro Bizerra, V., do Nascimento, J. G. A., Valério, R. B. R., de Sousa Junior, P. G., ... & dos Santos, J. C. S. (2023). Sustainable bioethanol production from first-and second-generation sugar-based feedstocks: advanced bibliometric analysis. *Bioresource Technology Reports, 23*, 101543. https://doi.org/10.1016/j.biteb.2023.101543

Gupta, S., Fernandes, A., Lopes, A., Grasa, L., & Salafranca, J. (2024). Photo-fermentative bacteria used for hydrogen production. *Applied Sciences, 14*, 1191. https://doi.org/10.3390/app14031191

Gurung, A., & Oh, S. -E. (2015). Rice straw as a potential biomass for generation of bioelectrical energy using Microbial Fuel Cells (MFCs). *Energy Sources, Part A: Recovery, Utilization, and Environmental Effects, 37*(24), 2625–2631. https://doi.org/10.1080/15567036.2012.728678

Han, W., & Chen, W. Y. (2024). Survival of biomass and waste power generation: a global overview. *Science of The Total Environment, 940*, 173593. https://doi.org/10.1016/j.scitotenv.2024.173593

Haq, I. U., Qaisar, K., Nawaz, A., Akram, F., Mukhtar, H., Zohu, X., Xu, Y., Mumtaz, M. W., Rashid, U., Ghani, W. A. W. A. K., & Choong, T. S. Y. (2021). Advances in valorization of lignocellulosic biomass towards energy generation. *Catalysts, 11*, 309. https://doi.org/10.3390/catal11030309

Heleniak, T. (2021). The future of the Arctic populations. *Polar Geography, 44*(2), 136–152.

Hitam, C. N. C., Jalil, A. A. (2023). A review on biohydrogen production through photo-fermentation of lignocellulosic biomass. *Biomass Conversion and Biorefinery, 13*, 8465–8483. https://doi.org/10.1007/s13399-020-01140-y

Hu, C. K., Bai, F. W., & An, L. J. (2005). Effect of flocculence of a self-flocculating yeast on its tolerance to ethanol and the mechanism. *Sheng wu gong cheng xue bao= Chinese Journal of Biotechnology, 21*(1), 123–128. https://pubmed.ncbi.nlm.nih.gov/15859341/

International Civil Aviation Organization (ICAO). (n.d.). Conversion processes. Available online: https://www.icao.int/environmental-protection/GFAAF/Pages/Conversion processes.aspx (accessed on March 7, 2024).

International Energy Agency (IEA). (2023). *World Energy Outlook 2023*. https://www.iea.org/reports/world-energy-outlook-2023 (License: CC BY 4.0, Annex A: CC BY NC SA 4.0).

International Renewable Energy Agency (IRENA). (2018). *Renewable Power Generation Costs in 2017* (Report). International Renewable Energy Agency.

Ivanenko, A. A., Laikova, A. A., Zhuravleva, E. A., Shekhurdina, S. V., Vishnyakova, A. V., Kovalev, A. A., Kovalev, D. A., Trchounian, K. A., & Litti, Y. V. (2024). Biological production of hydrogen: from basic principles to the latest advances in process improvement. *International Journal of Hydrogen Energy, 55*, 740–755. https://doi.org/10.1016/j.ijhydene.2023.11.179

Jara-Cobos, L., Abril-González, M., & Pinos-Vélez, V. (2023). Production of hydrogen from lignocellulosic biomass: a review of technologies. *Catalysts, 13*(4), 766. https://doi.org/10.3390/catal13040766

Jatoi, A. S., Abbasi, S. A., Hashmi, Z., Shah, A. K., Alam, M. S., Bhatti, Z. A., ... & Iqbal, A. (2021). Recent trends and future perspectives of lignocellulose biomass for biofuel production: a comprehensive review. *Biomass Conversion and Biorefinery*, 1–13.

Jin, Y., Zhang, L., Yi, Z., Fang, Y., & Zhao, H. (2022). Waste-to-energy: biobutanol production from cellulosic residue of sweet potato by Clostridia acetobutylicum. *Environmental Engineering Research, 27*(5).

Jung, S., Lee, J., Moon, D. H., Kim, K. H., & Kwon, E. E. (2021). Upgrading biogas into syngas through dry reforming. *Renewable and Sustainable Energy Reviews, 143*. https://doi.org/10.1016/j.rser.2021.110949

Justicia, J., Heras, F., Moreno, I., Baeza, J. A., Calvo, L., Fermoso, J., & Gilarranz, M. A. (2024). Understanding the relationship between catalytic pyrolysis conditions and hydrogen production by aqueous phase reforming of the water-soluble fractions of bio-oils. *Energy Conversion and Management, 320*, 118999. https://doi.org/10.1016/j.enconman.2024.118999

Jwad, M. K. M. M. K. M., & Al-Haidari, S. H. J. S. H. J. (2017). The effect of flocculation on the ethanolic production in some various Yeast. *Ibn AL-Haitham Journal for Pure and Applied Science, 21*(4), 19–30. https://mracpc.uobaghdad.edu.iq/jih/j/article/view/1424/.

Kalayasiri, P., Jeyashoke, N., & Krisnangkura, K. (1996). Survey of seed oils for use as diesel fuels. *Journal of the American Oil Chemists' Society, 73*, 471–474. https://doi.org/0.1007/bf02523921

Kallarakkal, K. P., Muthukumar, K., Alagarsamy, A., Pugazhendhi, A., & Mohamed, S. N. (2021). Enhancement of biobutanol production using mixotrophic culture of Oscillatoria sp. in cheese whey water. *Fuel, 284*, 119008.

Kamini, N. R., & Gunasekaran, P. (1987). Simultaneous ethanol production from lactose by Kluyveromyces fragilis and Zymomonas mobilis. *Current Microbiology, 16*, 153–157. https://doi.org/10.1007/BF01568395.

Kant Bhatia, S., Palai, A. K., Kumar, A., Kant Bhatia, R., Kumar Patel, A., Kumar Thakur, V., & Yang, Y. H. (2021). Trends in renewable energy production employing biomass-based biochar. *Bioresource Technology, 340*. https://doi.org/10.1016/j.biortech.2021.125644

Karthick, C., & Nanthagopal, K. (2021). A comprehensive review on ecological approaches of waste to wealth strategies for production of sustainable biobutanol and its suitability in automotive applications. *Energy Conversion and Management, 239*, 114219.

Karthikeyan, P. K., Bandulasena, H. C. H., Radu, T. (2024). A comparative analysis of pretreatment technologies for enhanced biogas production from anaerobic digestion of lignocellulosic waste. *Industrial Crops & Products, 215*, 118591. https://doi.org/10.1016/j.indcrop.2024.118591

Katam, K., Saisriyoot, M., Roddecha, S., Pornwongthong, P., & Sriariyanun, M. (2023). Conversion of lignocellulose residue obtained from biorefinery stream to electricity by microbial fuel cell. *E3S Web Conf*erence, *428*, 01011. https://doi.org/10.1051/e3sconf/202342801011

Kumar, K., Jadhav, S. M., & Moholkar, V. S. (2024). Acetone-Butanol-Ethanol (ABE) fermentation with clostridial co-cultures for enhanced biobutanol production. *Process Safety and Environmental Protection, 185*, 277–285.

Kumar, S., Awasthi, M. K., Kumar, V., & Awasthi, S. K. (2020). Lignocellulosic biomass for biobutanol production: A comprehensive review. *Journal of Cleaner Production*, 255, 120092.

Kwapinski, W. (2019). Char production technology. In: *Char and Carbon Materials Derived from Biomass* (pp. 39–68). https://doi.org/10.1016/B978-0-12-814893-8.00002-X

Lachos-Perez, D., Martins-Vieira, J. C., Missau, J., Anshu, K., Siakpebru, O. K., Thengane, S. K., Morais, A. R. C., Tanabe, E. H., & Bertuol, D. A. (2023). Review on biomass pyrolysis with a focus on bio-oil upgrading techniques. *Analytica, 4*, 182–205. https://doi.org/10.3390/analytica4020015

Laird, D. A., Brown, R. C., Amonette, J. E., & Lehmann, J. (2009). Review of the pyrolysis platform for coproducing bio-oil and biochar. *Biofuels, Bioproducts and Biorefining, 3*(5). https://doi.org/10.1002/bbb.169

Lee, J., Kim, S., You, S., & Park, Y. K. (2023). Bioenergy generation from thermochemical conversion of lignocellulosic biomass-based integrated renewable energy systems. *Renewable and Sustainable Energy Reviews, 178*, 113240.

Lee, Y., & Nirmalakhandan, N. (2011). Electricity production in membrane-less microbial fuel cell fed with livestock organic solid waste. *Bioresource Technology, 102*(10), 5831–5835. https://doi.org/10.1016/j.biortech.2011.02.090

Lepage, T., Kammoun, M., Schmetz, Q., & Richel, A. (2021). Biomass-to-hydrogen: a review of main routes production, processes evaluation and techno-economical assessment. *Biomass and Bioenergy, 144*, 105920. https://doi.org/10.1016/j.biombioe.2020.105920

Li, G., Yang, Y., Yu, Q., Ma, Q., Lam, S. S., Chen, X., He, Y., Ge, S., Sonne, C., & Peng, W. (2024). Application of nanotechnology in hydrogen production from biomass: a critical review. *Advanced Composites and Hybrid Materials, 7*, 17. https://doi.org/10.1007/s42114-023-00815-0

Li, M., Wang, T., Chen, X., & Ma, X. (2023). Conversion study from lignocellulosic biomass and electric energy to H2 and chemicals. *International Journal of Hydrogen Energy, 48*, 21004–21017. https://doi.org/10.1016/j.ijhydene.2022.09.191

Li, S., Xu, S., Liu, S., Yang, C., & Lu, Q. (2004). Fast pyrolysis of biomass in free-fall reactor for hydrogen-rich gas. *Fuel Processing Technology, 85*, 1201–1211.

Liang, J., Nabi, M., Zhang, P., Zhang, G., Cai, Y., Wang, Q., Zhou, Z., & Ding, Y. (2020). Promising biological conversion of lignocellulosic biomass to renewable energy with rumen microorganisms: a comprehensive review. *Renewable and Sustainable Energy Reviews*, 134, 110335.

Lim, M. K., Marty, L., & Feng, Q. (2023). Economic opportunities and challenges in biojet production: a literature review and analysis. *Biomass and Bioenergy, 170*, 106727. https://doi.org/10.1016/j.biombioe.2023.106727

Liu, B., & Zhao, Z. (2007). Biodiesel production by direct methanolysis of oleaginous microbial biomass. *Journal of Chemical Technology & Biotechnology: International Research in Process, Environmental & Clean Technology, 82*(8), 775–780. https://doi.org/10.1002/jctb.1744.

Liu, J., Liu, G., Zhang, W., Li, Z., Xing, F., & Tang, L. (2022). Application potential analysis of biochar as a carbon capture material in cementitious composites: a review. *Construction and Building Materials, 350*. https://doi.org/10.1016/j.conbuildmat.2022.128715

Liu, W., Liu, C, Gogoi, P., & Deng, Y. (2020). Overview of biomass conversion to electricity and hydrogen and recent developments in low-temperature electrochemical approaches. *Engineering, 6*(12), 1351–1363. https://doi.org/10.1016/j.eng.2020.02.021

Luo, H., Shi, Y., Xie, F., Zhou, T., Gao, L., Yang, R., & Wang, Z. (2023). Efficient co-production of fermentable sugars and biobutanol from corn stover based on a novel butyric acid pretreatment strategy. *Industrial Crops and Products, 191*, 115976.

Lv, P., Yuan, Z., Wu, C., Ma, L., Chen, Y., & Tsubaki, N. (2007). Bio-syngas production from biomass catalytic gasification. *Energy Conversion and Management, 48*(4). https://doi.org/10.1016/j.enconman.2006.10.014

Ma, G., Chen, Y., & Ndegwa, P. (2021). Association between methane yield and microbiota abundance in the anaerobic digestion process: a meta-regression. *Renewable and Sustainable Energy Reviews, 135*, 110212. https://doi.org/10.1016/j.rser.2020.110212

Machineni, L., Deepanraj, B., Chew, K. W., & Rao, A. G. (2023). Biohydrogen production from lignocellulosic feedstock: abiotic and biotic methods. *Renewable and Sustainable Energy Reviews, 182*, 113344. https://doi.org/10.1016/j.rser.2023.113344

Maitlo, G., Ali, I., Mangi, K. H., Ali, S., Maitlo, H. A., Unar, I. N., & Pirzada, A. M. (2022). Thermochemical conversion of biomass for syngas production: current status and future trends. *Sustainability (Switzerland), 14*(5). https://doi.org/10.3390/su14052596

Malik, F. R., Yuan, H.-B., Moran, J. C., & Tippayawong, N. (2023). Overview of hydrogen production technologies for fuel cell utilization. *Engineering Science and Technology, an International Journal, 43*, 101452. https://doi.org/10.1016/j.jestch.2023.101452

Mathiyazhagan, M., & Ganapathi, A. (2011). Factors affecting biodiesel production. *Research in Plant Biology, 1*(2), 1–5.

Maytorena, V. M., & Buentello-Montoya, D. A. (2024). Worldwide developments and challenges for solar pyrolysis. *Heliyon, 10*(15), e35464. https://doi.org/10.1016/j.heliyon.2024.e35464

Melendez, J. R., Mátyás, B., Hena, S., Lowy, D. A., & El Salous, A. (2022). Perspectives in the production of bioethanol: a review of sustainable methods, technologies, and bioprocesses. *Renewable and Sustainable Energy Reviews, 160*, 112260. https://doi.org/10.1016/j.rser.2022.112260

Mirabi, M., Karrabi, M., & Shahnavaz, B. (2024). Anaerobic co-digestion of lignocellulosic/lipidic wastes with cattle manure: investigating biogas production and methane yield. *Fuel, 366*, 131286. https://doi.org/10.1016/j.fuel.2024.131286

Miran, W., Nawaz, M., Jang, J., & Lee, D.S. (2016). Conversion of orange peel waste biomass to bioelectricity using a mediator-less microbial fuel cell. *Science of The Total Environment, 547*, 197–205. https://doi.org/10.1016/j.scitotenv.2016.01.004

Mishra, K., Siwal, S. S., Saini, A. K., & Thakur, V. K. (2023). Recent update on gasification and pyrolysis processes of lignocellulosic and algal biomass for hydrogen production. *Fuel, 332*, 126169. https://doi.org/10.1016/j.fuel.2022.126169

Miskin, V. S., & Hinge, G. A. (2024). An iterative comprehensive evaluation of pretreatment methods, comparative biomass analysis, and sustainability assessment for optimizing anaerobic biogas production from lignocellulosic biomass. *Innovative Infrastructure Solutions, 9*(172), 1–15. https://doi.org/10.1007/s41062-024-01476-8

Mohd Zaini Makhtar, M., & Tajarudin, H. A. (2020). Electricity generation using membraneless microbial fuel cell powered by sludge supplemented with lignocellulosic waste. *International Journal of Energy Research, 44*, 3260–3265. https://doi.org/10.1002/er.5151

Moradian, J. M., Fang, Z., & Yong, Y.C. (2021). Recent advances on biomass-fueled microbial fuel cell. *Bioresources and Bioprocessing, 8*, 14. https://doi.org/10.1186/s40643-021-00365-7

Moradian, J. -M., Xu, Z. -A., Shi, Y. -T., Fang, Z., & Yong, Y. -C. (2020). Efficient biohydrogen and bioelectricity production from xylose by microbial fuel cell with newly isolated yeast of Cystobasidium slooffiae. *International Journal of Energy Research, 44*, 325–333. https://doi.org/10.1002/er.4922

Muharja, M., Darmayanti, R. F., Fachri, B. A., Palupi, B., Rahmawati, I., Rizkiana, M. F., ... & Halim, A. (2023). Biobutanol production from cocoa pod husk through a sequential green method: Depectination, delignification, enzymatic hydrolysis, and extractive fermentation. *Bioresource Technology Reports, 21*, 101298.

Nassir, T. H., & Al-Sahlany, S. (2021). Bioethanol production from agricultural wastes by Zymomonas mobilis is used in vinegar production. *Journal of Microbiology, Biotechnology and Food Sciences, 11*(3), e3709–e3709. https://office2.jmbfs.org/index.php/JMBFS/article/view/3709/353.

Offei, F., Mensah, M., Kemausuor, F., & Thygesen, A. (2019). A biorefinery approach to bioethanol and bioelectricity co-production from tropical seaweeds. *Journal of Applied Phycology, 31*, 3899–3913. https://doi.org/10.1007/s10811-019-01887-6

Okolie, J. A., Mukherjee, A., Nanda, S., Dalai, A. K., & Kozinski, J. A. (2021). Next-generation biofuels and platform biochemicals from lignocellulosic biomass. *International Journal of Energy Research, 45*, 14145–14169. https://doi.org/10.1002/er.6697

Okolie, O., Kumar, A., Edwards, C., Lawton, L. A., Oke, A., McDonald, S., Thakur, V. K., & Njuguna, J. (2023). Bio-based sustainable polymers and materials: from processing to biodegradation. *Journal of Composites Science, 7*(6), 213. https://doi.org/10.3390/jcs7060213.

Olatunji, K. O., Ahmed, N. A., & Ogunkunle, O. (2021). Optimization of biogas yield from lignocellulosic materials with different pretreatment methods: a review. *Biotechnology for Biofuels, 14*(159), 1–34. https://doi.org/10.1186/s13068-021-02012-x

Olortiga Asencios, Y. J., & Assaf, E. M. (2023). Green hydrogen Production from the oxidative reform of clean biogas over Ni/MgO-Nb_2O_5 catalysts. *Journal of Science & Sustainable Engineering, 1*(1). https://doi.org/10.34024/jsse.2023.v1.15904

Osman, A. I., Mehta, N., Elgarahy, A. M., et al. (2021). Conversion of biomass to biofuels and life cycle assessment: a review. *Environmental Chemistry Letters, 19*, 4075–4118. https://doi.org/10.1007/s10311-021-01273-0

Ottoni, C., do Valle Trotta, C., Martins, G., Matos, J. Maiorano, A. E., Brito, A. G., & Peizoto, L. (2023). In Situ trametes versicolor laccase biocathode performance assessment in dual-chamber microbial fuel cells. *Bioenergy Research, 16*, 2616–2624. https://doi.org/10.1007/s12155-023-10594-7

Pal, M., & Sharma, R.K. (2020). Development of wheat straw based catholyte for power generation in microbial fuel cell. *Biomass Bioenergy, 138*, 105591. https://doi.org/10.1016/j.biombioe.2020.105591

Panoutsou, C., Germer, S., Karka, P., Papadokostantakis, S., Kroyan, Y., Wojcieszyk, M., Maniatis, K., Marchand, P., & Landalv, I. (2021). Advanced biofuels to decarbonise European transport by 2030: Markets, challenges, and policies that impact their successful market uptake. *Energy Strategy Reviews, 34*, 100633. https://doi.org/10.1016/j.esr.2021.100633

Patel, M., Zhang, X., & Kumar, A. (2016). Techno-economic and life cycle assessment on lignocellulosic biomass thermochemical conversion technologies: a review. *Renewable and Sustainable Energy Reviews, 53*, 1486–1499. https://doi.org/10.1016/j.rser.2015.09.070

Peters, M. A., Alves, C. T., & Onwudili, J. A. (2023). A review of current and emerging production technologies for biomass-derived sustainable aviation fuels. *Energies, 16*, 6100. https://doi.org/10.3390/en16166100.

Poddar, B. J., Nakhate, S. P., Gupta, R. K., Chavan, A. R., Singh, A. K., Khardenavis, A. A., & Purohit, H. J. (2022). A comprehensive review on the pretreatment of lignocellulosic wastes for improved biogas production by anaerobic digestion. *International Journal of Environmental Science and Technology, 19*, 3429–3456. https://doi.org/10.1007/s13762-021-03248-8

Ramos, L. P., Kucek, K. T., Domingos, A. K., & Wilhelm, H. M. (2003). Biodiesel. *Revista Biotecnologia Ciência & Desenvolvimento, 31*, 29.

Rangel, M. C., Mayer, F. M., Carvalho, M. S., Saboia, G., & Andrade, A. M. (2023). Selecting catalysts for pyrolysis of lignocellulosic biomass. *Biomass, 3*(1), 31–63. https://doi.org/10.3390/biomass3010003

Rebequi, T., Pio, Y., Penteado, C. F. A., da Silva Martins, L. H., Diniz, A. A. R., Komesu, A., & Penteado, E. D. (2023). Waste treatment and sustainable bioelectricity generation using microbial fuel cell. *Journal of Science & Sustainable Engineering, 1*(1). https://doi.org/10.34024/jsse.2023.v1.15460

Rethinasabapathy, M., Lee, J. H., Roh, K. C., Kang, S. -M., Oh, S. Y., Park, B., Lee, G. -W., Cha, Y. L., & Huh, Y. S. l (2020). Silver grass-derived activated carbon with coexisting micro-, mesoand macropores as excellent bioanodes for microbial colonization and power generation in sustainable microbial fuel cells. *Bioresource Technology, 300*, 122646. https://doi.org/10.1016/j.biortech.2019.122646

Roy, A., & Pramanik, S. (2024). A review of the hydrogen fuel path to emission reduction in the surface transport industry. *International Journal of Hydrogen Energy, 49*(Part D), 792–821. https://doi.org/10.1016/j.ijhydene.2023.07.010

Ruckenstein, E., & Hu, Y. H. (1999). Methane partial oxidation over NiO/MgO solid solution catalysts. *Applied Catalysis A: General, 183*(1). https://doi.org/10.1016/S0926-860X(99)00047-2

Sánchez-Borrego, F. J., Álvarez-Mateos, P., & García-Martín, J. F. (2021). Biodiesel and other value-added products from bio-oil obtained from agrifood waste. *Processes, 9*, 797. https://doi.org/10.3390/pr9050797

Sangrulkar, P., Gupta, S., & Kandasubramanian, B. (2023). Advancements in biochar-based electrodes for improved performance of microbial fuel cells. *Bioresource Technology Reports, 24*. https://doi.org/10.1016/j.biteb.2023.101684

Sani, A., Salva, N., Pandit, S., Mathuriya, A. S., Gupta, P. K., Khanna, N., Babu, R. P., & Kumar, S. (2021). Recent advances in bioelectricity generation through the simultaneous valorization of lignocellulosic biomass and wastewater treatment in microbial fuel cell. *Sustainable Energy Technologies and Assessments, 48*, 101572. https://doi.org/10.1016/j.seta.2021.101572

Scheffers, W. A. (1987). Alcoholic fermentation. *Stud Mycol, 30*, 321–332. https://studiesinmycology.org/sim/Sim30/fulltext/index.html?i=25

Serrano, V. A., Fajardo, C. A. G., & Castro, K. T. (2024). A review on biobutanol: eco-friendly fuel of the future. *History, Current Advances, and Trends.*

Shah, N., & Rehan, T. (2014). Bioethanol production from biomass. *Journal Chemistry and Biochemistry, 2*(2), 161–167.

Sharma, M., Salama, E., Thakur, N., Alghamdi, H., Jeon, B. -H., & Li, X. (2023a). Advances in the biomass valorization in bioelectrochemical systems: A sustainable approach for microbial-aided electricity and hydrogen production. *Chemical Engineering Journal, 465*, 142546. https://doi.org/10.1016/j.cej.2023.142546

Sharma, R. K., Nazari, M. A., Haydary, J., Singh, T. P., & Mandal, S. (2023b). A review on advanced processes of biohydrogen generation from lignocellulosic biomass with special emphasis on thermochemical conversion. *Energies, 16*(17), 6349. https://doi.org/10.3390/en16176349

Shen, X., Kommalapati, R. R., & Huque, Z. (2015). The comparative life cycle assessment of power generation from lignocellulosic biomass. *Sustainability, 7*, 12974–12987. https://doi.org/10.3390/su71012974

Kang, S., Li, X., Fan, J., & Chang, J. (2013). Hydrothermal conversion of lignin: A review. Renewable and Sustainable Energy Reviews, 27, 546-558. https://doi.org/10.1016/j.rser.2013.07.013

Shrivastava, A., & Sharma, R. K. (2023). Conversion of lignocellulosic biomass: Production of bioethanol and bioelectricity using wheat straw hydrolysate in electrochemical bioreactor. *Heliyon, 9*(1), e12951. https://doi.org/10.1016/j.heliyon.2023.e12951

Shrivastava, A., & Sharma, R.K. (2022). Lignocellulosic biomass based microbial fuel cells: Performance and applications. *Journal of Cleaner Production, 361*, 132269. https://doi.org/10.1016/j.jclepro.2022.132269

Singh, N., Singhania, R. R., Nigam, P. S., Dong, C.-D., Patel, A. K., & Puri, M. (2022). Global status of lignocellulosic biorefinery: Challenges and perspectives. *Bioresource Technology, 344*(Part B), 126415. https://doi.org/10.1016/j.biortech.2021.126415

Singh, S., Pandey, G., Rath, G. K., Veluswamy, H. P., & Molokitina, N. (2023). Life cycle assessment of biomass-based hydrogen production technologies: a review. *International Journal of Green Energy*. https://doi.org/10.1080/15435075.2023.2245453

Sivaramakrishnan, R., Shanmugam, S., Sekar, M., Mathimani, T., Incharoensakdi, A., Kim, S. -H., Parthiban, A., Geo, V. E., Brindhadevi, K., & Pugazhendhi, A. (2021). Insights on biological hydrogen production routes and potential microorganisms for high hydrogen yield. *Fuel, 291*, 120136. https://doi.org/10.1016/j.fuel.2021.120136

Sohi, S. P., Krull, E., Lopez-Capel, E., & Bol, R. (2010). A review of biochar and its use and function in soil. *Advances in Agronomy, 105*(1). https://doi.org/10.1016/S0065-2113(10)05002-9

Song, J., Tan, Z., Nong, Z., Li, Y., Huang, J., Yi, J., Ma, J., Huo, Q., Lu, Z., Su, C., & Ai, C. (2024). Easily-prepared macroporous-mesoporous carbon from palm fiber as a sustainable anode for high-performance microbial fuel cells. *Fuel, 360,* 130514. https://doi.org/10.1016/j.fuel.2023.130514

Su, G., Ong, H. C., Gan, Y. Y., Chen, W. H., Chong, C. T., & Ok, Y. S. (2022). Co-pyrolysis of microalgae and other biomass wastes for the production of high-quality bio-oil: Progress and prospective. *Bioresource Technology, 344*, 126096. https://doi.org/10.1016/j.biortech.2021.126096

Sun, H., Li, J., Cui, X., Stinner, W., Guo, J., & Dong, R. (2021). Enhancement mechanism of biogas potential from lignocellulosic substrates in the ensiling process via acid-based hydrolysis and biological degradation. *Journal of Cleaner Production, 319*, 128826. https://doi.org/10.1016/j.jclepro.2021.128826

Tamilselvan, R., & Selwynraj, A. I. (2024). Enhancing biogas generation from lignocellulosic biomass through biological pretreatment: Exploring the role of ruminant microbes and anaerobic fungi. *Anaerobe, 85*, 102815. https://doi.org/10.1016/j.anaerobe.2023.102815

Tekin, N., Karatay, S. E., & Dönmez, G. (2023). Optimization studies about efficient biobutanol production from industrial tea waste by Clostridium beijerinckii. *Fuel, 331*, 125763.

Tippayawong, N., Kinorn, J., & Thavornun, S. (2008). Yields and gaseous composition from slow pyrolysis of refuse-derived fuels. *Energy Sources Part A, 30*, 1572–1580.

Uddin, M. M., & Wright, M. M. (2022). Anaerobic digestion fundamentals, challenges, and technological advances. *Physical Sciences Reviews, 8*(9), 2819–2837. https://doi.org/10.1515/psr-2021-0068

Umer, M., Brandoni, C., Jaffar, M., Hewitt, N. J., Dunlop, P., Zhang, K., & Huang, Y. (2024), An experimental investigation of hydrogen production through biomass electrolysis. *Processes, 12*(1), 112. https://doi.org/10.3390/pr12010112

Uthandi, S., Kaliyaperumal, A., Srinivasan, N., Thangavelu, K., Muniraj, I. K., Zhan, X., ... & Gupta, V. K. (2022). Microbial biodiesel production from lignocellulosic biomass: new insights and future challenges. *Critical Reviews in Environmental Science and Technology, 52*(12), 2197–2225.

Vamsi Krishna, K., Bharathi, N., George Shiju, S., Alagesan Paari, K., & Malaviya, A. (2022). An updated review on advancement in fermentative production strategies for biobutanol using Clostridium spp. *Environmental Science and Pollution Research, 29*(32), 47988–48019.

van Dyk, S., & Saddler, J. (2021). IEA Bioenergy: Progress in commercialization of biojet/sustainable aviation fuels (SAF): Technologies, potential and challenges.

Vela-García, N., Bolonio, D., García-Martínez, M. -J., Ortega, M. F., & Canoira, L. (2022). Chapter 2 - Thermochemical conversion of agricultural waste to biojet fuel. In: A. Yousuf & C. Gonzalez-Fernandez (Eds.), *Sustainable Alternatives for Aviation Fuels*. Elsevier. https://doi.org/10.1016/B978-0-323-85715-4.00002-1

Velvizhi, G., Jacqueline, P. J., Shetti, N. P., Latha, K., Mohanakrishna, G., & Aminabhavi, T. M. (2023). Emerging trends and advances in valorization of lignocellulosic biomass to biofuels. *Journal of Environmental Management, 345*, 118527.

Wang, F., Ouyang, D., Zhou, Z., Page, S. J., Liu, D., & Zhao, X. (2021a). Lignocellulosic biomass as sustainable feedstock and materials for power generation and energy storage. *Journal of Energy Chemistry, 57*, 247–280. https://doi.org/10.1016/j.jechem.2020.08.060

Wang, H., Peng, X., Zhang, H., Yang, S., & Li, H. (2021b). Microorganisms-promoted biodiesel production from biomass: a review. *Energy Conversion and Management: X, 12*, 100137. https://doi.org/10.1016/j.ecmx.2021.100137

Wang, Y., Akbarzadeh, A., Chong, L., Du, J., Tahir, N., & Awasthi, M.K. (2022a). Catalytic pyrolysis of lignocellulosic biomass for bio-oil production: a review. *Chemosphere, 297*, 134181. https://doi.org/10.1016/j.chemosphere.2022.134181.

Wang, Y., Huang, L., Zhang, T., & Wang, Q. (2022b). Hydrogen-rich syngas production from biomass pyrolysis and catalytic reforming using biochar-based catalysts. *Fuel, 313*. https://doi.org/10.1016/j.fuel.2021.123006

Wang, Y. -Z., Shen, Y., Gao, L., Liao, Z. -H., Sun, J. -Z., & Yong, Y. -C. (2017). Improving the extracellular electron transfer of Shewanella oneidensis MR-1 for enhanced bioelectricity production from biomass hydrolysate. *RSC Advances, 7*, 30488–30494. https://doi.org/10.1039/C7RA04106C

Why, E. S. K., Ong, H. C., Lee, H. V., Gan, Y. Y., Chen, W. H., & Chong, C. T. (2019). Renewable aviation fuel by advanced hydroprocessing of biomass: challenges and perspective. *Energy Conversion and Management, 199*, 112015. https://doi.org/10.1016/j.enconman.2019.112015

Yoshimura, Y.,.Nakashima, K., Kato, M., Inoue, K., Okazaki, F., Soyama, H., Kawasaki, S.(2018). Electricity generation from rice bran by a microbial fuel cell and the influence of hydrodynamic cavitation pretreatment. *ACS Omega, 3*(11), 15267–15271. https://doi.org/10.1021/acsomega.8b02077

Zerva, C., Karakoulia, S. A., Kalogiannis, K. G., Margellou, A., Iliopoulou, E. F., Lappas, A. A., Papayannakos, N., & Triantafyllidis, K. S. (2021). Hydrodeoxygenation of phenol and biomass fast pyrolysis oil (bio-oil) over Ni/WO3-ZrO2 catalyst. *Catalysis Today, 366*, 57–67. https://doi.org/10.1016/j.cattod.2020.06.018

Zetty-Arenas, A. M., Tovar, L. P., Alves, R. F., Mariano, A. P., van Gulik, W., Maciel Filho, R., & Freitas, S. (2021). Co-fermentation of sugarcane bagasse hydrolysate and molasses by Clostridium saccharoperbutylacetonicum: effect on sugar consumption and butanol production. *Industrial Crops and Products, 167*, 113512.

Zhang, S., Zhu, X., Zhou, S., Shang, H., Luo, J., & Tsang, D. C. W. (2018). Hydrothermal carbonization for hydrochar production and its application. *Biochar from Biomass and Waste: Fundamentals and Applications*. https://doi.org/10.1016/B978-0-12-811729-3.00015-7

Zhang, X., Zhou, H., & Cheng, Y. (2021). Enzymatic hydrolysis of lignocellulosic biomass: a review. *Biomass and Bioenergy, 150*, 106100.

Zhao, C., Xu, Q., Gu, Y., Nie, X., & Shan, R. (2023). Review of advances in the utilization of biochar-derived catalysts for biodiesel production. *ACS Omega, 8*(9). https://doi.org/10.1021/acsomega.2c07909

Zhao, X., Liu, W., Deng, Y., & Zhu, J. Y. (2017). Low-temperature microbial and direct conversion of lignocellulosic biomass to electricity: Advances and challenges. *Renewable and Sustainable Energy Reviews, 71*, 268–282. https://doi.org/10.1016/j.rser.2016.12.055

Zulqarnain, Mohd Yusoff, M. H., Ayoub, M., Ramzan, N., Nazir, M. H., Zahid, I., ... & Butt, T. A. (2021). Overview of feedstocks for sustainable biodiesel production and implementation of the biodiesel program in Pakistan. *ACS Omega, 6*(29), 19099–19114. https://doi.org/10.1021/acsomega.1c02402

3 Present Status and Future Scope of Lignocellulosic Biomass-based Biofuel Production

Moumita Bishai

3.1 INTRODUCTION TO LIGNOCELLULOSIC BIOMASS

Lignocellulosic biomass is defined as entirely plant-based organic compounds that contain cellulose, hemicellulose, and lignin. These complex carbohydrate-based polymers provide structural support and protection to plants. They are renewable and sustainable, with plentiful in number for manufacturing biofuels, biochemicals and other bioproducts (Ashokkumar et al., 2022; Usmani et al., 2021). They are primarily consisting of cellulose (40%–50%), a polysaccharide composed of glucose units, providing structural strength; Hemicellulose (20%–30%), a polysaccharide composed of various sugars (xylose, arabinose, galactose), acting as a bonding agent and lignin (15%–30%), a complex polymer binding cellulose and hemicellulose, providing rigidity and water resistance (Mujtaba et al., 2023; Bhardwaj et al., 2019). There are various sources from where these biomasses were obtained. They can be produced from farm leftovers (corn stover, wheat straw), forestry remnants (wood chips, sawdust), specialized energy crops (switchgrass, miscanthus), municipality solid trash, and industrial waste discharges (Abolore et al., 2024; Makela et al., 2014; Adeniyi and Fagbenro, 2023; Zhang et al., 2018). Lignocellulosic biomasses found to be crucial for various biofuel production such as bioethanol, biobutanol, biomethane, etc; for biopower generation, for biorefinery development, for carbon sequestration and for climate change mitigation (Mujtaba et al., 2023; Devi et al., 2022; Gunasekaran et al., 2021). Rationale to use lignocellulosic biomass in biofuel sector arises from the fact that it has various advantages such as its huge abundance and origin from renewable resources which lowers the cost of feedstock (Vinod et al., 2023; Manyi-loh et al., 2023). It also ensures consistent supply. It maintains the carbon-neutral balance of the ecosystem and its sustainability is also adjusted. Hence, reduces dependency on fossil fuels. It also diversified the energy mix and created employment while directly effecting the rural economic development and encourages investment in biorefineries and infrastructure development. Finally, it also improved the soil health through residue return and contribute highly for the

DOI: 10.1201/9781003571223-3

potential carbon sequestration (Abalore et al., 2024; Vinod et al., 2023; Manyi-loh et al., 2023). Although it explored various advantages, there are also some intricacies related to Lignocellulosic biomasses. They are considered to be having a very complex composition and recalcitrance. Also, use of different pretreatment techniques, integration, optimization, fermentation, and strain development along with their purification increase the production and transportation costs and storage issues (Bhandari and Khosravi, 2023). Furthermore, there remains a consistent competition with food crops for land and resources and all the above processes of scalability and commercialization challenges are also the vital. There also remains challenges related to public acceptance and social licensing. Lignocellulosic biomasses also face some environmental challenges such as habitat destruction, nutrient depletion, soil degradation, greenhouse gas emissions during production, and waste management and disposal. If optimization of such encounters were carried out in a proper way, then such biomasses would be a valuable feedstock in immediate horizon. The global availability, regional distribution, and potentials of lignocellulosic biomasses in different sectors proves that they are abundantly available worldwide, with an estimated annual potential of approximately 7.5 billion metric tons. Geographically, it showed high availability in Asia Pacific region (Haberzettl et al., 2021; Ojo, 2023). This vast, renewable resource has immense potential for bioenergy, biochemicals, bioplastics, animal feed, and soil amendments, providing an environmentally friendly substitute to fossil fuels while preventing climate change (Figures 3.1–3.3) as described by Rather et al. in 2022, Priya et al. in 2023, and Farghali et al. in 2023.

The availability of lignocellulosic biomass is impacted by various factors, which can either enhance or constrain its potential. Understanding these factors is crucial for optimization of biomass production and utilization. Increasing population drives agricultural production, leading to more residues available for biomass. Changes

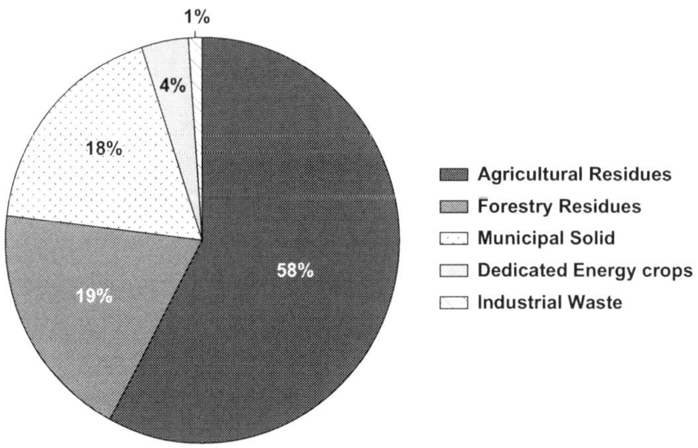

FIGURE 3.1 Global availability of lignocellulosic biomass.

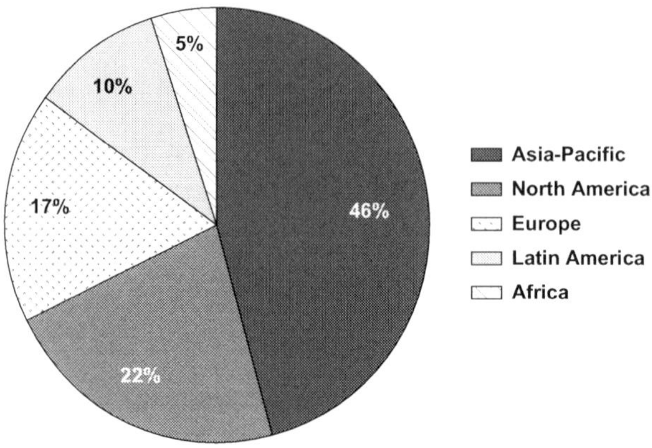

FIGURE 3.2 Regional distribution of lignocellulosic biomass.

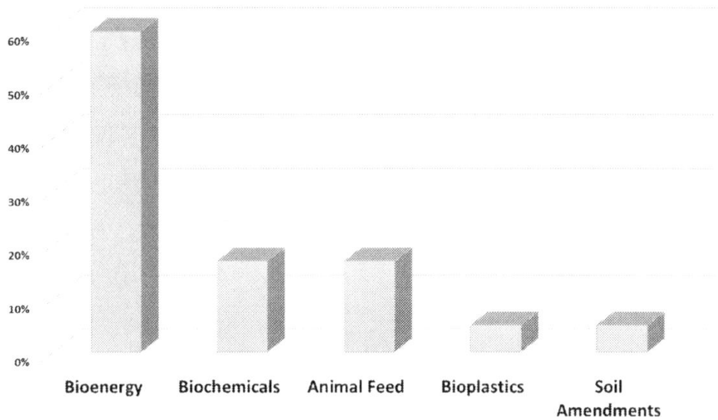

FIGURE 3.3 Potential of lignocellulosic biomass.

in land use and waste management practices affect biomass availability. Efficient farming methods, crop selection, and residue management influence biomass yield. Sustainable forestry practices ensure consistent wood residue supply. Effective waste collection, sorting, and recycling impact biomass availability. Altered weather patterns, temperature, and precipitation levels also affect biomass production and quality (Davis et al., 2018). Hence, to ensure consistent lignocellulosic biomass availability sustainable agricultural and forestry practices need to be implemented, efficient waste management systems need to be developed, climate-resilient agriculture, and forestry should be encouraged and policies promoting biomass production and utilization must be developed and supported.

3.2 CURRENT STATUS OF LIGNOCELLULOSIC BIOFUEL PRODUCTION

3.2.1 Overview of Conversion Technologies (Biochemical, Thermochemical)

Lignocellulosic biofuel production involves converting biomass into fuels, chemicals, and power. Two primary conversion technologies are biochemical and thermochemical conversion. Biochemical conversion utilizes microorganisms and enzymes to break down biomass into fermentable sugars. The key processes are: pretreatment (mechanical, chemical, or biological), hydrolysis (enzymatic or acid), fermentation (microbial or yeast), and distillation and purification to form products such as bioethanol, biobutanol, biodiesel, bioplastics, and so on. However, thermochemical conversion involves heat, pressure, and catalysts to transform biomass into fuels and chemicals. Here, the key processes are: pyrolysis (fast or slow), gasification, liquefaction, and hydroprocessing which produces products such as bio-oil, synthetic diesel, bio-gasoline, biochar, and so on (Khan and Zhang, 2023; Patel and Singh, 2023a).

The biochemical conversion found to be highly selective, using low energy and costly enzymes and found to be a slow process. In the other hand, thermochemical conversion is a fast, efficient, high energy using process where expensive catalyst was used (Singh and Gupta, 2023).

Nowadays, emerging trends shows that lignocellulosic biofuel can be produced using hybrid approaches (combining biochemical and thermochemical), using advanced biocatalysts and enzymes, using process integration and optimization and using algae-based lignocellulosic conversion which showed a high possibility in this field (Ershov et al. 2016; Singh and Gupta, 2023). Hybrid conversion improved the yield of bio-oil much higher as shown by Kumar et al. (2019). This was carried out by altering the biofuel production pathway. Effective conversion technologies are crucial for commercializing lignocellulosic biofuels. Ongoing research and development aim to overcome challenges, enhance efficiency, and reduce costs.

3.2.2. Commercial-scale Production: Success Stories and Challenges

Despite challenges of feedstock availability, high operating cost, conversion efficiency, scalability, and public acknowledgment, several companies had achieved positive note on commercial-scale lignocellulosic biofuel production. Some of them are as follows:

1. POET-DSM, located in Iowa, USA, is a pioneering facility producing 20 million gallons per year of cellulosic ethanol from corn stover, the agricultural residues left after corn harvesting. This facility utilizes advanced processing technologies, including pretreatment and enzymatic hydrolysis, to convert the lignocellulosic biomass into fermentable sugars, which are then fermented into ethanol using specialized microorganisms. By utilizing corn stover, POET-DSM not only helps reduce waste but also enhances

sustainability in agriculture, contributing to lower greenhouse gas emissions compared to fossil fuels. Overall, POET-DSM exemplifies the potential of cellulosic ethanol as a renewable energy solution, promoting both environmental sustainability and economic growth (https://poet.com/pr/first-commercial-scale-cellulosic-plant).

2. Red Rock Biofuels, located in California, USA, is an innovative facility producing 15 million gallons per year of renewable jet fuel from forest lignocellulose residues. The facility focuses on converting leftover biomass from forestry operations, such as tree limbs and other wood waste, into sustainable aviation fuel. Utilizing advanced thermochemical processes, Red Rock Biofuels efficiently breaks down lignocellulosic materials to produce high-quality renewable jet fuel, which helps reduce greenhouse gas emissions compared to traditional fossil fuels. This initiative not only addresses waste management in forestry but also contributes to the growing demand for sustainable aviation fuels, supporting both environmental sustainability and energy diversification in the aviation sector (Bakari et al., 2024).
3. Emerald Biofuels' at Cincinnati, Ohio, USA, converts lignocellulosic agricultural biomass into renewable diesel. Emerald Biofuels, located in Cincinnati, Ohio, USA, specializes in converting lignocellulosic agricultural biomass into renewable diesel. This facility utilizes advanced processing technologies to efficiently transform agricultural residues, such as corn stover and other plant materials, into high-quality diesel fuel. By focusing on biomass feedstocks, Emerald Biofuels contributes to reducing greenhouse gas emissions and enhancing energy sustainability. The renewable diesel produced is a drop-in fuel, compatible with existing diesel engines, making it a practical alternative to fossil fuels. This initiative helps diversify energy sources and promote environmental stewardship in the transportation sector (Rosales Calderon et al., 2024). Several other studies available for production of biomethane and other biofuel from lignocellulosic biomass (Stanley et al., 2022). These case studies exemplify the diverse range of biofuel production methods and feedstocks being utilized globally. From algae and waste materials to agricultural residues, these initiatives demonstrate the potential for sustainable energy solutions that contribute to environmental stewardship and energy security. Each project showcases innovative technologies and practices that address unique challenges in the biofuel industry.

3.2.3 Feedstock Sourcing and Supply Chain Management

Effective feedstock sourcing and supply chain management are crucial for lignocellulosic biofuel production. Strategies including diversifying feedstocks such as agricultural residues and municipal waste along with local sourcing from nearby locations to minimize transportation costs. The fermentation of lignocellulosic feedstock varies according on the raw materials and microorganisms used. The significance of effective fuel conversion, suitable biomass pretreatment, and effective supply chain management in the expansion of biomass power production was emphasized. Critical parameters for producing fuel gas with the ideal composition for internal combustion engines or turbines have been investigated by the author (Albashabsheh et al., 2020). Mujtaba et al. (2023) reviewed the conversion of lignocellulosic biomass from different agricultural waste to circular economy products, including biofuels, biocomposites, and bioplastics.

3.2.4 Economic and Environmental Assessments

Economic and environmental assessments highlight the potential benefits of lignocellulosic biofuels. Lignocellulosic biofuels offer a sustainable alternative to fossil fuels. Production and consumption of such biofuels reduces significantly greenhouse gas emissions. Also, the lignocellulosic biofuel industry has the potential to generate employment opportunities in rural areas. If produced domestically, biofuels might lead to a reduction in gasoline imports. As the production and use of biofuels increase, we are decreasing our need on imported fossil fuels. As green fuels, biofuels are more cost-effective and ecologically beneficial ways to satisfy energy demands in a sustainable manner (Sahoo et al., 2022; Rai et al., 2022). Rai and his research team described that the first-generation biofuels face limitations due to competition with food production and limited raw material supply. However, a viable substitute is provided by second-generation lignocellulosic biofuels, which are made from inedible biomass. Such a substitute would be economical and ecologically beneficial. Additionally, using lignocellulosic feedstock and agricultural waste residues may be a preferable choice in the near future to prevent the detrimental effects of first-generation biofuel production on food sources. Overall, lignocellulosic biofuel production has made significant strides, but continued innovation and investment need to be maintained to overcome existing challenges and achieve widespread adoption of the process.

3.3 PRETREATMENT AND PROCESSING TECHNOLOGIES OF LIGNOCELLULOSIC BIOFUEL

Lignocellulosic biofuels has emerged as a promising alternative to fossil fuels, offering a sustainable solution to mitigate climate change. However, the complex structure of lignocellulosic biomass requires efficient pretreatment and processing technologies to convert it into biofuels (Khan and Zhang, 2023). Various physical, chemical, biological, and advanced processing technologies were employed for lignocellulosic biofuel production.

3.3.1 Physical Pretreatment

(a) **Mechanical milling**: This method involves physically grinding of the biomass to reduce particle size and increase surface area. While effective in enhancing enzyme access, it requires significant energy input and may not sufficiently disrupt lignin structures (Khan and Zhang, 2023). Size of the particle plays a vital role in high yield of biofuel. The effectiveness of enzymatic hydrolysis is increased by smaller particles because they provide enzymes easier access to cellulose and hemicellulose. While the equipment and energy costs can be high, the process is generally straightforward and does not involve the use of chemicals. On the other hand, Mechanical milling can be energy-intensive, which offset some of the economic benefits. Also, though it effectively reduces particle size, milling may not sufficiently disrupt the lignin structure, which can inhibit enzyme access and overall conversion efficiency. Equipment can experience significant wear, leading to higher maintenance costs (Kumar et al., 2018). Mechanical millings were often combined with other pretreatment methods to enhance overall efficacy (Zhang et al., 2013).

(b) **Steam explosion**: Biomass undergoes fast decompression after being exposed to high-pressure steam in this process. As a result, the material becomes fragmented, which increases porosity and improves cellulose and hemicellulose accessibility. Additionally, it solubilizes some hemicellulose, increasing the efficiency of enzymatic hydrolysis later on (Zhang et al., 2013; Kumar et al., 2018). By increasing porosity, the method improves enzyme accessibility to cellulose and hemicellulose. The procedure solubilizes some hemicellulose, which may either be fermented directly or converted into useful by-products. Additionally, some inhibitory chemicals that might impede further fermentation can be reduced by employing steam explosion. On the other hand, a large initial capital expenditure is required due to the requirement for a high-pressure system during the process. Steam generation and pressure maintenance necessitate substantial energy input, despite the fact that it necessitates less energy than mechanical milling. Consequently, the degradation of sugars, particularly pentoses, can be influenced by elevated temperatures, which may have an impact on overall yields. Steam explosion is frequently implemented in the pretreatment of a variety of lignocellulosic materials, including agricultural residues (corn stover, sugarcane bagasse) and forestry residues (wood shavings, sawdust). It is frequently employed in conjunction with fermentation and enzymatic hydrolysis processes (Kumar et al., 2018).

In addition to these methods, other methods were employed, including microwave-assisted pretreatment, chilling, pyrolysis, and extrusion. This further demonstrated that the selection of pretreatment methods is entirely dependent on the type and composition of the biomass used in biofuel production (Kumari and Singh, 2018).

3.3.2 Chemical Pretreatment

(a) **Acid hydrolysis**: Using strong acids, such as sulfuric acid, this process hydrolyzes hemicellulose to produce fermentable sugars. Although efficient, it can result in the production of acetic acid and furfural, two inhibitors that have a detrimental effect on fermentation (Mansouri et al., 2016). Hemicellulose is broken down into fermentable sugars and lignin is depolymerized by acid hydrolysis (Kuglarz et al., 2018). Research revealed that while diluted acid pretreatment can produce notable sugar yields, high acid concentrations can also produce inhibitory byproducts such as hydroxymethylfurfural (HMF) and furfural (Mansouri et al., 2016). Sulfuric acid, for example, demonstrated a maximum bioethanol output of 90% (Kuglarz et al., 2018; Kumari and Singh, 2018).

(b) **Alkaline pretreatment**: Alkaline solutions (e.g., sodium hydroxide) are used to solubilize lignin and increase cellulose accessibility. This method can yield higher sugar concentrations and fewer inhibitors compared to acid pretreatment (Rezania et al., 2020). The alkalinity leads to saponification and depolymerization of lignin, facilitating the breakdown of the biomass structure (Kumar et al., 2022). Alkaline pretreatment can effectively enhance cellulose accessibility and produce fewer inhibitory compounds compared to acid pretreatment. However, it often requires high chemical concentrations and can lead to higher operational costs (Rezania et al., 2020; Molaverdi et al., 2019). Different biomasses such as sugarcane bagasse, sugarcane straw, Eucalyptus residue, Wheat and Rice straw, and so on, when treated with sodium carbonate, sodium hydroxide, and so on, showed high bioethanol yield upto 83 g/L as described by Rezania et al. in 2020.

(c) **Organo solvent pretreatment**: Utilizing organic solvents (such as ethanol or acetone) to dissolve lignin and hemicellulose, enhances cellulose accessibility while allowing for the recovery of valuable by-products (Rezania et al., 2020). This method effectively breaks down the lignin structure while preserving cellulose integrity (Fang, 2013). Organosolv pretreatment such as chlorine base and iron based organosolv can yield high-quality cellulose and high yield of biofuel with minimal degradation of sugars, but the recovery and recycling of solvents found to be the present challenge (Wang et al., 2017; Rezania et al., 2020, Kumari and Singh, 2018).

(d) **Ionic liquid pretreatment**: Ionic liquids can effectively dissolve cellulose and disrupt lignin structures at relatively low temperatures, facilitating easier enzymatic access. Although promising, this method is still under research for large-scale application for biofuel production (Zhang et al., 2015; Wang et al., 2017; Kumari and Singh, 2018).

3.3.3 Combined Pretreatment Methods

Integrating various pretreatment methods can enhance biomass deconstruction more effectively than individual treatments.

Physical and chemical pretreatment: Combination of steam explosion (a physical method) with dilute acid treatment (chemical method) can improve the accessibility of cellulose by breaking down hemicellulose and lignin (Rezania et al., 2020). This synergistic effect often results in higher sugar yields.

Sequential pretreatment strategies: Using a combination of alkaline and organosolv treatments can effectively solubilize lignin while preserving cellulose structure (Kumar et al., 2022). Such strategies enhance the quality of the resulting sugars and reduce the formation of inhibitory compounds.

3.3.4 Biological Pretreatment

Biological pretreatment employs fungi and bacteria to degrade lignin and hemicellulose. White-rot fungi, such as *Phanerochaete chrysosporium*, *Trametes hirsute*, white-rot fungus *Irpex lacteus*, and so on, produce ligninolytic enzymes capable of breaking down lignin and enhancing cellulose accessibility (Rezania et al., 2020). This approach is environmentally friendly and often involves lower energy inputs but requires longer processing times (Mäkelä et al., 2014).

3.3.5 Enzymatic Hydrolysis

After pretreatment, hydrolysis by enzymatic means is used to convert available cellulose and hemicellulose into fermentable sugars (Singh and Gupta, 2023; Makela et al., 2014):

Cellulases: A cocktail of cellulolytic enzymes, including endoglucanases, exoglucanases, and β-glucosidases, is utilized to break down cellulose into glucose (Singh and Gupta, 2023).

Hemicellulases: Enzymes such as xylanases and arabinofuranosidases are employed to hydrolyze hemicellulose into pentoses and hexoses, thus enhancing the overall sugar yield for fermentation. They are degraded by anaerobes (Bacteroidetes and Firmicutes especially *Clostridia* (Makela et al., 2014).

3.3.6 Fermentation

The sugars obtained from enzymatic hydrolysis can be fermented by various microorganisms, including bacteria and yeast. *Saccharomyces cerevisiae* is the most commonly used yeast for ethanol production, while engineered strains capable of fermenting pentoses have been developed to improve the utilization of hemi-cellulosic sugars (Rezania et al., 2020).

3.3.7 Integrated Processing Approaches

Recent advancements focus on integrated processing approaches that combine pretreatment, enzymatic hydrolysis and fermentation into a single step, known as consolidated bioprocessing (CBP). This method reduces costs and enhances efficiency by minimizing the time and resources required for separate processing stages (Lynd et al., 2017; Zhang et al., 2018b). Use of nanocatalysts enhances enzymatic hydrolysis and fermentation efficiency (Almeida et al., 2018). Membrane separation technology recovers and recycles enzymes, water, and other valuable compounds. Ultrasound-assisted processing improves enzymatic hydrolysis and fermentation efficiency (Wang et al., 2017). Microwave-assisted processing enhances biomass pretreatment and hydrolysis (Almeida et al., 2018). Emerging technologies, such as ionic liquid pretreatment and microwave-assisted processing, offer new avenues for enhancing biomass conversion efficiency (Wang et al., 2017). Ionic liquids can effectively dissolve lignin and facilitate cellulose accessibility but require further research to mitigate cost issues (Huang et al., 2018; Wang et al., 2017). Microwave-assisted processing offers rapid heating and can enhance sugar yields compared to traditional methods (Almeida et al., 2018; Huang, Z., et al., 2018). The development of effective pretreatment technologies is crucial for the successful conversion of lignocellulosic biomass into biofuels. Pretreatment and further processing technologies play a critical role in lignocellulosic biofuel production. Physical methods, such as mechanical milling and steam explosion, and so on, offer distinct advantages and challenges (Kumar et al., 2022; Kumar et al., 2018, Kumari and Singh, 2018). The integration of chemical and biological pretreatment strategies, along with emerging advanced technologies, holds promise for improving the efficiency and economic viability of lignocellulosic biofuel production. Future research should focus on optimizing these integrated methods and exploring synergistic strategies to maximize biofuel yields.

3.4 MICROBIAL AND ENZYMATIC CONVERSION OF LIGNOCELLULOSIC BIOFUEL

The process of converting of lignocellulosic matter into biofuels is a viable technique for producing renewable energy while minimizing the consequences of climate change. This method often includes fermentation by microbes, enzymatic hydrolysis, consolidated bioprocessing, and metabolic engineering, all of which help to improve the efficiency and viability of biofuel generation.

Microbial fermentation is an important stage in the breakdown of lignin-based biomass to biofuels, specifically ethanol (Patel and Singh, 2023a). This process can utilize various microorganisms, including bacteria and yeast. Certain bacteria, such as *Clostridium acetobutylicum* and *Enterobacter* species, are known for their ability to ferment sugars derived from lignocellulosic biomass into biofuels like butanol and ethanol (Guo et al., 2022). These bacteria can utilize a wide range of substrates, making them suitable for diverse lignocellulosic feedstocks (Zhang et al., 2018b). *Saccharomyces cerevisiae*, are widely used in bioethanol production due to their high fermentation efficiency and ability to tolerate high ethanol concentrations (Topaloglu et al., 2023). Recently, genetically engineered yeasts, capable of fermenting pentoses and hexoses, have been developed to improve the utilization of hemicellulosic sugars (Saha and Kennedy, 2020). Lignocellulosic biomass such as rice straw, wheat bran, different food waste improved the lipid yield, thus increases the biodiesel yield (Chintagunta et al., 2021).

Enzymatic hydrolysis, on the other hand, is the process of fragmenting down complex lignocellulosic molecules into fermentable sugars utilizing particular enzymes, primarily cellulases and xylanases. Cellulases are a class of enzymes that degrade cellulose into glucose. They are made by a variety of fungus and bacteria. The crystalline structure of cellulose can impair cellulase efficacy, necessitating the utilization of a cocktail of enzymes for optimal conversion (Chintagunta et al., 2021). Xylanases degrade hemicellulose, converting xylan into xylose and other sugars. Cellulases and xylanases working together can greatly increase the whole output of fermentable sugars from lignocellulosic material, resulting in increased biofuel production (Bhardwaj et al., 2019).

Consolidated bioprocessing (CBP) is an innovative approach that integrates enzymatic hydrolysis and microbial fermentation into a single step (Zhang et al., 2018b). This method can significantly reduce production costs and increase efficiency by minimizing the need for separate hydrolysis and fermentation processes. In CBP, engineered microorganisms are employed that can produce enzymes and ferment sugars simultaneously. For instance, some strains of *Saccharomyces cerevisiae* and *Zymomonas mobilis* have been developed to express cellulolytic enzymes, enabling them to directly utilize cellulose for ethanol production (Lynd et al., 2017; Dixit et al., 2023). This integrated approach holds the potential to streamline biofuel production, reducing both time and resource expenditure.

Combining enzymatic hydrolysis and fermentation processes in one phase, referred to as simultaneous saccharification and fermentation (SSF), may additionally enhance efficiency. In SSF, microorganisms and enzymes collaborate to turn cellulose into sugars, which are subsequently fermented into biofuels in one vessel. This technique lowers the likelihood of sugar inhibition while increasing total productivity (Yoon et al., 2014). CBP combines enzyme synthesis, substrate hydrolysis, and fermentation in a single process by using engineered microbes capable of generating enzymes and biofuels. This strategy lowers production expenses and streamlines the process (Zhang et al., 2018b).

Metabolic engineering is instrumental in optimizing microbial fermentation processes, significantly enhancing their efficiency. By modifying the metabolic pathways of microorganisms, researchers can optimize the conversion of lignocellulosic sugars into biofuels. Through genetic engineering, pathways that convert sugars to desired biofuels can be enhanced. For instance, the introduction of additional genes or modification of existing ones can improve the yield of ethanol or butanol (Nielsen et al., 2013). Additionally, engineering microbes to increase their tolerance to inhibitors present in lignocellulosic hydrolysates, such as furfural and acetic acid, can further improve fermentation efficiency (Patel and Singh, 2023a). This allows for higher concentration substrates, ultimately enhancing overall biofuel yields.

The microbial and enzymatic conversion of lignocellulosic biomass into biofuels represents a multifaceted approach to developing sustainable energy solutions. By leveraging microbial fermentation, enzymatic hydrolysis, consolidated bioprocessing, and metabolic engineering, researchers continue to improve the efficiency and viability of biofuel production from lignocellulosic sources. Future research should focus on integrating these strategies to optimize the conversion processes further, addressing both economic and environmental challenges associated with biofuel production.

3.5 ARTIFICIAL INTELLIGENCE (AI) APPLICATIONS FOR LIGNOCELLULOSIC BIOFUEL

The growing interest in renewable energy sources, particularly biofuels derived from lignocellulosic biomass, has highlighted the need for innovative technologies to optimize production processes. Artificial intelligence (AI) has emerged as a transformative tool in this domain, providing capabilities for predictive modeling, machine learning, real-time monitoring and integration with Internet of Things (IoT) technologies (Mishra and Thakur, 2022). These applications not only enhance efficiency but also improve decision-making and resource management in biofuel production (Chaturvedi and Choudhary, 2023; Kumar and Kumar, 2023; Clauser et al., 2022).

3.5.1 Predictive Modeling for Process Optimization

Predictive modeling involves using AI algorithms to analyze historical data and predict future outcomes, making it a valuable tool for optimizing biofuel production processes. AI can model complex biochemical processes such as enzymatic hydrolysis and fermentation, allowing for the identification of optimal operating conditions (e.g., temperature, pH, enzyme concentrations) (Fernandez and Gonzalez, 2023). Using techniques such as regression analysis and neural networks, predictive models can simulate different scenarios, and recommend adjustments to improve sugar yields and fermentation efficiency (Zhang et al., 2020). These models can continuously assess real-time data from production processes, enabling operators to make informed decisions based on predicted outcomes. For example, by predicting when a process is likely to deviate from optimal conditions, interventions can be implemented proactively, minimizing downtime and resource waste (Cai et al., 2021; Clauser et al., 2022: Li and Chen, 2022).

3.5.2 Machine Learning for Feedstock Selection and Quality Control

Machine learning algorithms are instrumental in enhancing the selection and quality control of lignocellulosic feedstocks. Machine learning techniques, such as clustering and classification, can analyze various feedstock characteristics (e.g., lignin content, cellulose accessibility) to identify the most suitable biomass for specific conversion processes. This is crucial for ensuring high yields and efficient conversion rates (Dodo et al., 2022). AI can also automate quality assessment through image recognition and sensor data analysis, providing real-time feedback on feedstock quality. For instance, computer vision systems can analyze the physical properties of biomass (color, size, etc.) to determine its suitability for processing, thereby improving the consistency of feedstock used in biofuel production (Chaturvedi and Choudhary, 2023; Kumar and Kumar, 2023; Li and Chen, 2022; Zhang and Wang, 2022).

3.5.3 AI-Driven Monitoring and Control Systems

AI-driven monitoring and control systems enhance the operational efficiency of biofuel production facilities by facilitating real-time data analysis and decision-making. AI systems can collect and analyze data from various sensors deployed throughout the production process, monitoring parameters such as temperature, pressure, and chemical concentrations using real-time monitoring system. This data can be used to identify anomalies, allowing for quick adjustments to maintain optimal conditions (Kumar, 2020). Advanced control systems that incorporate AI algorithms can automate adjustments to the production process based on real-time data. For example, machine learning algorithms can optimize enzyme dosages in enzymatic hydrolysis, improving efficiency and reducing costs associated with enzyme procurement (Clauser et al., 2022; Zhang and Wang, 2022).

3.5.4 Integration of AI with IoT for Real-Time Optimization

The integration of AI with IoT technologies enables the creation of smart biofuel production systems capable of real-time optimization. IoT devices can collect vast amounts of data from various stages of the biofuel production process, including feedstock preparation, pretreatment, hydrolysis, and fermentation. This data serves as input for AI algorithms, allowing for comprehensive analysis and modeling of the entire production chain (Shaw et al., 2021; Ghosh and Sen, 2023). AI algorithms can analyze IoT-collected data to optimize production processes continuously. For instance, machine learning can predict optimal operational parameters based on real-time feedstock quality and environmental conditions. This dynamic approach allows for quick adaptations to changing conditions, maximizing efficiency and yield (Clauser et al., 2022). By combining AI and IoT, predictive maintenance enables machine learning models to analyze sensor data, forecasting equipment failures and minimizing downtime and maintenance expenses (Ghosh, 2016; Zhang and Wang, 2022).

Hence, application of artificial intelligence in the production of lignocellulosic biofuels is paving the way for more efficient, cost-effective, and sustainable biofuel production processes. Through predictive modeling, machine learning for feedstock selection, AI-driven monitoring systems, and integration with IoT technologies, AI can significantly enhance decision-making and operational efficiency in the biofuel sector. As research and development in this area continue to advance, the potential for AI to optimize Lignocellulosic biofuel production will grow, contributing to the broader transition toward renewable energy sources.

3.6 CHALLENGES AND LIMITATIONS

The production of biofuels from lignocellulosic biomass offers significant potential for renewable energy, but it is not without its challenges. Several factors affect the efficiency and feasibility of converting lignocellulosic materials into viable biofuels. This write-up examines the primary challenges and limitations in this domain.

3.6.1 Recalcitrance of Lignocellulosic Biomass

Lignocellulosic biomass is characterized by its complex structure, composed of cellulose, hemicellulose, and lignin. This composition results in recalcitrance, which is the resistance to degradation and conversion into fermentable sugars. The intertwined nature of cellulose, hemicellulose, and lignin makes it difficult for enzymes to access and break down cellulose and hemicellulose. This requires effective pretreatment methods, which can be energy-intensive and costly. Also, during pretreatment, the breakdown of lignin and hemicellulose can produce inhibitory compounds (like phenols and furfural) that adversely affect fermentation processes. This not only reduces yields but also complicates downstream processing (Abolore et al., 2024; Manyi-loh et al., 2023). Overcoming recalcitrance is crucial for improving biofuel yields. Research continues to focus on optimizing pretreatment and enzymatic hydrolysis processes. Different optimization process needs to be implemented for further improvement of the biofuel yield. Anaerobic co-digestion technology has emerged as a vital solution for efficiently managing organic waste and sewage sludge disposal. Also, some advanced modeling system development can be carried out, but significant advancements are still needed (Khan and Zhang, 2023; Kucharska et al., 2018).

3.6.2 Scalability and Cost-Effectiveness

Transitioning from laboratory-scale production to commercial-scale biofuel facilities poses significant challenges related to scalability and economic viability. The costs associated with biomass collection, transportation, pretreatment, and enzyme production can be prohibitively high. Economies of scale have not been fully realized, making lignocellulosic biofuels less competitive with fossil fuels (Broda et al., 2022; Zheng et al., 2022). Many advanced technologies, such as consolidated bioprocessing, remain in the research phase and have not yet been proven at scale. Sun et al. in 2023 showed enhanced volatile fatty acid production without methanogenesis inhibitors might be vital to reduce the cost and further effect the scalability for biofuel production. To optimize lignocellulose breakdown, alkali-based pretreatments are commonly supplemented with CO_2 or acid additives in the post-pretreatment stage for efficient neutralization as described by Kruger et al., 2022. Further addressing scalability and cost-effectiveness requires significant investment in research and infrastructure, as well as the development of innovative technologies that can reduce production costs while maintaining high efficiency.

3.6.3 Water and Energy Consumption

The production of lignocellulosic biofuels can be water and energy-intensive, which poses challenges for sustainable production. Many pretreatment and processing methods for biofuels, such as steam explosion or chemical treatments, require substantial amounts of water and energy. This can strain local resources, particularly in regions where water scarcity is an issue (Broda et al., 2022; Sharma et al., 2023).

Further, Yuan et al. in 2021 approached integrated biorefinery process for dual production of ethanol and biogas which might reduce the water consumption in a way lessen energy consumption. But, comprehensive assessments of the water and energy inputs across the entire production lifecycle are often lacking in most of the cases. Without this data, it is difficult to gauge the environmental sustainability of Lignocellulosic biofuels compared to traditional fossil fuels. Developing more efficient processes that minimize water and energy consumption is critical. The integration of renewable energy sources into the production process could help alleviate some of these concerns.

3.6.4 Policy and Regulatory Frameworks

The regulatory landscape plays a significant role in shaping the biofuel industry. Policies can either support or hinder the development and deployment of lignocellulosic biofuels. Many countries lack comprehensive and stable policies that support lignocellulosic biofuel development. Inconsistent regulations can deter investment and innovation (Dutta et al., 2023).. Policies that favor conventional biofuels (like corn ethanol) can disadvantage lignocellulosic biofuels, which are often perceived as less mature technologies. This creates a challenging market environment for new entrants. Creating supportive policy frameworks that encourage research, development, and commercialization of lignocellulosic biofuels is essential for fostering a robust biofuel industry. This includes incentives for investment and mechanisms to ensure fair competition (Broda et al., 2022; Dutta et al., 2023). A study by Usmani et al in 2023 showed development of framework named Comprehensive Biofuel Policy Analysis (CBPA) and its utilization by different major countries. Also, Saravanan et al in 2020 studied evolution of liquid biofuel policy in BRICKS nations. In India also various programs were performed for policy related to biofuels (Saravanan et al., 2020).

Lignocellulosic biofuel production faces a complex array of obstacles, spanning technological hurdles, economic viability, environmental concerns, and regulatory frameworks. Overcoming these challenges demands a collaborative approach, bringing together experts from research, industry, and policy-making to pioneer novel solutions that would boost efficiency, ensure sustainability, and enhance competitiveness of lignocellulosic biofuels, ultimately unlocking their full potential.

3.7 FUTURE SCOPE AND DIRECTIONS

The future of lignocellulosic biofuel production holds significant promise as advancements in technology and a growing commitment to sustainability reshape the energy landscape. Emerging concepts and practices will enhance efficiency, lower costs, and promote environmental benefits. This write-up explores key future directions for lignocellulosic biofuels, focusing on advanced biorefinery concepts, integration with renewable energy, new conversion technologies, emerging feedstocks, and sustainability assessments.

3.7.1 Advanced Biorefinery Concepts

The biorefinery concept involves the integrated processing of biomass into a spectrum of valuable products, including biofuels, biochemicals, and bioproducts. Future research will focus on optimizing biorefinery processes to increase the efficiency of converting lignocellulosic biomass into multiple products. Advanced separation techniques and integrated processing schemes will enhance the economic viability of biorefineries (Ozdenkçi et al., 2017; Ragauskas et al., 2014). Developing modular and decentralized biorefinery systems will facilitate localized production, reducing transportation costs, and enhancing energy security. These systems can adapt to varying feedstock availability and regional energy needs (Ragauskas et al., 2014). Emphasizing waste valorization and circular economy principles will enable biorefineries to minimize waste and create closed-loop systems, where by-products are repurposed within the process.

3.7.2 Integration with Other Renewable Energy Sources

Combining Lignocellulosic biofuel production with other renewable energy sources can enhance overall energy efficiency and reliability. Integrating biofuel production with solar, wind, and geothermal energy can optimize resource use and reduce reliance on fossil fuels. Hybrid systems can utilize excess energy generated from renewables to power biorefineries, further lowering operational costs (Ershov et al., 2016; Yuan et al., 2021). Research into energy storage technologies that can store excess energy from renewables for later use in biofuel production processes will enhance system flexibility and resilience. This will also enable better management of supply and demand (Zhou et al., 2021).

3.7.3 Development of New Conversion Technologies

Innovative conversion technologies are essential for improving the efficiency and effectiveness of Lignocellulosic biomass conversion. Ongoing research into next-generation enzymes and engineered microbial strains capable of efficiently breaking down lignocellulosic substrates will play a crucial role. Improved enzyme cocktails can enhance hydrolysis rates and reduce the need for extensive pretreatment (Lynd et al., 2017). Exploring gasification and pyrolysis such as plasma assisted pyrolysis as alternative conversion routes will provide opportunities to produce biofuels, biochar, and bio-oil from lignocellulosic biomass (Velvizhi et al., 2023). These technologies can also enable the utilization of lower-quality feedstocks, enhancing overall feedstock flexibility (Pena-Castro et al., 2023). The application of AI in optimizing conversion processes and enhancing real-time monitoring will streamline production and improve process control, leading to higher yields and lower costs (Zhang et al., 2020; Mishra and Thakur, 2022).

3.7.4 Emerging Feedstocks (Agricultural Residues, Municipal Waste)

Utilizing diverse feedstocks can enhance the sustainability and economic viability of lignocellulosic biofuels. Increasing the use of agricultural residues (e.g., straw, corn stover) as feedstocks can reduce competition with food crops and promote waste valorization. Research should focus on optimizing collection and processing techniques for these materials (Zhang et al., 2018). The conversion of municipal solid waste (MSW) into biofuels presents an opportunity to address waste management challenges while reducing energy. Developing efficient methods for separating and processing organic waste into biofuel will be a critical area of research (Pérez et al., 2017). Exploring algae as a complementary feedstock offers additional potential for biofuel production. Algae can utilize non-arable land and waste nutrients, contributing to a more sustainable biofuel landscape (Chisti, 2007; Lin and Lu, 2021).

3.7.5 SWOT Analysis

Conducting comprehensive assessments of the environmental impacts of biofuel production is crucial for ensuring sustainability. Future studies should integrate life cycle assessments (LCA) and SWOT analysis to account for all stages of biofuel production, from feedstock cultivation to end-use. This integrated approach will identify areas for improvement, optimize sustainability, and reveal opportunities for growth (Czyrnek-Delêtre et al., 2017; Patel and Singh, 2023b; Prasad et al., 2016). Specifically, SWOT analysis will help pinpoint:

- Strengths: efficient feedstock cultivation, low-carbon processing
- Weaknesses: high water usage, land-use concerns
- Opportunities: job creation, energy access, social equity
- Threats: market fluctuations, regulatory changes

Incorporating socio-economic considerations into sustainability analyses will provide a nuanced understanding of biofuel production's impacts on local communities and economies (Davis et al., 2018). Standardized sustainability certification frameworks will ensure lignocellulosic biofuels meet environmental and social criteria, bolstering consumer confidence and market acceptance.

3.8 CONCLUSION

The exploration of lignocellulosic biofuels represents a significant frontier in the quest for sustainable energy solutions. As the world increasingly turns its attention to renewable energy sources, lignocellulosic biomass—derived from agricultural residues, forestry by-products, and even municipal waste—offers a promising avenue for producing environmentally friendly fuels. Currently, the lignocellulosic biofuel industry is at a critical juncture, characterized by rapid technological advancements and an expanding body of research. While significant progress has been made in pretreatment, enzymatic hydrolysis, and fermentation technologies, the sector still faces challenges related to feedstock recalcitrance, scalability, and cost-effectiveness. The integration of advanced biorefinery concepts and innovative conversion technologies is essential to overcome these barriers and maximize the efficiency of lignocellulosic biofuel production. Looking to the future, the prospects for lignocellulosic biofuels are promising, particularly with the growing emphasis on sustainability and carbon neutrality. The increasing integration of lignocellulosic biofuel systems with other renewable energy sources and the exploration of new feedstocks are set to enhance the viability and attractiveness of biofuels. The industry's evolution will hinge on addressing existing challenges through research, investment, and collaboration across sectors.

Despite the advances in lignocellulosic biofuel production, several critical research gaps remain that present opportunities for further exploration. There is a need for more effective and economically viable pretreatment methods that reduce energy and resource consumption while maximizing sugar release. Research into novel enzymatic formulations and microbial strains that can improve hydrolysis and fermentation efficiency is also essential. Investigating the use of alternative feedstocks, such as algae and urban organic waste, can expand the resource base for biofuel production. Research into optimizing the collection, processing, and conversion of these feedstocks can significantly enhance sustainability. Comprehensive lifecycle assessments that encompass social, economic, and environmental impacts will provide insights into the true sustainability of lignocellulosic biofuels. Developing frameworks for continuous monitoring and evaluation will support informed decision-making. The integration of artificial intelligence in lignocellulosic biofuel production has the potential to revolutionize the sector. AI can enhance predictive modeling. Furthermore, AI-powered surveillance and management systems may perform modifications in real-time, significantly reducing operational costs and improving yield.

In summary, the future of lignocellulosic biofuel production is bright, driven by innovation, research, and the transformative capabilities of technology. By addressing existing challenges and capitalizing on emerging opportunities, particularly through the application of AI, the lignocellulosic biofuel industry can perform a pivotal character in the changeover to a further sustainable energy future.

REFERENCES

https://poet.com/pr/first-commercial-scale-cellulosic-plant First commercial-scale cellulosic ethanol plant in the U.S. opens for business.

Abolore, R.S., Jaiswal, S. and Jaiswal, A.K. (2024). Green and sustainable pretreatment methods for cellulose extraction from lignocellulosic biomass and its applications: A review. *Carbohydrate Polymer Technologies and Applications*, 7, 100396. doi: 10.1016/j.carpta.2023.100396.

Adeniyi, A.A. and Fagbenro, J.A. (2023). Enabling technologies for the production of second-generation biofuels from lignocellulosic biomass. *Renewable Energy*, 216, 444–457. doi: 10.1016/j.renene.2023.04.027.

Albashabsheh, N.T. and Stamm, J.L.H. (2020). Optimization of lignocellulosic biomass-to-biofuel supply chains with densification: Literature review. *Biomass and Bioenergy*, 144, 105888.

Almeida, J.R.M., Modig, T., Petersson, A., Hahn-Hägerdal, B., Liden, G., Gorwa-Grauslund, M.F. (2018). Microwave-assisted hydrolysis of lignocellulosic biomass: A review. *Renewable and Sustainable Energy Reviews*, 82, 2835–2851. doi: 10.1016/j.rser.2017.10.036.

Ashokkumar, V., Venkatkarthick, R., Jayashree, S., Chuetor, S., Dharmaraj, S., Kumar, G., Chen, W.-H. and Ngamcharussrivichai, C. (2022). Recent advances in lignocellulosic biomass for biofuels and value-added bioproducts: A critical review. *Bioresource Technology*, 344, Part B, 126195. doi: 10.1016/j.biortech.2021.126195.

Bakari, R., Asha, R., Hossein, M., Huang, X., Islam, N.F., Liew, R.K., Narayan, M., Lam, S. S. and Sarma, H. (2024). Converting food waste to biofuel: A sustainable energy solution for Sub-Saharan Africa. *Sustainable Chemistry for the Environment*, 7, 100126. doi: 10.1016/j.scenv.2024.100126.

Bhandari, G.R. and Khosravi, M. (2023). The role of predictive analytics in the optimization of biofuel production processes. *Journal of Cleaner Production*, 367, 132694. doi: 10.1016/j.jclepro.2022.132694.

Bhardwaj, N., Kumar, B. and Verma, P. (2019). A detailed overview of xylanases: An emerging biomolecule for current and future prospective. *Bioresources and Bioprocessing*, 6(1), 40. doi: 10.1186/s40643-019-0276-2.

Broda, M., Yelle, D.J. and Serwańska, K. (2022). Bioethanol production from lignocellulosic biomass—Challenges and solutions. *Molecules*, 27(24), 8717. doi: 10.3390/molecules27248717.

Cai, Y., Zheng, Z., Schäfer, F., Stinner, W., Yuan, X., Wang, H., Cui, Z. and Wang, X. (2021). Pretreatment of lignocellulosic biomass in anaerobic digestion: Achievements and challenges in Germany and China. *Journal of Cleaner Production*, 299, 126885. doi: 10.1016/j.jclepro.2021.126885.

Chaturvedi, S. and Choudhary, A. (2023). Machine learning applications in biofuel production: A comprehensive review. *Fuel Processing Technology*, 228, 107190. doi: 10.1016/j.fuproc.2023.107190.

Chintagunta, A.D., Zuccaro, G., Kumar, M., Kumar, S.P.J., Garlapati, V.K., Postemsky, P.D., Kumar, N.S.S., Chandel, A.K. and Simal-Gandara, J. (2021). Biodiesel production from lignocellulosic biomass using oleaginous microbes: Prospects for integrated biofuel production. *Frontiers in Microbiology*, 12, 658284. doi: 10.3389/fmicb.2021.658284.

Chisti, Y. (2007). Biodiesel from microalgae. *Biotechnology Advances*, 25(3), 294–306. doi: 10.1016/j.biotechadv.2007.02.001.

Clauser, N.M., Felissia, F.E., Area, M.C. and Vallejos, M.E. (2022). Integrating the new age of bioeconomy and industry 4.0 into biorefinery process design. *BioResources*, 17(3), 5510.

Czyrnek-Delêtre, M.M., Smyth, B.M. and Murphy, J.D. (2017). Beyond carbon and energy: The challenge in setting guidelines for life cycle assessment of biofuel systems. *Renewable Energy*, 105, 436–448. doi: 10.1016/j.renene.2016.11.043.

Davis, R.E., Grundl, N.J., Tao, L., Biddy, M.J., Tan, E.C., Beckham, G.T., Humbird, D., Thompson, D.N. and Roni, M.S. (2018). *Process Design and Economics for the Conversion of Lignocellulosic Biomass to Hydrocarbon Fuels and Coproducts: Biochemical Design Case Update; Biochemical Deconstruction and Conversion of Biomass to Fuels and Products via Integrated Biorefinery Pathways*. United States. https://doi.org/10.2172/1483234.

Devi, A., Bajar, S., Kour, H., Kothari, R., Pant, D. and Singh, A. (2022). Lignocellulosic biomass valorization for bioethanol production: A circular bioeconomy approach. *Bioenergy Research*, 15(4), 1820–1841. doi: 10.1007/s12155-022-10401-9.

Dixit, Y., Yadav, P., Sharma, A.K., Pandey, P. and Kuila, A. (2023). Multiplex genome editing to construct cellulase engineered *Saccharomyces cerevisiae* for ethanol production from cellulosic biomass. *Renewable and Sustainable Energy Reviews,* 187, 113772. doi: 10.1016/j.rser.2023.113772.

Dodo, U.A., Ashigwuike, E.C. and Abba, S.I. (2022). Machine learning models for biomass energy content prediction: A correlation-based optimal feature selection approach. *Bioresource Technology Reports*, 19, 101167. doi: 10.1016/j.biteb.2022.101167.

Dutta, S., Saravanabhupathy, S., Anusha, Rajak, R.C., Banerjee, R., Dikshit, P.K., Padigala, C. T., Das, A.K. and Kim, B.S. (2023). Recent developments in lignocellulosic biofuel production with nanotechnological intervention: An emphasis on ethanol. *Catalysts*, 13(11), 1439. doi: 10.3390/catal13111439.

Ershov, M.A., Grigoreva, E.V., Habibullin, I.F., Emelyanov, V.E. and Strekalina, D.M. (2016). Prospects of bioethanol fuels E30 and E85 application in Russia and technical requirements for their quality. *Renewable and Sustainable Energy Reviews*. 66, 228–232. 10.1016/j.rser.2016.07.054.

Fang, Z. (2013). *Pretreatment Techniques for Biofuels and Biorefineries* (p. 457). Berlin: Springer.

Farghali, M., Mohamed, I.M.A., Osman, A.I., Rooney and D.W. (2023). Seaweed for climate mitigation, wastewater treatment, bioenergy, bioplastic, biochar, food, pharmaceuticals, and cosmetics: A review. *Environmental Chemistry Letters*, 21, 97–152. doi: 10.1007/s10311-022-01520-y.

Fernandez, A. and Gonzalez, A. (2023). The potential of AI in the optimization of biomass pretreatment processes: A review. *Fuel*, 335, 126794. doi: 10.1016/j.fuel.2022.126794.

Ghosh, A. (2016). Systems and synthetic biology for the microbial production of biofuels. *Current Metabolomics*, *4*(1), 5–13.

Ghosh, S. and Sen, R. (2023). Integrating AI and IoT in biomass energy systems: Prospects and challenges. *Renewable Energy Focus*, 41, 113–125. doi: 10.1016/j.ref.2022.11.005.

Gunasekaran, M.P., Kumar, G., Karthikeyan, O.P., Varjani, S. and Banu, J.R. (2021). Lignocellulosic biomass as an optimistic feedstock for the production of biofuels as valuable energy source: Techno-economic analysis. *Environmental Technology and Innovation*, 24, 102080. doi: 10.1016/j.eti.2021.102080.

Guo, Y., Liu, Y., Guan, M., Tang, H., Wang, Z., Lin, L. and Pang, H. (2022). Production of butanol from lignocellulosic biomass: Recent advances, challenges, and prospects. *RSC Advances*, 12(29), 18848–18863. doi: 10.1039/d1ra09396g.

Haberzettl, J., Hilgert, P. and von Cossel, M. (2021). A critical review on lignocellulosic biomass yield modeling and the bioenergy potential from marginal land. *Agronomy*, 11(12), 2397.

Huang, Y.-F., Kuan, W.-H. and Chang, C.-Y. (2018). Effects of particle size, pretreatment, and catalysis on microwave pyrolysis of corn stover. *Energy*, 143, 696–703. doi: 10.1016/j.energy.2017.11.022.

Khan, M.A. and Zhang, S. (2023). Advancements in biomass pretreatment technologies: A machine learning perspective. *Biomass and Bioenergy*, 163, 106521. doi: 10.1016/j.biombioe.2022.106521.

Kruger, J.S., Dreiling, R.J., Wilcox, D.G., Ringsby, A.J., Noon, K.L., Amador, C.K., Brandner, D.G., Ramirez, K.J., Haugen, S.J., Klein, B.C., Davis, R., Hanes, R.J., Happs, R.M., Cleveland, N.S., Christensen, E.D., Miscall, J. and Beckham, G.T. (2022). Lignin alkaline oxidation using reversibly-soluble bases. *Green Chemistry*, 24(22), 8733–8741. doi: 10.1039/d2gc03333j.

Kucharska, K., Hołowacz, I., Konopacka-Łyskawa, D., Rybarczyk, P. and Kamiński, M. (2018). Key issues in modeling and optimization of lignocellulosic biomass fermentative conversion to gaseous biofuels. *Renewable Energy*, 129, Part A, 384–408. doi: 10.1016/j.renene.2018.06.018.

Kuglarz, M., Alvarado-Morales, M., Dąbkowska, K. and Angelidaki, I. (2018). Integrated production of cellulosic bioethanol and succinic acid from rapeseed straw after dilute-acid pretreatment. *Bioresource Technology*, 265, 191–199.

Kumar, A. and Kumar, S. (2023). Role of machine learning in optimizing biofuel production from lignocellulosic biomass: A review. *Journal of Cleaner Production*, 391, 135648. doi: 10.1016/j.jclepro.2022.135648

Kumar, G., Dharmaraja, J., Arvindnarayan, S., Shoban, S., Bakonyi, P., Saratale, G.D., Nemestóthy, N., Bélafi-Bakó, K., Yoon, J.J. and Kim, S.H. (2019). A comprehensive review on thermochemical, biological, biochemical and hybrid conversion methods of bio-derived lignocellulosic molecules into renewable fuels. *Fuel*, 251, 352–367.

Kumar, P., Barrett, D.M., Delwiche, M.J. and Stroeve, P. (2018). Methods for pretreatment of lignocellulosic biomass for efficient hydrolysis and biofuel production. *Industrial and Engineering Chemistry Research*, 48(8), 3713–3729. doi: 10.1021/ie801542g.

Kumar, R., Kim, T., Basak, B., Patil, S., Kim, H., Ahn, Y., Yadav, K., Pinto, M.C. and Jeon, B.-H. (2022). Emerging approaches in lignocellulosic biomass pretreatment and anaerobic bioprocesses for sustainable biofuels production. *Journal of Cleaner Production*, 333, 130180. doi: 10.1016/j.jclepro.2021.130180 .

Kumar, S. (2020). Estimation capabilities of biodiesel production from algae oil blend using adaptive neuro-fuzzy inference system (ANFIS). *Energy Sources, Part A: Recovery, Utilization, and Environmental Effects*, 42(7), 909–917.

Kumari, D. and Singh, R. (2018). Pretreatment of lignocellulosic wastes for biofuel production: A critical review. *Renewable and Sustainable Energy Reviews,* 90. 877–891. 10.1016/j.rser.2018.03.111.

Li, Y. and Chen, H. (2022). AI-driven biorefineries: Innovations in biofuel production and sustainability. *Biotechnology Advances*, 60, 107937. doi: 10.1016/j.biotechadv.2022.107937

Lin, C.-Y. and Lu, C. (2021). Development perspectives of promising lignocellulose feedstocks for production of advanced generation biofuels: A review. *Renewable and Sustainable Energy Reviews*, 136, 110445. doi: https://doi.org/10.1016/j.rser.2020.110445.

Lynd, L.R., Liang, X., Biddy, M.J., Allee, A., Cai, H., Foust, T. and Wyman, C.E. (2017). Cellulosic ethanol: Status and innovation. *Current Opinion in Biotechnology*, 45, 202–211.

Makela, M.R., Donofrio, N. and de Vries, R.P. (2014). Plant biomass degradation by fungi. *Fungal Genetics and Biology*, 72, 2–9.

Mansouri, A., Rihani, R., Laoufi, A.N. and Özkan, M. (2016). Production of bioethanol from a mixture of agricultural feedstocks: Biofuels characterization. *Fuel*, 185, 612–621.

Manyi-Loh, C.E. and Lues, R. (2023). Anaerobic digestion of lignocellulosic biomass: Substrate characteristics (challenge) and innovation. *Fermentation*, 9, 755.

Mishra, P. and Thakur, I. S. (2022). Artificial intelligence for the sustainable production of biofuels from lignocellulosic biomass: Current trends and future perspectives. *Energy Reports*, 8, 1724–1738. doi: 10.1016/j.egyr.2022.12.013.

Molaverdi, M., Karimi, K. and Mirmohamadsadeghi, S. (2019). Improvement of dry simultaneous saccharification and fermentation of rice straw to high concentration ethanol by sodium carbonate pretreatment. *Energy*, 167, 654–660.

Mujtaba, M., Fraceto, L.F., Fazeli, M., Mukherjee, S., Savassa, S.M., de Medeiros, G.A., Pereira, A. do E.S., Mancini, S.D., Lipponen, J. and Vilaplana, F. (2023). Lignocellulosic biomass from agricultural waste to the circular economy: A review with focus on biofuels, biocomposites, and bioplastics. *Journal of Cleaner Production*, 402, 136815. doi: 10.1016/j.jclepro.2023.136815.

Nielsen, J., Larsson, C., van Maris, A.J.A. and Pronk, J.T. (2013). Metabolic engineering of yeast for production of fuels and chemicals. *Current Opinion in Biotechnology*, 24(3), 398–404. doi: 10.1016/j.copbio.2013.03.023.

Ojo, A.O. (2023). An overview of lignocellulose and its biotechnological importance in high-value product production. *Fermentation*, 9, 990.

Ozdenkci, K., De Blasio, C., Muddassar, H.R., Melin, K., Oinas, P., Koskinen, J., Sarwar, G. and Järvinen, M. (2017). A novel biorefinery integration concept for lignocellulosic biomass. *Energy Conversion and Management*, 149, 974–987. doi: 10.1016/j.enconman.2017.04.034.

Patel, S. and Singh, V. (2023a). Microbial fermentation of lignocellulosic biomass: Advances and challenges. *Journal of Industrial Microbiology and Biotechnology*, 50(1), 45–67. doi: 10.1007/s10295-022-02869-2.

Patel, K. and Singh, S. K. (2023b). Environmental sustainability analysis of biofuels: A critical review of LCA studies. *Clean Technologies and Environmental Policy,* 25, 2489–2510. doi: 10.1007/s10098-023-02596-y.

Pena-Castro, J.M., Muñoz-Páez, K.M., Robledo-Narvaez, P.N. and Vázquez-Núñez and E. (2023). Engineering the metabolic landscape of microorganisms for lignocellulosic conversion. *Microorganisms*, 11(9), 2197. doi: 10.3390/microorganisms11092197.

Pérez-Pimienta, J., López-Ortega, M., Sánchez, A. (2017). Recent developments in Agave performance as a drought-tolerant biofuel feedstock: agronomics, characterization, and biorefining. *Biofuels*, 11. https://doi.org/10.1002/bbb.1776.

Prasad, A., Sotenko, M., Blenkinsopp, T. and Coles, S. (2016). Life cycle assessment of lignocellulosic biomass pretreatment methods in biofuel production. *The International Journal of Life Cycle Assessment*, 21, 44–50. doi: 10.1007/s11367-015-0985-5.

Priya, A.K., Alagumalai, A., Balaji, D. and Song, H. (2023). Bio-based agricultural products: A sustainable alternative to agrochemicals for promoting a circular economy. *RSC Sustainability*, 4, 746–762. doi: 10.1039/d3su00075c.

Ragauskas, A.J., Beckham, G.T., Biddy, M.J., Chandra, R., Chen, F., Davis, M.F. and Wyman, C.E. (2014). Lignin valorization: Improving lignin processing in the biorefinery. *Science*, 344(6185), 1246843.

Rai, A.K., Al Makishah, N.H., Wen, Z., Gupta, G., Pandit, S. and Prasad, R. (2022). Recent developments in lignocellulosic biofuels, a renewable source of bioenergy. *Fermentation*, 8(4), 161. doi: 10.3390/fermentation8040161.

Rather, R.A., Wani, A.W., Mumtaz, S., Padder, S.A., Khan, A.H., Almohana, A.I., Almojil, S.F., Alam, S.S. and Baba, T.R. (2022). Bioenergy: A foundation to environmental sustainability in a changing global climate scenario. *Journal of King Saud University - Science,* 34(1), 101734. doi: 10.1016/j.jksus.2021.101734.

Rezania, S., Oryani, B., Cho, J., Talaiekhozani, A., Sabbagh, F., Hashemi, B. and Mohammadi, A.A. (2020). Different pretreatment technologies of lignocellulosic biomass for bioethanol production: An overview. *Energy*, 199, 117457. doi: 10.1016/j.energy.2020.117457.

Rosales Calderon, O., Tao, L., Abdullah, Z., Talmadge, M., Milbrandt, A., Smolinski, S. and Moriarty, K. (2024). Sustainable aviation fuel state-of-industry report: Hydro processed esters and fatty acids pathway (NREL/TP-5100-87803). *National Renewable Energy Laboratory.*

Saha, B.C. and Kennedy, G.J. (2020). Production of xylitol from mixed sugars of xylose and arabinose without co-producing arabitol. *Biocatalysis and Agricultural Biotechnology,* 29, 101786.

Sahoo, G., Wani, A., Swamy, S.L. and Sharma, A. (2022). Environmental impact and economic benefits of biofuel production. In Bio-Clean Energy Technologies, Chowdhary, P., Khanna, N., Pandit, S., Kumar, R. (eds). 1, 349–378. Springer. doi: 10.1007/978-981-16-8090-8_16.

Saravanan, A.P., Pugazhendhi, A. and Mathimani, T. (2020). A comprehensive assessment of biofuel policies in the BRICS nations: Implementation, blending target, and gaps. *Fuel*, 272, 117635. doi: 10.1016/j.fuel.2020.117635.

Sharma, S., Tsai, M.-L., Sharma, V., Sun, P.-P., Nargotra, P., Bajaj, B.K., Chen, C.-W. and Dong, C.-D. (2023). Environment friendly pretreatment approaches for the bioconversion of lignocellulosic biomass into biofuels and value-added products. *Environments*, 10(1), 6. doi: 10.3390/environments10010006.

Shaw, R.N., Mendis, N., Mekhilef, S. and Ghosh, A. (Eds.). (2021). *AI and IOT in Renewable Energy*. Singapore: Springer.

Singh, R. and Gupta, A. (2023). Advancements in enzyme engineering for lignocellulosic biomass conversion: A review. *Bioresource Technology Reports*, 20, 100937. doi: 10.1016/j.biteb.2023.100937.

Stanley, J.T., Thanarasu, A., Kumar, P.S., Periyasamy, K., Raghunandhakumar, S., Periyaraman, P., Devaraj, K., Dhanasekaran, A. and Subramanian, S. (2022). Potential pre-treatment of lignocellulosic biomass for the enhancement of biomethane production through anaerobic digestion: A review. *Fuel*, 318, 123593. doi: 10.1016/j.fuel.2022.123593.

Sun, J., Zhang, L. and Loh, K.-C. (2023). Enhancing scalability and economic viability of lignocellulose-derived biofuels production through integrated pretreatment and methanogenesis arrest. *Bioresource Technology*, 389, 129790. doi: 10.1016/j.biortech.2023.129790.

Topaloglu, A., Esen, Ö., Turanlı-Yıldız, B., Arslan, M. and Çakar, Z.P. (2023). From *Saccharomyces cerevisiae* to ethanol: Unlocking the power of evolutionary engineering in metabolic engineering applications. *Journal of Fungi*, 9(10), 984. doi: 10.3390/jof9100984.

Usmani, R., Mohammad, A. and Ansari, S. (2023). Comprehensive biofuel policy analysis framework: A novel approach evaluating the policy influences. *Renewable and Sustainable Energy Reviews*, 183, 113403. doi: 10.1016/j.rser.2023.113403.

Usmani, Z., Sharma, M., Awasthi, A.K., Lukk, T., Tuohy, M.G., Gong, L., Nguyen-Tri, P., Goddard, A.D., Bill, R.M., Nayak, S.C. and Gupta, V.K. (2021). Lignocellulosic biorefineries: The current state of challenges and strategies for efficient commercialization. *Renewable and Sustainable Energy Reviews,* 148, 111258. doi: 10.1016/j.rser.2021.111258.

Velvizhi, G., Jennita Jacqueline, P., Shetti, N.P., Latha, K., Mohanakrishna, G. and Aminabhavi, T.M. (2023). Emerging trends and advances in valorization of lignocellulosic biomass to biofuels. *Journal of Environmental Management*, 345, 118527. doi: 10.1016/j.jenvman.2023.118527.

Vinod, A., Pulikkalparambil, H., Jagadeesh, P., Rangappa, S.M. and Siengchin, S. (2023). Recent advancements in lignocellulose biomass-based carbon fiber: Synthesis, properties, and applications. Heliyon, 9(3), e13614. doi: 10.1016/j.heliyon.2023.e13614.

Wang, F.L., Li, S., Sun, Y.X., Han, H.Y., Zhang, B.X., Hu, B.Z. and Hu, X.M. (2017). Ionic liquids as efficient pretreatment solvents for lignocellulosic biomass. *RSC advances*, 7(76), 47990–47998.

Yoon, S.Y., Han, S.H. and Shin, S.J. (2014). The effect of hemicelluloses and lignin on acid hydrolysis of cellulose. *Energy*, 77, 19–24.

Yuan, H.-W., Tan, L., Kida, K., Morimura, S., Sun, Z.-Y. and Tang, Y.-Q. (2021). Potential for reduced water consumption in biorefining of lignocellulosic biomass to bioethanol and biogas. *Journal of Bioscience and Bioengineering,* 131(5), 461–468. doi: 10.1016/j.jbiosc.2020.12.015.

Zhang, J., Zhang, W.X., Wu, Z.Y., Yang, J., Liu, Y.H., Zhong, X. and Deng, Y. (2013). A comparison of different dilute solution explosions pretreatment for conversion of distillers'grains into ethanol. *Preparative Biochemistry and Biotechnology,* 43(1), 1–21.

Zhang, P., Dong, S.J., Ma, H.H., Zhang, B.X., Wang, Y.F. and Hu, X.M. (2015). Fractionation of corn stover into cellulose, hemicellulose and lignin using a series of ionic liquids. *Industrial Crops and Products*, 76, 688–696.

Zhang, Y., Minaret, J., Yuan, Z., Dutta, A. and Xu, C. (2018). Mild hydrothermal liquefaction of high water content agricultural residue for bio-crude oil production: A parametric study. *Energies*, 11(11), 3129.

Zhang, Z., Duan, H., Zhang, Y., Guo, X., Yu, X., Zhang, X. and Cai, J. (2020). Investigation of kinetic compensation effect in lignocellulosic biomass torrefaction: Kinetic and thermodynamic analyses. *Energy*, 207, 118290.

Zhang, Y. and Wang, Y. (2022). Data-driven approaches in the sustainable production of biofuels from lignocellulosic materials: A review. *Sustainable Energy Technologies and Assessments*, 53, 102538. doi: 10.1016/j.seta.2022.102538.

Zheng, B., Yu, S., Chen, Z. and Huo, Y.-X. (2022). A consolidated review of commercial-scale high-value products from lignocellulosic biomass. *Frontiers in Microbiology,* 13, 933882. doi: 10.3389/fmicb.2022.933882.

Zhou, X., Xu, Y., Zhang, X., Xu, D., Linghu, Y., Guo, H., Wang, Z. and Chen, H. (2021). Large-scale underground seasonal thermal energy storage in China. *Journal of Energy Storage*, 33, 102026. doi: 10.1016/j.est.2020.102026.

4 An Overview of Artificial Intelligence and Machine Learning

Adwika Singh, Falguni Sharma, and Anshuman Shastri

4.1 INTRODUCTION

The rapid advancement of artificial intelligence (AI) and machine learning (ML) has ushered in a transformative era, redefining how machines process information, make decisions, and interact with their environment. The pivotal transformation occurred with AI, wherein robots transcended preprogrammed constraints to learn from data, identify patterns, and make autonomous judgments. Unlike traditional computing, which relies on rigid rule-based systems, AI enables machines to adapt, learn from data, and refine their performance over time. This fundamental shift has made AI an indispensable force across diverse industries, from healthcare and finance to transportation and sustainable energy.

Machine learning, a subset of AI, plays a crucial role in this evolution, leveraging data-driven models to recognize patterns, and improve efficiency without direct human intervention. From early rule-based systems to the emergence of deep learning architectures, AI has undergone significant milestones—each marked by technological breakthroughs, periods of stagnation, and renewed progress. Despite challenges such as computational limitations, ethical dilemmas, and data biases, ongoing research continues to push the boundaries of AI's potential.

This chapter explores the fundamental principles of AI and ML, their historical development, key achievements, and the challenges they face. By bridging theoretical foundations with real-world applications, it highlights AI's role in driving innovation, optimizing processes, and shaping the future of intelligent systems.

4.2 DEFINITION OF ARTIFICIAL INTELLIGENCE

For decades, machines have been engineered to augment human capabilities—calculators supplanted manual arithmetic, industrial robots assumed repetitive work, and search engines surpassed conventional research methodologies. Nonetheless, these breakthroughs depended on inflexible, rule-based directives, constraining machines to solely what they were explicitly programmed to execute. The advent of artificial intelligence (AI) signified a pivotal transformation, allowing systems to not

 DOI: 10.1201/9781003571223-4

only automate jobs but also to learn, reason, and adapt, reflecting elements of human intellect (Figure 4.1). Fundamentally, AI encompasses the examination of algorithms and computational frameworks that enable machines to address issues, identify patterns, and provide informed conclusions. In contrast to traditional software, which functions based on established logic and predetermined rules, AI systems derive insights from data, enhance their performance over time, and apply knowledge to novel circumstances. This flexibility, mostly propelled by Machine learning (ML) and deep learning methodologies, has established AI as a disruptive influence across several domains, including healthcare, finance, and scientific exploration. Artificial Intelligence is not a solitary technology but a multidisciplinary domain that integrates mathematics, computer science, neuroscience, and cognitive psychology. It is primarily classified into two categories: Narrow AI (Weak AI) and General AI (Strong AI). Narrow AI, which underpins the majority of contemporary applications such as speech recognition, recommendation systems, and autonomous vehicles, is engineered to do specific tasks but is devoid of general intelligence. Conversely, General AI denotes a hypothetical system possessing human-like cognitive faculties, capable of reasoning across various fields without specialized training for specific tasks. Although research advances the frontiers of AI, general AI remains conjectural, presenting one of the most significant difficulties in the domain. A key difference between AI and conventional programming is the method of decision-making. Traditional programs adhere to well specified rules, utilizing if–else situations, deterministic logic, and preordained replies. AI systems, especially those utilizing machine learning, acquire knowledge from data patterns, enhancing their decision-making capabilities over time. This capacity to adjust beyond clearly defined directives allows AI to traverse intricate, unpredictable contexts, hence revealing new opportunities for intelligent automation and issue resolution.

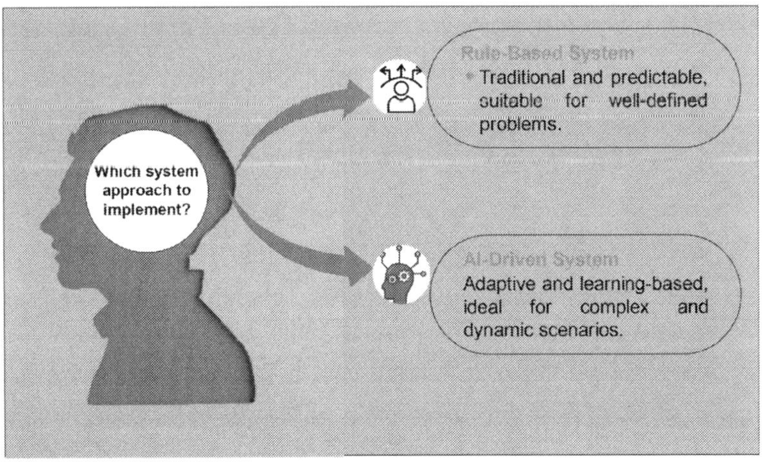

FIGURE 4.1 Comparison of rule-based systems and AI-driven systems in problem-solving.

4.2.1 Machine Learning and Its Connection to AI

Artificial intelligence (AI) seeks to develop systems that can learn, reason, and adapt; yet, what is the mechanism by which these systems "learn?" This is the juncture at which Machine learning (ML) becomes relevant. Machine learning is a branch of AI that allows machines to identify patterns and enhance task performance autonomously, without explicit programming. In contrast to conventional algorithms that adhere to strict directives, machine learning models discern patterns within data and enhance their predictions or conclusions progressively (Figure 4.2).

Machine learning fundamentally depends on statistical methods and mathematical models to analyze extensive datasets, reveal concealed patterns, and generate informed predictions. This process is influenced by human cognition, wherein experience informs decision-making. An ML model enhances its performance with increased data input, similar to the way individuals acquire knowledge through repeated exposure to material. A voice assistant, such as Siri or Google Assistant, does not depend on fixed rules for speech comprehension; rather, it assimilates extensive voice data to enhance speech recognition and respond more precisely to user commands, adapting to various accents, speech patterns, and languages over time.

The link between AI and ML is frequently characterized by AI delineating the objective, whereas ML supplies the methodology to attain it. Artificial intelligence spans a wide range of intelligent systems, while Machine learning serves as the catalyst for numerous breakthroughs within this domain. Early AI systems relied on

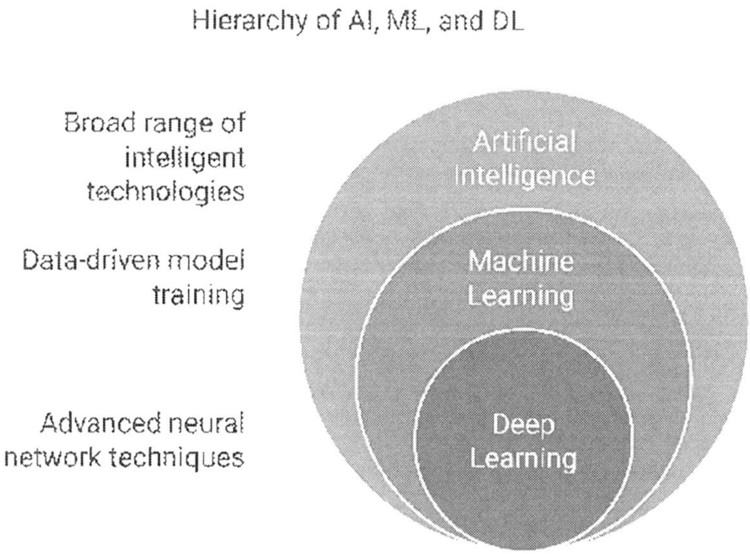

FIGURE 4.2 Hierarchical taxonomy illustrating the relationship between AI, machine learning, and deep learning.

manually constructed rules, but contemporary AI predominantly utilizes machine learning methodologies to address complexity, scalability, and flexibility.

4.2.2 Development and Key Achievements of AI and ML

The history of artificial intelligence (AI) and machine learning (ML) dates back to the inception of mechanical invention and intellectual endeavors. The concept of developing machines that emulate human reasoning dates back to ancient automata, with Greek engineers such as Hero of Alexandria constructing rudimentary mechanical devices. These primitive ideas initiated a protracted pursuit to develop machines capable of exhibiting intelligent behavior.

A significant advancement occurred in the 20th century due to the contributions of Alan Turing. His 1936 paper, On Computable Numbers, proposed the Turing Machine concept, formalizing computational principles. Subsequently, his Turing Test (1950) introduced a method to assess if a machine could demonstrate human-like intelligence, influencing AI research for decades. Approximately concurrently, Warren McCulloch and Walter Pitts (1943) presented the inaugural mathematical model of artificial neurons, establishing the groundwork for neural networks.

The formal establishment of AI as an academic field transpired in 1956 at the Dartmouth Summer Research Project on Artificial Intelligence. Pioneering academics such as John McCarthy, Marvin Minsky, Nathaniel Rochester, and Claude Shannon aimed to develop machines capable of advanced cognitive functions, including abstract reasoning and problem-solving. This era witnessed the emergence of early AI systems, such the Logic Theorist and the General Problem Solver, which proved that computers were capable of resolving mathematical proofs and logical dilemmas. Nonetheless, these systems depended significantly on manually encoded rules, constraining their capacity to adapt to real-world complexities. In the late 1960s, academics initiated investigations into alternative methodologies, culminating in the emergence of expert systems during the 1970s and 1980s. These systems utilized rule-based knowledge for specialised tasks in medical, finance, and engineering, attaining significant success in industrial applications. Nevertheless, expert systems necessitated considerable manual knowledge encoding, rendering them challenging to scale and sustain. The 1990s signified a transition from rule-based artificial intelligence to machine learning, wherein systems acquired knowledge from data instead of depending on predetermined reasoning. The resurgence of neural networks, especially due to Geoffrey Hinton's contributions to backpropagation, enhanced the adaptability of AI systems. Moreover, probabilistic reasoning techniques like Bayesian networks facilitated AI's management of uncertainty, leading to progress in speech recognition, robotics, and recommendation systems. The significant advancement of AI occurred in the 2010s, driven by deep learning, extensive datasets, robust GPUs, and enhanced neural architectures. Significant advancements, including AlexNet's triumph in the ImageNet competition (2012), DeepMind's AlphaGo (2016) overcoming human champions, and the emergence of transformer models such as BERT and GPT, showcased AI's unparalleled proficiency in vision, language, and decision-making. The integration of big data, cloud computing, and

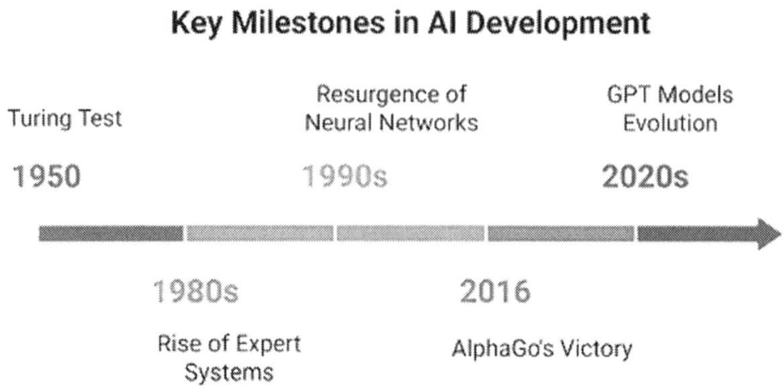

FIGURE 4.3 Timeline highlighting key advancements and milestones in artificial intelligence.

reinforcement learning has expedited advancements, resulting in improvements in autonomous systems, robotics, and generative AI (Figure 4.3).

Currently, AI research is advancing in self-supervised learning, reinforcement learning, and investigations into Artificial General Intelligence (AGI). Researchers are developing models that can learn effectively with limited supervision, adapt across several domains, and reason similarly to humans. Artificial intelligence has progressed from theoretical frameworks to practical implementations, revolutionizing sectors including healthcare, finance, and automation.

4.2.3 AI Dormancies and Revivals

Not withstanding the initial potential of artificial intelligence, advancements have not followed a linear trajectory. Historically, AI has undergone phases of swift progress succeeded by stasis, referred to as AI Winters—intervals characterized by reduced interest, funding reductions, and doubt over AI's promise. These losses frequently resulted from exaggerated expectations, technical constraints, and the failure of AI systems to fulfill their lofty commitments. Each AI Winter was ultimately succeeded by a revival, propelled by innovative breakthroughs, enhanced computational capabilities, and new learning paradigms. The inaugural significant AI Winter (1974–1980) emerged following the initial fervor of the 1950s and 1960s. Initial AI programs, including the General Problem Solver and SHRDLU, exhibited efficacy in controlled settings but encountered difficulties with real-world intricacies. Funding entities, including DARPA and the British government, retracted financial backing due to the absence of tangible outcomes, resulting in a notable deceleration in AI development. The discipline experienced a resurgence in the 1980s with the advent of expert systems, which utilized rule-based methodologies in specialized fields such as medicine and finance. This resulted in a surge of commercial AI applications; nevertheless, their dependence on manually written knowledge bases hindered scalability, ultimately leading to the second AI Winter (1987–1993). The decline of the expert systems market, along with a change in financing objectives, resulted in diminished investment and skepticism regarding the viability of AI.

The resurgence of AI in the 1990s was driven by the advent of machine learning and probabilistic reasoning. Researchers transitioned from inflexible, rule-based systems to data-driven methodologies, utilizing statistical and optimization tools. The comeback intensified with the reemergence of neural networks, notably Geoffrey Hinton's contributions to backpropagation, which allowed AI models to learn from data with enhanced adaptability. The pivotal moment occurred in the 2010s with the advent of deep learning. Improvements in technology, particularly GPUs, together the accessibility of extensive datasets, have allowed neural networks to surpass conventional AI techniques in speech recognition, computer vision, and language processing. Significant milestones, like as DeepMind's AlphaGo (2016) and the emergence of transformer-based models like GPT and BERT, showcased AI's enhanced capabilities, resulting in extensive investment and adoption.

Currently, artificial intelligence is advancing, although problems persist, encompassing issues related to ethical AI, data privacy, and explainability. Despite concerns regarding a potential AI Winter stemming from exaggerated expectations, current advancements in self-supervised learning, reinforcement learning, and neuromorphic computing indicate that AI has become increasingly robust. The cyclical history of AI highlights a fundamental lesson: progress may not be linear, yet innovation persistently expands the limits of AI's capabilities.

4.2.4 The Distinction between AI and Conventional Programming

The cyclical progression of AI, characterized by phases of swift advancement interspersed with regressions, underscores its essential distinction from conventional programming. In contrast to traditional software, which functions based on a static array of predetermined instructions, AI-driven systems are engineered to adapt, learn, and enhance over time. The transition from rule-based automation to data-driven intelligence signifies a fundamental difference between AI and conventional programming. In conventional programming, developers explicitly define every rule and condition that the system must adhere to. A typical fraud detection tool may utilize a series of established criteria, such as identifying transactions above a specified threshold or originating from atypical places. Although efficient for straightforward, clearly delineated scenarios, such rule-based systems encounter difficulties with intricate, dynamic patterns. Artificial intelligence, especially machine learning, provides a more versatile and responsive methodology. An AI-driven fraud detection system analyses extensive transaction data, uncovers concealed relationships, and enhances its predictions based on emerging trends, rather than depending on manually created rules. This adaptability enables AI to manage uncertainty and fluctuation, which traditional programming struggles to accommodate. Tasks such image identification, natural language processing, and predictive analytics necessitate systems to handle ambiguous, noisy, or missing data—challenges in which AI markedly surpasses traditional methods. Rule-based systems fail when confronted with edge situations or unfamiliar inputs, whereas AI models extrapolate

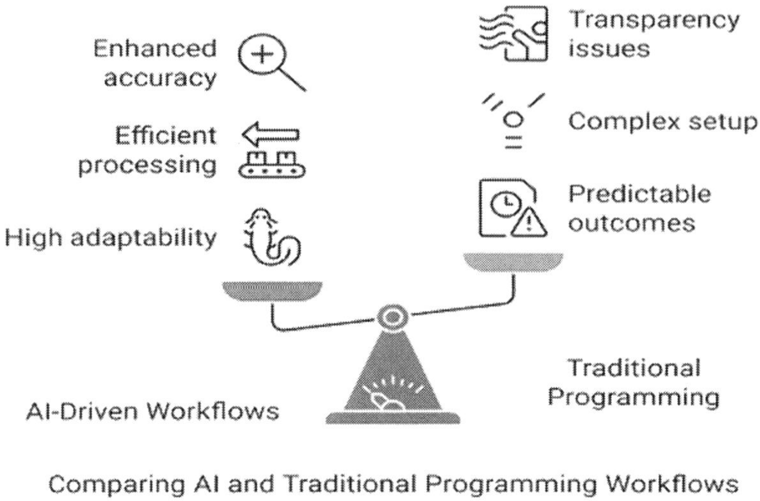

FIGURE 4.4 Comparative analysis illustrating the distinction between AI-driven workflows and traditional programming methodologies.

from prior experiences, rendering them significantly more robust in dynamic contexts. Furthermore, AI-driven systems demonstrate exceptional scalability and ongoing enhancement. Conventional software remains fixed unless updated manually, whereas AI models possess the capability to self-optimize through the assimilation of fresh data. This attribute is especially significant in fields such as healthcare diagnostics, autonomous vehicles, and recommendation systems, where changing patterns require continuous adaptation.

Notwithstanding these benefits, AI serves as an augmentation of traditional programming rather than a substitute for it (Figure 4.4). Numerous applications continue to necessitate deterministic logic, security assurances, and human supervision, rendering traditional software development essential. Rather than supplanting established approaches, AI augments them—broadening the realm of computational possibilities and unveiling new horizons of intelligent automation.

4.2.5 Essential Elements of AI and ML Systems

The increasing influence of AI and ML across sectors is propelled by many essential elements that collaborate to facilitate intelligent decision-making. In contrast to conventional software that adheres to predetermined instructions, AI systems depend on data, algorithms, and computational resources to learn, adapt, and execute intricate tasks. Comprehending these essential elements offers insight into the operation and development of AI models. Data is the fundamental element at the heart of every AI system, driving learning, and prediction. AI models necessitate extensive quantities of both organized and unstructured data, derived from text, photos, sensor inputs, or real-time interactions. The caliber, variety, and quantity of data directly affect model precision and generalization. Data pretreatment methods, including cleaning, normalization, and augmentation, are crucial for facilitating effective learning. Algorithms and models constitute the intelligence underlying AI systems. This encompasses supervised and unsupervised learning methods, as well as intricate deep learning frameworks such as convolutional neural networks (CNNs) for image identification and transformers for natural language processing. The selection of an algorithm is contingent upon the specific nature of the task, be it classification, regression, clustering, or reinforcement learning. Computational power constitutes another essential element. The proliferation of AI has been propelled by advancements in high-performance computing (HPC), graphics processing units (GPUs), and specialized AI technology such as tensor processing units (TPUs). These facilitate the effective training of deep neural networks necessitating extensive parallel computation. Cloud-based AI services significantly improve scalability, enabling organizations to implement and optimize models without requiring expensive on-site infrastructure. Furthermore, the assessment and enhancement of models are essential for sustaining AI efficacy. Methods such as cross-validation, hyperparameter optimization, and regularization enhance model resilience and mitigate overfitting. Ongoing surveillance guarantees the dependability of AI systems, particularly when utilized in practical applications where circumstances evolve over time. Ultimately, ethics, interpretability, and security increasingly influence AI development. Explainable AI (XAI) methodologies enhance model transparency, while fairness and bias mitigation approaches guarantee ethical implementation. Security protocols, including adversarial robustness and privacy-preserving methods, protect AI models from manipulation and data breaches.

Collectively, these elements constitute the foundation of AI and ML systems, facilitating information processing, predictive analytics, and automation across several domains, including healthcare, finance, and autonomous systems. As artificial intelligence progresses, developments in these critical domains will shape the forthcoming generation of intelligent technologies.

4.3 FUNDAMENTAL PRINCIPLES AND MECHANISMS IN ARTIFICIAL INTELLIGENCE, MACHINE LEARNING, AND DEEP LEARNING

4.3.1 Comprehending Artificial Intelligence, Machine Learning, and Deep Learning

Artificial Intelligence (AI), Machine Learning (ML), and Deep Learning (DL) are closely related fields that have transformed how machines process information, learn from data, and make decisions. While AI serves as the overarching concept of creating intelligent systems, ML and DL define specific approaches to achieving intelligence through learning from experience. Understanding the distinctions and interconnections between these domains is essential for grasping their real-world impact and future potential.

4.3.2 Defining Artificial Intelligence, Machine Learning, and Deep Learning

Artificial intelligence (AI) denotes the creation of systems that can do tasks usually necessitating human intelligence. These tasks encompass thinking, problem-solving, pattern identification, language comprehension, and decision-making. Artificial Intelligence encompasses a wide array of methodologies, ranging from rule-based expert systems to sophisticated learning-based models. Machine learning (ML) is a branch of AI dedicated to creating algorithms that enable machines to discern patterns from data and enhance their performance autonomously over time, without explicit programming. In contrast to conventional programming, which relies on explicit rules to determine behavior, machine learning models extrapolate from precedents, rendering them flexible in the face of novel and unencountered data. Deep learning (DL) is a distinct subset of Machine learning (ML) that use artificial neural networks with numerous layers to extract features and learn representations from extensive datasets. Deep learning models, inspired by the architecture of the human brain, excel at tasks that require the analysis of intricate patterns, including image and audio recognition, natural language processing, and autonomous systems.

Although AI, ML, and DL are interrelated, their differences reside in the approaches employed. Artificial intelligence (AI) is the broad domain, machine learning (ML) is a subset of AI concentrating on data-driven learning, and deep learning (DL) is a further specialization of ML utilizing neural networks for sophisticated pattern identification and decision-making.

4.3.3 Interrelation of AI, ML, and DL/the Position of Deep Learning Within AI and ML

Artificial intelligence, machine learning, and deep learning are interrelated in a hierarchical structure, where each domain enhances the functionalities of its predecessor. Artificial intelligence constitutes the overarching category that includes all endeavors aimed at developing intelligent machines, utilizing methods such as logical reasoning, expert systems, or learning-based techniques. Machine learning is an essential element of artificial intelligence, allowing robots to enhance their performance via experience instead of depending exclusively on predetermined rules.

Deep learning is a specialized subset of machine learning that utilizes intricate neural network structures. Conventional machine learning models frequently necessitate manual feature extraction, wherein engineers identify essential features for the model's processing. Conversely, deep learning models autonomously acquire hierarchical features, enabling them to excel at processing high-dimensional and unstructured input, including photos and natural language.

The advent of deep learning has markedly propelled AI, resulting in substantial advancements in areas such as computer vision, speech synthesis, and autonomous decision-making. Nonetheless, machine learning approaches beyond deep learning, like decision trees, support vector machines, and probabilistic models, continue to be advantageous for numerous AI applications.

4.3.4 Determining the Appropriate Application of Machine Learning, Deep Learning, or Traditional Programming

The selection among traditional programming, machine learning, and deep learning is contingent upon the problem's characteristics, the data at hand, and the computational resources accessible. Conventional programming: Rule-based programming is appropriate for situations when specific rules can be established. For instance, banking systems employ hardcoded logic for transaction processing, whereas traffic control systems depend on established rules for signal alterations. These methodologies are effective for structured, well defined issues but falter in the presence of uncertainty, variability, or extensive data sets. Machine learning: ML is advantageous for analyzing intricate patterns or when it is challenging to establish explicit rules. Spam detection in emails, fraud detection in financial transactions, and recommendation systems in e-commerce platforms all leverage machine learning, since these systems discern patterns from prior data and adjust to new inputs. Deep learning: DL is most efficacious when processing extensive volumes of unstructured data, including images, audio, and text. It is frequently utilized in facial recognition, speech-to-text transcription, and autonomous vehicle technologies. Nonetheless, deep learning models necessitate substantial computer resources and big datasets to function effectively.

Traditional programming is pertinent for organized problem-solving, machine learning excels in predictive analytics and data-driven decision-making, while deep learning is optimal for high-dimensional, complicated problems necessitating deep feature extraction. Choosing the correct methodology guarantees efficiency, precision, and scalability in AI-driven systems.

4.3.5 Classifications of AI, ML, and DL Methodologies

Artificial intelligence, machine learning, and deep learning comprise various specialized methodologies, each tailored to distinct jobs and obstacles. The categorization of AI is predicated on its skills and functionalities, differentiating between systems with restricted, general, or superintelligent competencies. Machine learning is classified based on its learning framework, including supervised, unsupervised, and reinforcement learning, which dictate how models analyze and adjust to input. Deep learning, a branch of machine learning, utilizes many neural network topologies, each tailored to address different complexities in pattern recognition and decision-making. Comprehending these classifications offers a more lucid perspective on the development and application of intelligent systems across several fields. Provide an in-depth analysis utilizing diverse categorization criteria, including diagrams and flowcharts, culminating in the classification of AI types.

4.3.5.1 Categories of Artificial Intelligence

Artificial intelligence can be extensively classified according to its capabilities and functionalities. The categorization of AI differentiates between systems that execute certain, specialized functions and those that aspire to emulate human cognitive

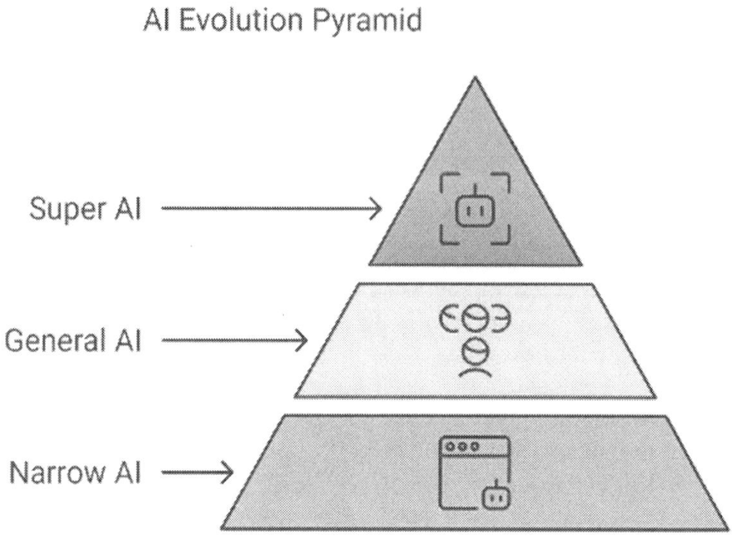

FIGURE 4.5 AI-evolution pyramid, illustrating the hierarchical levels of artificial intelligence.

capabilities. The initial category, Narrow AI (Weak AI), includes the majority of contemporary AI applications. These systems are engineered to execute certain functions, such as picture recognition, language translation, and recommendation systems, although they are devoid of general thinking capabilities. Virtual assistants, spam filters, and autonomous vehicles act inside this realm, proficiently executing specific tasks while being confined to their established parameters. General AI (Strong AI) denotes the hypothetical phase in which machines demonstrate human-like intellect, comprehension, reasoning, and learning across various domains without the necessity for task-specific training. Although this is a continuous research endeavor, progress in cognitive modeling and reinforcement learning is advancing AI toward enhanced generalized problem-solving capabilities (Figure 4.5).

Beyond General AI exists the notion of Artificial Superintelligence (ASI), a hypothetical phase in which machines exceed human intelligence across all domains, encompassing creativity, decision-making, and problem-solving. The discourse around ASI, characterized by debate and speculation, has stimulated conversations regarding AI ethics, governance, and alignment with human values.

4.3.5.2 Categories of Machine Learning

Machine learning systems are categorized according to their data processing methods and their capacity for improvement over time. These learning paradigms delineate how models recognize patterns, modify their parameters, and enhance their predictions (Figure 4.6).

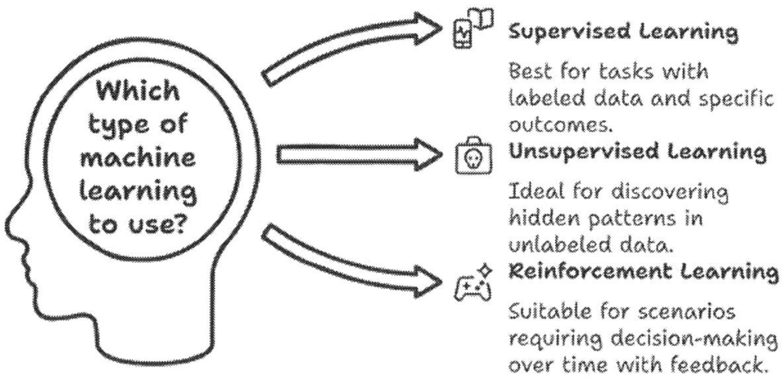

FIGURE 4.6 Classification of machine learning paradigms based on data availability and learning objectives.

Supervised learning depends on labeled data, wherein input–output pairs direct the model's learning process. The system correlates inputs with established outputs, facilitating applications like spam identification, medical diagnosis, and financial forecasting. Algorithms such as decision trees, support vector machines, and neural networks function within this domain, utilizing previous data to enhance predicted precision.

Unsupervised learning functions without labeled data, uncovering concealed patterns and correlations within datasets. Clustering algorithms, like k-means and hierarchical clustering, categorize data into significant groups, while dimensionality reduction methods, such as principal component analysis (PCA), facilitate the simplification of intricate datasets. Unsupervised learning applications encompass customer segmentation, anomaly detection, and feature extraction in image processing.

Reinforcement learning is acquiring knowledge via trial and error, wherein an agent engages with an environment and refines its actions depending on rewards and penalties. This framework is essential for autonomous systems, robotics, and game-playing AI, shown by DeepMind's AlphaGo. Reinforcement learning models employ techniques like Q-learning and policy gradient methods to enhance their decision making procedures.

4.3.5.3 Categories of Deep Learning Architectures

Deep learning, a branch of machine learning, utilizes neural network topologies to analyze intricate patterns in data. These designs differ in form and application, facilitating advancements in natural language processing, computer vision, and autonomous decision-making.

Feedforward Neural Networks (FNNs) constitute the most basic type of deep learning models, characterized by unidirectional data flow from input to output. These networks are employed for fundamental classification and regression tasks but may encounter difficulties with sequential dependencies in data.

Convolutional Neural Networks (CNNs) are specifically designed for the processing of images and videos. Convolutional Neural Networks (CNNs) utilize convolutional layers to identify spatial hierarchies in images, facilitating applications such as facial recognition, medical imaging, and autonomous navigation. Recurrent Neural Networks (RNNs) are proficient in processing sequential data, rendering them suitable for applications such as speech recognition, language modeling, and time-series forecasting. Variants such as Long Short-Term Memory (LSTM) and Gated Recurrent Units (GRUs) address challenges such disappearing gradients, enabling networks to preserve long-term relationships (Figure 4.7). Transformer models, such as BERT and GPT architectures, have transformed natural language processing. These models utilize self-attention mechanisms to concurrently analyze complete sequences, markedly enhancing tasks such as translation, text production, and contextual comprehension. Generative Adversarial Networks (GANs) utilize two opposing neural networks—a generator and a discriminator—to produce highly realistic synthetic data. Generative Adversarial Networks (GANs) are extensively employed in image synthesis, deepfake production, and artistic content development.

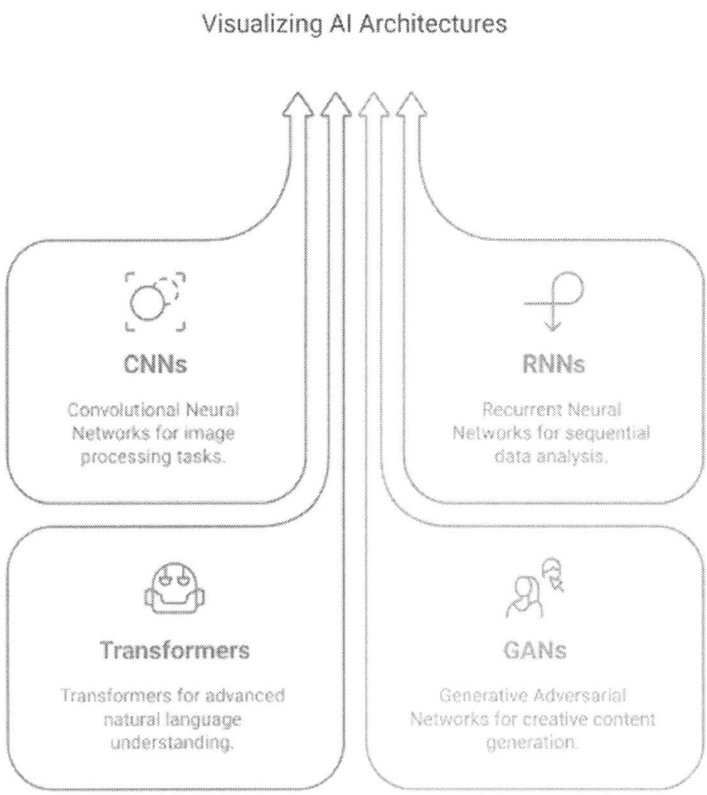

FIGURE 4.7 Key AI architectures—CNNs for image processing, RNNs for sequential data, transformers for NLP, and GAN for content generation.

Each deep learning architecture is tailored to tackle particular difficulties, enhancing AI capabilities across many fields. As research advances, hybrid models and innovative designs persist in expanding the capabilities of intelligent systems.

4.4 THE WORKFLOW FOR AI, ML, AND DL DEVELOPMENT

Developing AI, Machine Learning (ML), and Deep Learning (DL) models follows a structured workflow that ensures accuracy, efficiency, and scalability. This workflow involves multiple stages, from data collection to model deployment, with each step playing a crucial role in building a reliable AI-driven system. Proper implementation of these stages ensures that the model generalizes well, remains robust, and performs optimally in real-world applications. The key stages of AI/ML/DL developments are as follows.

4.4.1 Data Acquisition

Data constitutes the fundamental basis for artificial intelligence and machine learning systems. The quality, quantity, and diversity of data profoundly affect the accuracy and reliability of a model. Data may originate from diverse sources, encompassing organized databases, unstructured text, sensor outputs, pictures, and real-time streams. Recommendation systems depend on user interaction data, whereas self-driving cars employ sensor and video inputs to traverse their surroundings.

Data collecting may entail manual gathering, automatic scraping, or interaction with APIs and data pipelines, contingent upon the application (Figure 4.8). Challenges at this stage encompass addressing incomplete data, maintaining compliance with privacy regulations, and efficiently managing extensive databases.

4.4.2 Data Preparation

Raw data frequently exhibits noise, inconsistency, and incompleteness, rendering preprocessing a crucial phase prior to its integration into an AI or ML model. Preprocessing methodologies differ based on the data type and the employed learning paradigm. Preprocessing structured data involves addressing missing values, eliminating duplicates, normalizing numerical features, and encoding categorical variables. Conversely, text-based data necessitates tokenization, elimination of stop words, and either stemming or lemmatization. Image preprocessing may encompass scaling, normalization, and augmentation procedures to enhance model resilience.

An adequately preprocessed dataset enhances model performance by mitigating biases, augmenting data integrity, and ensuring the model acquires knowledge from significant patterns instead of extraneous noise.

4.4.3 Feature Engineering (Machine Learning)/ Feature Extraction (Deep Learning)

After data cleansing, the subsequent stage involves converting it into a format that optimizes model performance. In machine learning, feature engineering entails the

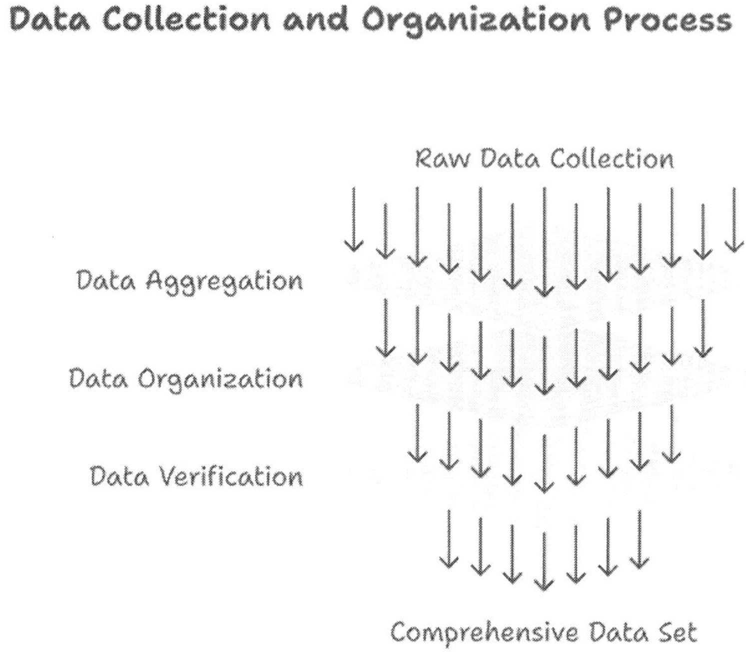

FIGURE 4.8 Data collection and organization process highlighting aggregation, organization, and verification for a comprehensive dataset.

selection, creation, or transformation of pertinent variables that enhance forecast accuracy. This may encompass polynomial characteristics, domain-specific aggregations, and dimensionality reduction methods such as Principal Component Analysis (PCA). In deep learning, neural networks typically manage feature extraction. Deep learning models autonomously acquire hierarchical representations from unprocessed data, eliminating the need for manual feature selection. Convolutional Neural Networks (CNNs) identify edges and textures in images, whereas transformer models in Natural Language Processing (NLP) discern contextual relationships among words. Feature engineering and extraction significantly influence a model's generalization capacity, rendering this phase essential for performance enhancement.

4.4.4 Selection of Models

Selecting the appropriate model is contingent upon the problem's characteristics, data accessibility, and computing limitations. In machine learning, algorithms such as decision trees, support vector machines, and ensemble approaches (e.g., random forests, gradient boosting) are frequently employed for classification and regression tasks. Choosing an adequate architecture is essential for deep learning. Convolutional Neural Networks (CNNs) are optimal for image-related activities, Recurrent Neural Networks (RNNs) or transformers are superior in sequence-oriented applications, and reinforcement learning models are proficient in dynamic decision-making contexts. Considerations including interpretability, scalability, and training complexity influence model selection. Basic models may be enough for organized datasets with few features, whereas deep learning models are beneficial for high-dimensional and unstructured data.

4.4.5 Instruction and Calibration

Upon selection of a model, it enters a training phase, during which it acquires knowledge from data by modifying its internal parameters. The training procedure entails supplying input data, calculating loss functions, and optimizing model weights through algorithms such as gradient descent. Hyperparameter tuning is an essential aspect of training, involving the adjustment of parameters like as learning rate, batch size, and network depth to enhance performance. Methods including grid search, random search, and Bayesian optimization facilitate the identification of optimal hyperparameter combinations. Deep learning training frequently necessitates substantial computer resources, especially for big models. Hardware accelerators such as GPUs and TPUs markedly enhance this process, allowing models to efficiently process extensive data sets.

4.4.6 Evaluation of the Model

The model's performance is evaluated post-training to confirm its generalization to novel data. Evaluation metrics are contingent upon the nature of the problem—classification tasks utilize accuracy, precision, recall, and F1-score, whereas regression models depend on metrics such as mean squared error (MSE) and R-squared. Precision and memory are crucial in classification issues characterized by class imbalance. Precision quantifies the proportion of retrieved results that are relevant, whereas recall assesses the proportion of actual relevant items that were successfully retrieved. The diagram demonstrates that precision is the ratio of true positives to the total number of retrieved items (true positives plus false positives), while recall is the ratio of true positives to the total number of relevant elements (true positives plus false negatives). The F1-score balances precision with recall, rendering it an essential metric when both are critical.

Cross-validation methods, including k-fold cross-validation, reduce overfitting by evaluating the model on several subsets of the dataset. In deep learning, validation loss and accuracy curves inform model enhancements, while sophisticated methods such as adversarial testing and robustness analysis evaluate adaptability to unforeseen circumstances.

4.5 ESSENTIAL TECHNOLOGIES AND INSTRUMENTS IN ARTIFICIAL INTELLIGENCE, MACHINE LEARNING, AND DEEP LEARNING

Progress in AI, ML, and DL is primarily propelled by robust technologies that facilitate effective model creation and implementation. Specialised libraries, deep learning frameworks, high-performance computational resources, and extensive datasets are the foundation of contemporary AI applications. Every component is essential for facilitating deployment, enhancing scalability, and expediting research advancements.

4.5.1 Artificial Intelligence/Machine Learning Libraries

The advancement of machine learning depends on comprehensive libraries that offer pre-existing algorithms, data processing tools, and optimization methods. These libraries facilitate the model-building process, enabling academics and engineers to concentrate on problem-solving rather than low-level implementation. Scikit-learn has emerged as a prevalent tool for traditional machine learning applications, including an array of classification, regression, and clustering techniques. Likewise, gradient boosting frameworks such as XGBoost and LightGBM have become prominent due to their efficacy in managing structured data and extensive datasets. Statsmodels is a significant asset in statistical modeling, especially in econometrics and hypothesis testing. These libraries function as fundamental components for the development and enhancement of machine learning models across several fields.

4.5.2 Frameworks for Deep Learning

Deep learning, necessitating extensive computations, and complex neural network designs, has been significantly expedited by specialized frameworks. PyTorch has been a favored option among researchers because of its adaptability and dynamic computing structure, facilitating real-time debugging and experimentation. TensorFlow, created by Google, is extensively utilized for both research and large-scale deep learning applications, providing comprehensive support for distributed training and deployment. Keras functions as a high-level API within TensorFlow, offering an intuitive interface for constructing and training deep learning models with reduced complexity. These frameworks accommodate several neural network topologies, including convolutional networks employed in image recognition and transformers that have transformed natural language processing.

4.5.3 Computational Resources

The escalating intricacy of AI models has heightened the necessity for high-performance computational resources. In contrast to conventional computing, AI workloads frequently necessitate parallel processing capabilities, rendering specialized hardware essential. Graphics processing units (GPUs) are fundamental to deep learning, expediting model training by quickly managing extensive matrix computations. NVIDIA's CUDA-capable GPUs and Tensor Cores have established benchmarks for AI acceleration in the industry. Alongside GPUs, Google's Tensor Processing Units (TPUs) present an alternative tailored exclusively for TensorFlow tasks, delivering accelerated training and inference durations for deep learning models. Cloud computing services, such Amazon Web Services (AWS), Google Cloud, and Microsoft Azure, have enhanced accessibility to robust AI infrastructure by providing scalable virtual machines equipped with pre-installed deep learning libraries. The enhancements in processing capacity have facilitated the training of models comprising billions of parameters, propelling AI research and applications to unparalleled heights.

4.5.4 Datasets and Pretrained Models

The efficacy of AI and ML models is significantly contingent upon the quality and accessibility of data. Extensive datasets form the basis for developing resilient models, offering varied instances that assist algorithms in generalizing to novel situations. In computer vision, datasets like ImageNet and COCO have established benchmarks for object recognition and image segmentation tasks. Collections such as Common Crawl and Wikipedia dumps have propelled the advancement of language models proficient in comprehending and producing human-like content for natural language processing. In addition to raw datasets, pretrained models have transformed AI through transfer learning, a method that allows models trained on extensive datasets to be refined for specific tasks with minimum supplementary data. This method has considerably lowered the time and computational resources required to create high-performance AI systems. BERT and GPT have revolutionized natural language comprehension, while designs like as ResNet and EfficientNet have enhanced picture recognition proficiency. Utilizing open-source datasets and pretrained networks, AI practitioners can develop robust models without the necessity for massive labeled data or substantial computational resources.

The collaboration of advanced software, robust hardware, and vast datasets has accelerated the rapid advancement of AI. These technologies persist in their evolution, rendering AI systems more accessible, efficient, and adept at addressing progressively intricate real-world difficulties.

4.6 CHALLENGES AND CONSIDERATIONS IN ARTIFICIAL INTELLIGENCE, MACHINE LEARNING, AND DEEP LEARNING

Notwithstanding the significant advancements in artificial intelligence, machine learning, and deep learning, numerous problems persist that impede their extensive acceptance and efficacy. These obstacles encompass data-related issues, ethical considerations, and technical constraints. Confronting these challenges is essential for guaranteeing the appropriate and efficient advancement of AI systems.

4.6.1 Challenges Pertaining to Data

Data constitutes the cornerstone of AI and ML models; nonetheless, acquiring high-quality datasets continues to be a formidable task. A key difficulty is the availability of data—numerous sectors lack adequate labeled data to train precise algorithms. When data is available, it frequently exhibits unstructured, noisy, or biased characteristics, resulting in erroneous forecasts and unreliable decision-making. Data privacy and security provide further considerations. Organizations must manage legal limits when collecting and utilizing user data due to rules such as the General Data

Protection Regulation (GDPR) and the California Consumer Privacy Act (CCPA). Ensuring adherence to these standards while preserving model efficacy necessitates strategies such as differential privacy and federated learning; yet, these methods involve further complexity. A further difficulty is data imbalance, characterized by the under-representation of specific categories within a dataset. This may result in biased models that excel with majority classes but underperform with minority groups. Methods like as oversampling, undersampling, and synthetic data generation are employed to address these challenges; nonetheless, they may not consistently maintain the authentic statistical characteristics of the data.

4.6.2 Ethical Considerations in Artificial Intelligence

As AI systems grow more embedded in daily life, ethical questions have gained urgency. A significant issue is algorithmic bias, when models inadvertently perpetuate socioeconomic imbalances. Facial recognition technology exhibit elevated error rates for those with darker skin tones, prompting concerns over equity and prejudice. Bias in AI frequently arises from skewed training data or prejudiced feature selection, necessitating the integration of fairness-aware learning methodologies to alleviate these hazards. Transparency and explainability represent significant ethical challenges. Numerous deep learning models operate as "black boxes," complicating the interpretation of their decision-making mechanisms. In essential sectors like healthcare, banking, and criminal justice, where accountability is crucial, the absence of explainability generates trust and regulatory concerns. Research in explainable AI (XAI) aims to create methodologies that elucidate model behavior, facilitating the auditing and comprehension of judgments. An emerging ethical dilemma pertains to the proliferation of deepfakes—artificially generated images, movies, and sounds produced by deep learning methodologies. Although deepfake technology has valid uses in entertainment and content creation, it has also been used to produce deceptive media, enable identity theft, and influence public perception. The capacity of deepfake algorithms to generate very realistic but wholly fictitious content poses a significant risk to misinformation management, digital security, and individual privacy. Addressing the detrimental effects of deepfakes necessitates progress in AI-based detection techniques, digital watermarking, and regulatory policies to mitigate their abuse.

Furthermore, the possible exploitation of AI beyond deepfakes. Unregulated autonomous weapons, extensive surveillance, and AI-facilitated cyberattacks present considerable dangers. Formulating global governance frameworks, ethical AI protocols, and accountability mechanisms is crucial to guarantee the responsible development and deployment of AI technologies.

4.6.3 Technical Difficulties

In addition to data and ethical issues, AI development encounters other technical constraints. A significant difficulty is computational efficiency. Training extensive deep learning models demands substantial computational power and memory, frequently requiring costly hardware like GPUs and TPUs. Although hardware accelerators have enhanced performance, the growing size and complexity of models, particularly in natural language processing, persistently challenge computational boundaries, raising concerns regarding energy usage and sustainability.

Generalization constitutes another critical concern. AI algorithms frequently excel with training data but encounter difficulties when faced with novel situations. Overfitting, in which a model retains specific patterns rather than acquiring generalizable characteristics, is a prevalent issue. Methods like as regularization, dropout, and transfer learning assist in alleviating this problem; however, guaranteeing strong generalization continues to be a prominent research focus.

Moreover, the implementation of AI in real-world scenarios has distinct obstacles owing to the fluidity of data settings. In contrast to conventional software, which adheres to established protocols, machine learning models must adjust to evolving circumstances. This is particularly vital in domains such as fraud detection, autonomous driving, and cybersecurity, where patterns change over time. Researchers are investigating continual learning and adaptive AI methodologies to tackle these difficulties; yet, these technologies remain in their nascent phases.

Notwithstanding these obstacles, AI, ML, and DL persist in their evolution, propelled by continuous research and technical advancement. Addressing these difficulties is crucial for developing AI systems that are not just potent but also ethical, dependable, and adaptable for practical applications.

4.7 THREE APPLICATIONS OF AI, ML, AND DL ACROSS VARIOUS INDUSTRIES

Having developed a fundamental comprehension of AI, ML, and DL—including their progression, elements, processes, and obstacles—it is now essential to investigate their practical ramifications. These technologies have infiltrated almost every sector, transforming conventional procedures, improving decision-making, and facilitating solutions previously deemed unattainable. This section examines the ways in which artificial intelligence, machine learning, and deep learning are transforming various industries, fostering innovation, and tackling real-world issues.

4.7.1 Artificial Intelligence in Healthcare

Artificial Intelligence (AI) is transforming healthcare through early diagnosis, customized treatment strategies, and enhanced operational efficiency. AI-driven solutions are improving clinical outcomes and minimizing human errors in medical imaging and medication research. Machine learning algorithms scrutinize extensive medical data, identifying patterns that frequently evade human skill. One of the most revolutionary applications is in medical imaging, where AI models trained on radiological data aid in identifying abnormalities such as tumors, fractures, and organ anomalies with exceptional precision. AI-driven systems, such as Google's DeepMind, have exhibited expertise in diagnosing ocular illnesses using retinal scans, frequently exceeding the capabilities of human ophthalmologists. In addition to diagnostics, AI enhances personalized medicine by evaluating genetic, clinical, and lifestyle information to customize treatment strategies. AI-driven solutions such as IBM Watson Health analyze extensive medical literature to suggest effective cancer therapies according to a patient's specific biomarkers.

Artificial intelligence significantly contributes to operational efficiency in hospitals. Predictive analytics enhance patient admissions, maximizing bed utilization and alleviating emergency room overcrowding. Moreover, AI chatbots and virtual health assistants, including Babylon Health and Ada Health, offer first consultations, alleviating the workload of healthcare professionals and enhancing accessibility in rural regions. Notwithstanding these gains, AI in healthcare encounters hurdles including data protection, ethical dilemmas, and the requirement for human supervision. Regulatory frameworks, such as the FDA's AI recommendations, seek to guarantee that AI applications are transparent, safe, and advantageous for healthcare practitioners and patients alike.

4.7.2 Artificial Intelligence in Manufacturing and Industry 4.0

The incorporation of AI in manufacturing, commonly known as Industry 4.0, is revolutionizing production methods via automation, predictive maintenance, and supply chain optimization. AI-driven technologies provide real-time surveillance, minimizing downtime and augmenting productivity. A primary application of AI in manufacturing is predictive maintenance. Machine learning systems evaluate sensor data from industrial machinery to anticipate faults before to their occurrence, hence minimizing expensive downtime. Corporations such as General Electric (GE) utilize AI-driven predictive maintenance to enhance turbine and jet engine performance, thereby averting unforeseen breakdowns. AI-driven robots and computer vision technologies enhance quality control by identifying flaws with superior accuracy compared to human inspectors. AI-driven visual inspection systems in car manufacturing may detect small cracks in engine components, hence assuring elevated safety standards.

AI has made considerable advancements in supply chain optimization. AI-driven demand forecasting tools evaluate market trends, meteorological conditions, and geopolitical influences to enhance inventory management. Corporations such as Amazon and Tesla employ artificial intelligence to optimize logistics, hence decreasing delays and minimizing waste. Although AI-driven automation enhances productivity, it simultaneously engenders apprehensions around job displacement. The future of AI in manufacturing will likely entail a collaborative approach, wherein AI enhances human capabilities instead of completely displacing workers. Human-AI collaboration, facilitated by cobots (collaborative robots), is anticipated to characterize the forthcoming phase of Industry 4.0, promoting safer and more efficient work environments.

4.7.3 Artificial Intelligence in Autonomous Systems and Transportation

AI-driven autonomous systems are transforming the transportation industry by improving safety, efficiency, and sustainability. AI-driven developments, ranging from autonomous vehicles to sophisticated traffic control systems, are minimizing human error and enhancing mobility solutions. A notable advancement is in autonomous vehicles (AVs). Companies such as Tesla, Waymo, and NVIDIA are utilizing AI-driven deep learning models to facilitate autonomous driving capabilities. These cars utilize computer vision, LiDAR, and sensor fusion to traverse intricate urban landscapes, reducing accidents attributable to human fatigue and distraction. Artificial intelligence significantly contributes to traffic management by utilizing intelligent technologies to analyze real-time traffic data, hence optimizing signal timings and alleviating congestion. Cities such as Singapore and Los Angeles utilize AI-based traffic monitoring systems that adapt traffic light sequences according to vehicle flow patterns, enhancing urban mobility. In aviation, artificial intelligence improves flight operations via predictive maintenance and the optimization of air traffic control. AI-driven solutions, exemplified by Airbus' Skywise, evaluate aircraft health data to predict component failures, thereby facilitating smoother and safer flights. Likewise, AI-powered air traffic control systems facilitate the management of rising air traffic volumes, minimizing delays, and enhancing fuel efficiency.

Notwithstanding these breakthroughs, obstacles like as governmental approval, ethical dilemmas concerning accountability in autonomous mishaps, and cybersecurity risks persist as substantial impediments. Resolving these difficulties is essential for the extensive implementation of AI-driven transportation systems.

4.7.4 Artificial Intelligence in Education

Artificial intelligence is revolutionizing education through the personalization of learning experiences, the automation of administrative activities, and the improvement of accessibility. AI-driven solutions empower instructors to customize classes according to specific student requirements, enhancing the efficacy of learning environments. A principal application of AI in education is adaptive learning platforms. Platforms such as Coursera, Duolingo, and Carnegie Learning employ AI algorithms to evaluate student progress and modify content accordingly, guaranteeing that learners receive tailored training based on their strengths and shortcomings.

Automated grading systems utilizing natural language processing (NLP) optimize the evaluation process, enabling educators to concentrate on teaching rather than administrative duties. AI-powered essay assessment technologies, exemplified by those employed by ETS in standardized examinations, analyze grammar, coherence, and content pertinence, delivering immediate feedback to students. Artificial intelligence also improves accessibility in schooling. Speech-to-text and text-to-speech programs facilitate enhanced engagement with educational resources for students with disabilities. AI-driven translation systems enhance cross-linguistic communication, aiding non-native speakers in comprehending intricate topics. Notwithstanding these advantages, issues such as data privacy, disparities in digital access, and potential biases in AI-generated content must be resolved to guarantee fair access to AI-enhanced educational solutions.

4.7.5 Artificial Intelligence in Sustainable Energy and Biofuels

Artificial intelligence is crucial in enhancing sustainable energy solutions and biofuel production through the optimization of energy management, the augmentation of resource efficiency, and the acceleration of scientific research (Figure 4.9). In renewable energy, artificial intelligence improves power grid stability and demand prediction. AI-driven predictive analytics assist utility firms in harmonizing electricity supply and demand, including renewable sources like solar and wind with traditional energy infrastructures. Google's DeepMind use artificial intelligence to enhance wind energy forecasts, thereby improving the dependability of renewable energy production.

In biofuel production, artificial intelligence expedites research by examining intricate molecular events. Machine learning algorithms facilitate the identification of appropriate microbial strains for biofuel synthesis, hence decreasing the time and expense associated with biofuel development. MIT researchers have utilized AI-driven simulations to optimize enzymatic processes, thereby increasing the efficiency of bioethanol and biodiesel synthesis.

Artificial intelligence contributes to energy-efficient production and the reduction of carbon footprints. AI-powered smart sensors monitor energy use in industrial environments, pinpointing potential for waste reduction. Corporations such as Siemens utilize AI-powered energy management systems to enhance energy efficiency in smart factories, hence advancing sustainability objectives. Challenges in AI-driven sustainable energy solutions including data reliability, interaction with legacy systems, and regulatory obstacles. Nevertheless, ongoing developments in AI,

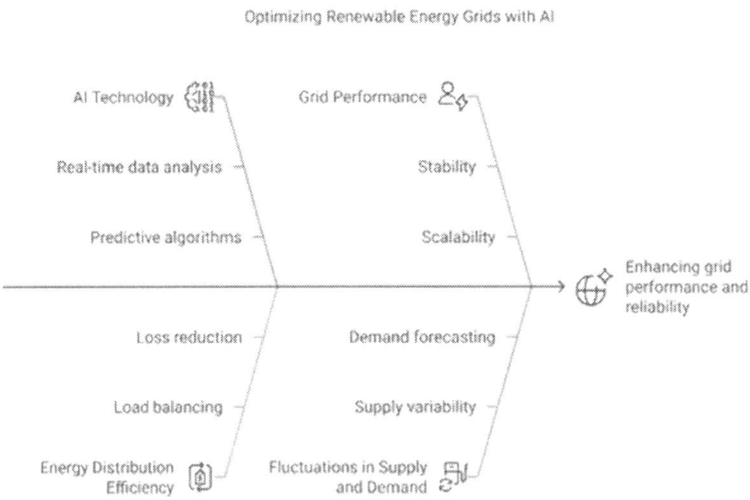

FIGURE 4.9 AI enhances renewable grid performance through predictive analytics and real-time optimization.

along with policy endorsement, will propel the shift towards cleaner, more sustainable energy alternatives.

4.8 CONCLUSION

Artificial Intelligence (AI) and Machine Learning (ML) are transformative technologies reshaping industries and society. Their applications, from healthcare diagnostics to personalized entertainment, demonstrate their immense potential to solve complex problems, enhance decision-making, and drive innovation. However, the journey to fully harnessing AI and ML is not without challenges. Ethical considerations, data quality, computational requirements, and workforce readiness are pivotal issues that need addressing for sustainable and responsible deployment.

5 Artificial Intelligence and Its Application in Biofuel Production

Shivani Sahu and Arindam Kuila

5.1 INTRODUCTION

In the coming years, there is expected to be a significant growth in both the global population and the need for energy (Boriah et al., 2008). Berahas et al. (2019) predict numerous estimates that the world's fossil fuel supplies—including coal, oil, and gas—will run out within the next century due to the exponential growth in population. According to Milano et al. (2016), the world's population has shifted its focus to biofuels, an environmentally beneficial alternative energy source, as a result of the scarcity of fossil fuels, increased energy consumption, and environmental risks. Brodin et al. (2017b) state the number of benefits of using biofuels made from trash, particularly lignocellulosic feedstocks. Energy crops, forestry waste, agricultural wastes, and other plant-based materials are some examples of lignocellulosic biomass, which is abundant and renewable and offers a reliable and affordable feedstock for the generation of biofuels. Furthermore, Martínez (2021), the manufacture of biofuels from biomass-based feedstocks, is consistent with the circular economy's tenets of resource efficiency, environmental effect minimization, and waste material repurposing. Liew et al. (2014) proposed that many thermochemical and biochemical conversion processes yield many forms of biofuels, including ethanol, methanol, biogas, bio-syngas, biodiesel, bio-oil, charcoal, biohydrogen, and Fischer–Tropsch liquids. Nowadays, Gupta et al. (2021) referred to the majority of biomass-based biofuels produced commercially as being produced using biochemical processes, then maintaining and optimizing effective parameters. The task is to determine the relationship between the outputs of the biofuel production process and process parameters in an ideal range to guarantee high productivity and quality. Amini et al. (2016) gave underlying nonlinear relations between the responses and the parameters, as well as the abundance of process parameters, which make factor effect prediction unfeasible even in the presence of experiments. For this reason, highly accurate experimental modeling techniques, such as machine learning-based prediction and artificial intelligence (AI) techniques, are helpful in overcoming the challenges posed by traditional computing techniques and experimental methods.

 DOI: 10.1201/9781003571223-5

5.1.1 Artificial Intelligence

The past few decades have seen significant investments in artificial intelligence, which have accelerated its growth. It all began with the development of various models that could conduct tests and evaluate theories on the workings of various mental processes (Haenlein & Kaplan, 2019). The two distinct models that form the foundation of AI systems are machine learning (ML) and deep learning (DL). Machine learning is a process where a model learns to recognize patterns in a given quantity of data so that it may automatically adapt to new data without the need for human intervention. When it comes to deep learning, the model is meant to run patterns in order to anticipate future events and be able to take in a lot of unstructured data. According to Helei et al. (2021b), the applications of AI technologies, such as machine learning, have become more and more common in numerous scientific and engineering research fields outside of the energy sector. The scientists also state that by employing specific ML technologies, like artificial neural networks (ANN), some of the key characteristics in CO_2 capture procedures can be successfully predicted. Machine learning algorithms use a technique named inductive learning to learn how to accomplish a specific goal function by sending in training data. The dataset has to be made more broadly so that the whole domain is covered and the machine learning model is able to predict values that it has never seen before.

Ghaderi et al. (2019) work together using empirical and experimental approaches to provide helpful insights and hasten the advancement of fermentation-based cellulosic biomass biofuel production. They suggested that great efforts have been made in biofuel research and development to overcome these obstacles. According to Caspeta et al. (2014), biofuels made from lignocelluloses are more plentiful than first-generation alcohol and produce fewer greenhouse gas emissions while also not competing with food crops; they are seen as a desirable replacement for petroleum products. The surplus lignocellulosic biomass available is most typically employed by combustion of biomass to generate power or heat, pyrolysis of biomass, or creation of second-generation bioethanol by hydrolysis of biomass. Technology that is technically feasible, economically viable, and environmentally sustainable must be available to properly utilize lignocellulosic biomass for the manufacture of bioethanol. The hydrolysis process is the costliest part of turning lignocellulosic material into bioethanol. It is possible to perform hydrolysis chemically or enzymatically. Because the enzymatic route produces fewer inhibitors and has less of an impact on the environment, it is preferred. To create a process that is economically viable, it is necessary to optimize a number of other factors, including the type of pretreatment, substrate loading, substrate size, substrate and product inhibition, and substrate feeding strategy.

Qdais et al. (2010) put forward the better understanding of the process inputs for ideal yield, and production rate will result from the modeling and optimization of biofuel production processes. Optimizing the steps involved in the production of these biofuels in order to increase yields is the primary objective of modeling.

In the production process of biofuels (Kana et al., 2012, Mohamed et al., 2013), a variety of modeling techniques have been used. The results have demonstrated that modeling and optimization can increase biofuel yields. For example, Ghosh et al. (2012) employed the response surface methodology (RSM) in a single-stage photo fermentation using the photosynthetic bacteria *Rhodobacter capsulatus* to maximize biohydrogen production on inputs of glucose concentration, fixed nitrogen, and light intensity. Their findings demonstrated that there was a significant interactive effect between these parameters and nitrogenase activity and biohydrogen yield. The optimum production in this case of biohydrogen was 85% greater than the previous record. Tools for artificial intelligence have become a viable approach to bioprocess modeling and optimization. These include fuzzy logic, ant algorithms, particle swarm optimization, artificial neural networks (ANN), and genetic algorithms (GA) (Zhang et al., 2010). All of these are thought to be appropriate for use in case designing of bioprocesses for the research and development (Haider et al., 2007; Garlapati & Banerjee, 2010). Nagata and Chu (2003) described ANN as a mathematical representation of how the human brain functions neurologically. By simulating the network structure of inter-connected nerve cells arithmetically, they mimic the learning process found in the brain. Y. Huang et al. (2007) suggest in their review literature that the main driver behind the widespread use of ANNs in research and development is the quick advancement of algorithms and information technologies. Tan et al. (2006) additionally carry out a study in which the combustion process was predicted using an artificial neural network (ANN) technique, and the parameters (combustion temperature, feed flow rate, catalyst, etc.) were optimized using a genetic algorithm approach.

5.2 ROLE OF ARTIFICIAL INTELLIGENCE IN BIOFUEL PRODUCTION

The need for alternative energy sources has become more pressing in light of the growing population, environmental degradation, and concerns about climate change. To solve these issues (Biofuels, n.d.), biofuels derived from waste materials have been embraced as an alternative to the petroleum-based precursors for numerous industrial uses. According to Okolie et al. (2022), biofuels can be generated from biomass through integrated processes, thermochemical conversion, and biological conversion. Microorganisms and bacteria are used in biological processes such as anaerobic digestion and fermentation to effectively convert biomass to biofuels (Okolie et al., 2022). In contrast, biomass is broken down by thermochemical processes like pyrolysis and gasification, which employ heat and chemical energy. It should be noted that the majority of techniques and processes utilized in the manufacturing of biofuels are intricate and nonlinear. Therefore, in accordance with Aghbashlo et al. (2021), for their design and subsequent optimization, a quick and precise modeling technique is required. When it comes to modeling such extremely complicated systems, data-driven machine learning (ML) and artificial intelligence (AI) techniques have proven to be more predictive than traditional approaches. In recent times, data-driven machine learning techniques have optimized biofuel systems, thereby adhering to environmental, social, and governance norms, resolving environmental issues, and promoting sustainable practices within the sector.

Thus, this study examines the state of the art and future directions of AI/ML applications in the field of biofuel technology. It should be noted that a thorough search of the literature was done using a variety of databases, including Research Gate, Google Scholar, and Scopus PubMed. Artificial intelligence (AI) is a broad notion that involves machines or computers performing tasks that are normally associated with human intelligence. These include solving puzzles, comprehending spoken language, seeing patterns, and drawing lessons from the past. Conversely, machine learning (ML) is a subset of artificial intelligence (AI) that is a particular method or methodology that enables computers to learn from data. The sophistication of AI technologies has increased to the point where academics are starting to pay more attention to them, especially in the field of bioenergy. Considering the significance of artificial intelligence (AI) in the particular domain of biofuel production, many AI methods for improving current systems and resolving issues with bioenergy generation have been extensively explored. In order to produce biofuel efficiently from natural resources, be environmentally conscious, and make the most use of available resources, artificial intelligence (AI) offers numerous sophisticated processes (Williams et al., 2015).

Both AI and ML work together using empirical and experimental approaches to provide helpful insights and hasten the advancement of fermentation-based cellulosic biomass biofuel production. Even though AI and ML have been used extensively in the field of biofuel production, there is still a lack of thorough reviews on the importance of these potent techniques in the fermentation-based biofuel production of cellulosic biomass in the studies that are now available. According to Del Rio-Chanona et al. (2018), such approaches may be able to accurately predict the results, using algorithms or models that are trained using specific inputs and employ mathematical techniques to predict the outputs. Furthermore, Mutlu and Yucel (2018) added that as soon as new data is obtained, the results are instantly improved. Because ML algorithms can predict product concentration, their use in biorefineries has increased dramatically in recent times. ML creates new opportunities for bioenergy system forecasts. In addition, many researchers worked on integrating AI into bioenergy to boost the efficiency of bioenergy technology and obtain the highest possible energy output. Artificial intelligence (AI) served as the newest technology, in accordance with Meena et al. (2021), that has significantly improved processes like control and management in the biomass industry's biofuel, bioenergy, and other biobased production, illustrated in Figure 5.1. The goal of AI is to develop systems and procedures that might be able to do tasks like problem-solving and learning that are often associated with human intellect. Researchers (Cinar et al., 2021) in this field take into account the application of AI in light of the demands of contemporary industries with respect to matters such as minimizing production costs, managing energy consumption, improving efficiency, and the need for artificial intelligence processes in the various phases and procedures used to generate biofuel in various sectors. Hence, it can be concluded that AI-based tools and solutions have the potential to have a big impact on environmental quality–related

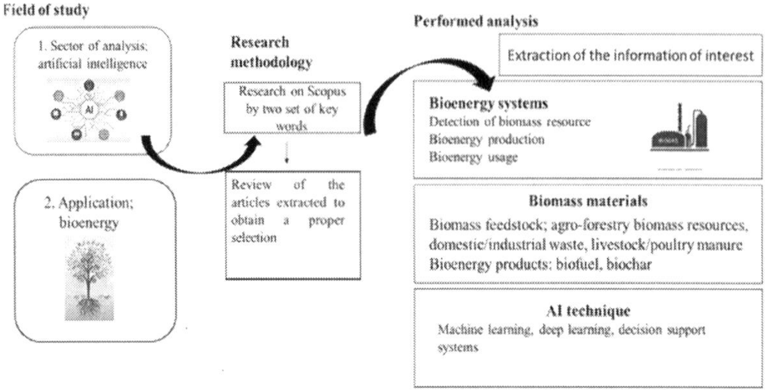

FIGURE 5.1 Biofuel production using artificial intelligence.

factors like reducing pollutant emissions, controlling waste disposal, maximizing energy use, and resource management.

5.3 AI IN FEEDSTOCK CHARACTERIZATION AND SELECTION

Biofuel has gained attention as a possible energy source with advantages for society, the economy, and the environment in recent decades. Biofuel is produced from biomass, or organic material, which is cultivated in fields and forests. A thriving biofuel sector would improve energy security, lower energy and fuel prices, and decrease oil imports, all of which would be advantageous to society. The usage of biofuel has been suggested as an alternative to traditional liquid fuels due to its ability to lower life cycle emissions and the consequences they have, such as climate change. Sharma et al. (2023) reviewed the biomass supply chain design and modeling, particularly for mathematical programming, and gave an overview of biomass feedstock and biofuel generation as depicted in Figure 5.2. Jha et al. (2017) proposed that the goal of the current study is to investigate how artificial intelligence (AI) approaches can be used to improve biofuel production processes and control supply and demand. Biofuel made from biomass is dependable and helps lower atmospheric CO_2 levels when used in place of traditional fuels; it has attracted increased attention in the energy, transportation, and residential sectors. The degree of characterization detail for the feedstock and effluent is critical to the building of realistic reactor models. The aforementioned suggests the need for reactor modeling approaches in addition to feedstock reconstruction. While research in these subjects is plentiful, few methods use artificial intelligence. Entropy maximizing techniques were successfully applied in feedstock reconstruction of different petroleum fractions by Hudebine and Verstraete (2004), Verstraete et al. (2004), and subsequently Van Geem et al. (2007). Because of this, biofuel made from oil crops and animal fat is becoming increasingly popular as an alternative to conventional petroleum fuels. In particular,

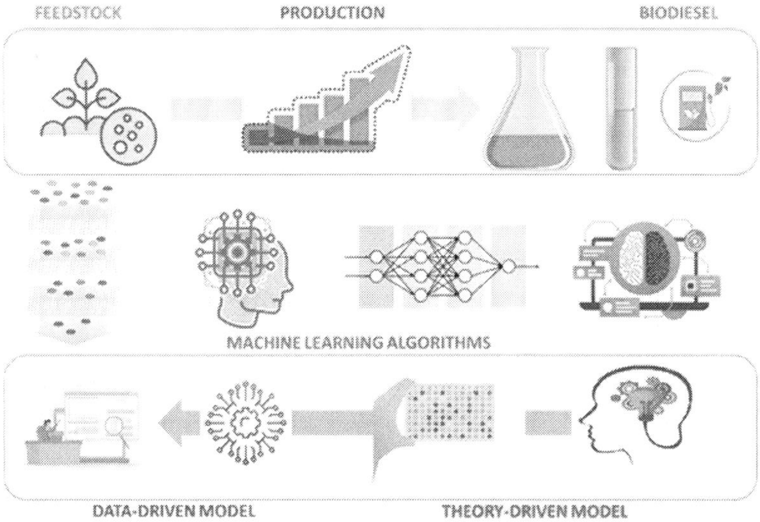

FIGURE 5.2 Overview of machine learning algorithms on biofuel data.

fuels derived from biomass feedstock are now more environmentally friendly and sustainable (Xing et al., 2021). They have a number of benefits over petroleum fuels, such as a more varied regional distribution of feedstock, biodegradability, increased potential for reducing greenhouse gas emissions, and viability. It is viewed in Guo and Song (2019) and Güleç et al. (2021) that the application of biomasses in terms of process application and product standardization is complicated by the variations in their physical and chemical properties. Ruiz et al. (2013), Şimşek et al. (2020), Heidari et al. (2019), and Cherad et al. (2016) stated that a range of characterization and experimental studies are necessary to determine the usability of different forms of biomass feedstocks for their prospective application in energy conversion processes like gasification, liquefaction, hydrothermal processing, pyrolysis, and combustion.

Ambaye et al. (2021) mentioned that to investigate the viability of using biofuel derived from innovative feedstock and to establish it as a commonplace fuel for transportation, more study is required. Securing a steady supply of raw materials and managing their high costs are problems for biofuels, particularly biodiesel. Furthermore, a substantial amount of experimental data is required to verify fuel quality. Sharma et al. (2023) state that it is difficult to get large amounts of gasoline, especially from new sources, for testing engines with various characteristics. This limitation is especially noticeable for second-generation biofuels, as novel sources and methods are currently being investigated. Additionally, they (Sharma et al., 2023) also emphasize the value of life cycle assessments in improving knowledge and reducing the indirect consequences of carbon emissions and land use. As the growing trend in the use of artificial intelligence and machine learning predicts the fuel properties, screening, and choosing the most promising feedstocks, it also optimizes the fuel production processes. These applications also minimize the need for substantial experimental data and extensive testing. Recently, there has been a lot of interest in using deep learning algorithms to discover and optimize feedstock and production methods for biofuel. Awogbemi and Von Kallon (2023) highlighted the lack of thorough information covering a broad variety of geographic regions and feedstock types in their study on the selection of feedstock. There are still questions about the deep learning models' generalization and transferability when choosing feedstock for biofuel. According to R. Wang et al. (2024), models that are trained on certain datasets might not be as resilient when applied to different geographic areas or feedstock kinds. More research is needed to develop domain adaptation strategies and transfer learning techniques that enable deep learning models to adapt to novel datasets and environments without compromising prediction accuracy. To properly capture the wide range and complex characteristics of feedstocks, comprehensive statistics must be gathered and arranged collaboratively.

5.4 PREDICTIVE MODELING FOR BIOMASS YIELD

The most abundant renewable energy source at the moment, bioenergy, has enormous promise for resolving the world's energy and climate change problems. According to J. Guo and Baghban (2017), the contemporary industry has created a wide range of technologies to transform biomass into solid, liquid, and gaseous energy products. The advantages of bioenergy, which include improving energy security, regional agriculture growth (Maltsoglou et al., 2013), rural economic development (Demirbas & Demirbas, 2007), and the efficient use of natural resources (Khishtandar, 2019), have been extensively covered in the current studies. However, the difficulties associated with supply chains, biorefineries, and biomass feedstock continue to restrict the widespread use of bioenergy systems (Asadullah, 2014; Nguyen et al., 2015). The infrastructure for the production of bioenergy must transition to high-tech processes in light of the growing need for energy.

In order to comprehend (Sewsynker-Sukai et al., 2016; Fischer et al., 2017), analyze, and forecast biological processes, mathematical and statistical models are vital. They are also required for the optimization of crucial parameters that enhance process performance. In order to achieve maximum efficiency, a better understanding of the process expenditures will be facilitated by modeling and optimizing the biofuel manufacturing processes. Improving efficiency is the primary goal of modeling, which involves optimizing the production processes. In the recent past, Rivera et al. (2010) stated, there has been an increasing use of artificial intelligence and machine learning techniques, such as genetic algorithms (GA) and artificial neural networks (ANN), for complex biotechnological problems in order to achieve high operational performance. Such techniques are typically involved in modeling and optimization of complex phenomena, including various process variables. It works better when examining data that show nonlinear interactions between several process parameters or when deterministic equation formation is too challenging. Kraszkiewicz et al. (2019) have proven useful in creating biofuel models even in the absence of previous knowledge about the kinetics and metabolic fluxes that take place in cells and their environments. Moreover, ANNs rely solely on data and do not have prior knowledge of the events that govern the process. The main feature of ANNs that makes them attractive as modeling tools is their ability to learn and generalize, along with their capacity to handle assigned data in a unique way. These features include high parallelism, error and noise acceptance, and linearity.

Mashhadimoslem et al. (2021) used machine learning algorithms to anticipate different biomass synthesis by applying pyrolysis temperatures and CO_2 adsorption performance techniques. They found that the artificial neural network's (ANN's) ability to predict the CO_2 adsorption and Brunauer–Emmett–Teller (BET) of different biomass precursors was satisfactory. The study by He et al. (2021) and Shafizadeh et al. (2022) shows the great potential of machine learning techniques for stimulating the biomass hydrothermal liquefaction by employing kinetics datasets. Lu et al. (2022) used machine learning and computational fluid dynamics (CFD) simulation to quickly pyrolyze four different biomass feedstocks of biofuels, as illustrated in Table 5.1. Studies using machine learning assessed the yields of syngas, biochar, and bio-oil depending on different input features as shown in Figure 5.3. Numerous creative methods in this area produced noteworthy outcomes and greatly raised their efficiency. For instance, Sun et al. (2016) combined the experimental data with the machine learning pathways to examine the relationships between pyrolysis conditions and bioproducts. According to Safaei-Farouji et al. (2022) and Parsaeian et al. (2022), machine learning models enhanced our understanding of biomass production while reducing the need for an experimental method to evaluate results. Additionally, two helpful techniques to lower complexity and improve prediction are partial dependence analysis and feature reduction (Schmidhuber, 2014; Mashhadimoslem & Ghaemi, 2022). Reviewing the studies revealed the substantial association between the biomass content and the bio-oil component, which is worth noticing. Analyzing all the machine learning processes and outcomes, Rahimi et al. (2023) conclude that, when combined with nonlinear datasets produced from biomass-derived precursor composition and pyrolysis conditions, they provide a dependable means of performing comparative analyses of multiple ML approaches. The majority of scientific methods rely on yield production, optimization, and partial dependency on input features. To assess the capability modeling of machine learning and feature inputs, respectively, generalizability and sensitivity analysis are utilized. In order to increase the yields of bioproducts, a genetic algorithm was also used to assess the optimization of pyrolysis parameters. Based on pyrolysis settings, four different types of biomass were covertly improved (Table 5.1).

5.5 AI IN BIOMASS PRE-PROCESSING

When working with large datasets, it is imperative to remove significant deviations from the remainder of the datasets. Taleb et al. (2015) propose data preprocessing techniques such as data cleansing, deduplication, compression filtering, and format conversion to achieve this. Large data preprocessing can provide a number of difficulties, which will affect data processing. One of the most important things to keep in mind when doing preprocessing is data quality. Data quality can be increased by

TABLE 5.1
A Comparative Analysis of Machine Learning Models Applied to Forecast the Final Output of Bioproducts, Including Biochar, Syngas, and Bio-oil

Bioproduct	Utilized Model	Features and Coverage of the Recent Studies
Biochar	Multilayer perceptron (MLP)	Determine the amount of biochar produced under various pyrolysis conditions and using different types of biomass
	Random Forest (RF) and Decision Tree (DT)	To model the process of co-pyrolyzing coal and biomass, random forest and decision tree algorithms were employed
	Extreme gradient boosting (XGB)	The study aimed to predict the biochar yield from algae by analyzing the impact of pyrolysis temperature, heating rate, and time. Different biomass types and microwave pyrolysis settings (power, duration, weight, and absorber) were also considered
Biochar and Bio-oil	Extreme gradient boosting (XGB)	Machine learning techniques were utilized to forecast the production biochar and bio-oil from biomass pyrolysis
Syngas	Tree-based Pipeline Optimization Tool (TPOT) and Gaussian process regression (GPR)	A tree-based machine model was used to estimate the production of syngas
	Support vector machine (SVM) and Random Forest (RF)	Machine learning was used to predict syngas yield, identify optimal pyrolysis conditions, analyze biomass properties, and simplify the model for better performance
Bio-oil	Random Forest (RF)	Machine learning models were employed to predict bio-oil yields from various biomass sources, considering the properties of the raw materials and the pyrolysis process
	Multiple linear regression (MLR), Random Forest (RF), Support vector machine (SVM), and Decision Tree (DT)	Bio-oil yield was estimated using a machine learning model that incorporated biomass composition and pyrolysis conditions and pyrolysis conditions as temperature, heating rate, and particle size. Genetic algorithms were used to identify the most relevant features for accurate predictions
	Gaussian process regression (GPR) and Random Forest (RF)	Machine learning techniques were applied to predict bio-oil production from algal hydrothermal liquefaction, optimizing the process through both forward and reverse engineering approaches
	Random Forest (RF) and Multiple Linear Regression (MLR)	The study analyzes the relationship between biomass properties and pyrolysis conditions to determine their impact on bio-oil yield

Source: Adapted from Rahimi et al. (2023).

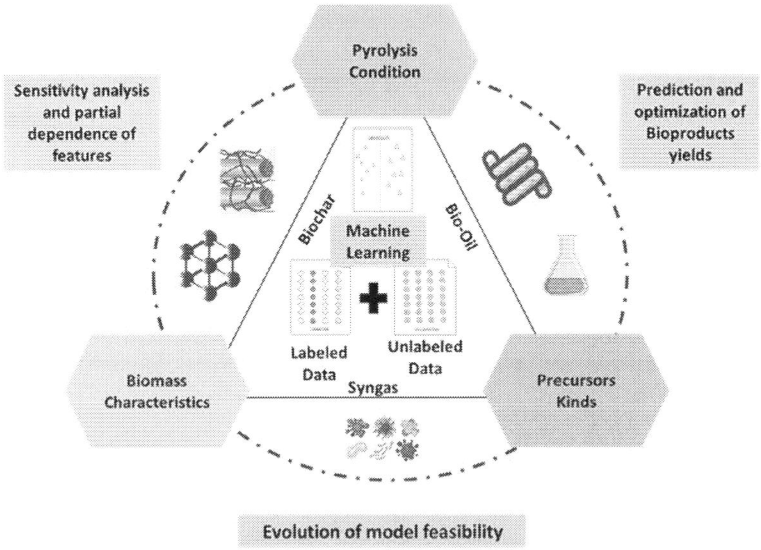

FIGURE 5.3 Machine learning models predicting biofuels (biochar, syngas, bio-oil).

cleaning the raw data utilizing a variety of methods and procedures. Moreover, by locating the low-quality data and rebuilding it, one can achieve increased data quality. Rinnan et al. (2009) proposed various preprocessing and multivariate calibration techniques. The material and its physical and chemical qualities determine which approaches to use in combination or separately. Although preprocessing biomass may lower handling costs, as stated by Nolan et al. (2010), Sokhansanj and Hess (2009), it raises the feedstock's price. Preprocessing biomass facilitates feedstock handling. For example, the process of pelletization, which results in drying and densification, can raise the feedstock's price by up to 50% when compared to the raw material without processing. Although costs remain the essential issue, several preprocessing stages might be critical depending on the type of biomass and local conditions. For example, it's important to regulate the moisture level since biomass that has more than 15% to 20% moisture is deemed aerobically unstable, which might pose issues during storage and transportation. According to Rentizelas et al. (2008), for anaerobic storage, biomass containing more than 20% moisture must be either dried to below the threshold or sealed. In spite of the relative dryness of the biomass, storage itself may provide challenges. Dry matter loss, quality deterioration (biomass with lower biofuel output), self-ignition safety risks, fungal spore dispersion health risks, and other issues might result from prolonged storage. Preprocessing is used by Rinnan et al. (2009), who examine how crucial it is to obtain acceptable findings. The study demonstrates how it is applied to the analysis of NIR data using various classification models. The NIR spectroscopy, according to Pasquini (2003), serves as the basis for the method's speed, accuracy, and low sample preparation requirements, which pave the way for the evolution of a wide range of analytical techniques and the method's replacement by other traditional analytical methods. The preprocessing data is separated into two sections: first, reference-dependent technology, which is used to orthogonalize data using a reference; and second, reference-independent technology, which is used for exploratory studies in which there is no reference value. When comparing the raw and preprocessed NIR datasets used to determine the moisture content of marzipan (Rinnan et al., 2009), minor differences were found; however, preprocessing techniques were suggested to be a considerable improvement (Kim et al., 2017; Leblon et al., 2013). Savitzky-Golay (SG) preprocessing is a popular approach for preprocessing NIR spectrum data. The SG approach makes use of time-domain technologies to let the user draw a trend line. The primary goal of the SG approach is to smooth out the raw data, which increases the model's precision. In order for the method to function, the input data must first be limited by a low-degree polynomial and then placed at equal intervals. In order to optimize a dataset, various numbers of necessary smoothing points were employed. When using SG to smooth the data by Chen et al. (2019), there are several processes involved, including figuring out the polynomial degree, polynomial derivative order, and number of smoothing points. For NIR spectroscopic data, multiplicative scatter correction (MSC), proposed by Mishra and Lohumi (2021), is another often-used preprocessing method. By adjusting the spectral data's offset and scaling, it normalizes the data. Nonetheless, it is capable of distinguishing between the spectral data's scattering signal and the chemical component's informative absorbance. By doing this, the spectrum difference resulting from surface intensity and particle size—which is devoid of any chemical or physical information—is eliminated. It is simple to read because the modified spectral is comparable to the original spectral. Because NIR spectrum data tends to amplify the spectral absorption information, this preprocessing technique is useful.

5.6 REAL-TIME MONITORING AND CONTROL USING AI

Svendsen et al. (2014) proposed that real-time reactor state measurements provide more control and process optimization than offline monitoring; real-time bioprocess monitoring is a hotly researched topic. It is viewed by K. Schügerl (2001) that due to a lack of appropriate and reasonably priced monitoring tools, real-time process monitoring is typically only somewhat possible in large-scale biotechnology processes. The majority of monitoring applications have been created for laboratory use. Numerous publications exist regarding the benefits, drawbacks, and availability of various monitoring approaches; however, as noted by Sonnleitner (2012), real-time large-scale monitoring with sophisticated sensors is rarely carried out. This is due to the paucity of studies on the possible advantages of these techniques. In order to provide guidance for choosing the best monitoring technique under particular circumstances, Kiviharju et al. (2008) analyzed several monitoring techniques based on particular needs for biomass monitoring. Several articles review certain monitoring devices; these are mentioned as separate entities in the papers. Ulber et al. (2003) investigated in situ microscopy, NIR spectroscopy, and fluorescence spectroscopy, while Marison et al. (2012) provided a thorough overview of near-infrared spectroscopy (NIR), mid-infrared spectroscopy (MIR), Raman spectroscopy, dielectric spectroscopy (DS), and biocalorimetry. However, because the effectiveness of these techniques for certain monitoring goals was not contrasted, the reader was not provided enough information to justify choosing a particular alternative.

According to Kadam et al. (2004), biomass, which primarily consists of plant material, is converted into cellulose and hemicellulose-based biofuels by fixing sugars in structures that are entangled with lignin. Cellulose-based biofuel production is currently too expensive to compete with non-cellulosic biofuels (Macrelli et al., 2012), where the carbon source is derived from relatively easy-to-find and basic sources like corn or sugar beet. Since non-cellulosic biofuels have been produced for over 20 years, Tyner (2008) states that a great deal of expertise has been gained in managing and constructing plants for the production of non-cellulosic biofuels. As a result, the technology is now mature. Therefore, Stephen et al. (2011) emphasized that, in order to ensure the process operates under optimal conditions and to compensate for the limited understanding of this technology, monitoring is crucial in the manufacture of cellulosic ethanol. Increasing the process's profit margin is one of the objectives of tracking the fermentation of cellulose to ethanol. Achieving a high ethanol yield and operating the process non-sterilely can lead to greater profit. It is crucial to monitor the process for contamination in order to prevent substrate loss and be able to halt the process as soon as contamination is found. The risk of contamination generally increases with the procedure duration. Therefore, in order to implement a real-time strategy to enhance process performance, monitoring is important of inhibitory components in the reactor and feed. New control and optimization methods come with a cost, not only in terms of implementation and equipment but also in terms of operator training. Therefore, the advantages must be shown to be greater than the operating benefits of the current process. For the first time, the use of combination strategies for advanced monitoring is discussed. The majority of model approaches given by Rathore et al. (2022) provide an estimate by employing multi-parameter techniques. Nevertheless, new developments in liquid chromatography and membrane modeling with mass spectrometry benchmarking, as well as machine learning models quantifying real-time control processes of bioreactors at pilot and commercial scales, enable improved on-site sensor feedback. The true difficulty in utilizing machine learning in the bioprocess industry in real time, though, lies in choosing the appropriate algorithm and model to create, as each model has its own restrictions and its predictive power changes depending on the desired outcome. The study by *Aghbashlo et al. (2021)* clearly links monitoring apparatus and their combinations with specific process objectives for the manufacture of cellulosic ethanol, going beyond earlier reviews. Furthermore, their paper examines the potential benefits of these technologies through a case study involving cellulosic ethanol production.

Since different measurements will add more direct data, increasing the number of monitoring methods to three or four enhances the added value. It is important to consider the extent to which supplementary monitoring techniques augment the overall information acquired through the combination of both hardware and software sensors. This is because soft sensors frequently possess the ability to analyze reactor operations from simpler measurement techniques, like off-gas composition measurement. More precise information on the present reactor status should be provided via hardware sensors. This is only true, though, if the sensors are accurate, impervious to interference, and able to measure every component of interest. Fast control also benefits from quick reaction times, but the methods shouldn't be overly costly. For measurement methods of biomass concentration to be useful at all, contaminations must be identified and viable cells must be distinguished from non-viable ones. Soft sensors are required for most procedures in order to calibrate and transform the measured data into useful information. Different techniques have different levels of complexity when it comes to data processing, calibration, and maintenance. This can be significant because factories are frequently located in remote locations where access to experts is not always guaranteed.

5.7 TECHNICAL AND COMPUTATION CHALLENGES

A variety of beneficial and profitable goods are produced from biobased resources by chemical technologies and the chemical process sector. These products are used in agrochemicals, transportation fuels, health products, and personal care items. The rapidly growing subject of science and engineering is creating instruments to identify, measure, and balance human needs and resource constraints in order to maximize benefits to both people and environment. The chemical industry's greatest difficulty is developing new chemical processes while keeping old design goals in mind and integrating sustainability and environmental goals. The industry is faced with a unique challenge as biofuels, a potentially sustainable and cleaner alternative to conventional fuels, are developing quickly. This presents an opportunity to understand the technical and computational challenges of designing sustainable supply chains by taking into account social, ecological, and economic factors at the same time. Many process technologies, according to Shi et al. (2012), can handle combinations of different types of biomasses; most processes necessitate a certain level of consistency in the feedstock they take. Muller-Feuga et al. (1998) shows that the continuous production is possible with artificial lighting, but it requires a lot more energy. The main goal of creating a fuel that is affordable is sometimes defeated by the fact that artificial illumination typically comes from fossil fuels, which also increases the carbon footprint of the system. It is crucial to evaluate the environmental impact of chemicals that may represent a concern by Wiloso and Heijungs (2013). The term "footprint" refers to the concentration of a chemical used or released into the environment as a result of human activity, which may have an impact on global sustainability. Many life cycle evaluations have been conducted on the production of biobased fuels. When doing such assessments, a lot of aspects are taken into consideration, especially knowing how land use changes, how much carbon is sequestered during the biomass development phase, and how credits from co-products like chemicals, heat, and power are used. Comparing biological conversion pathways to thermochemical conversion choices, this is particularly true. Any biorefinery's design and construction require access to industrial amounts of lignocellulosic biomass, and ensuring a steady supply of this raw material is still a major obstacle to the widespread use of lignocellulosic biofuels. Building a biorefinery requires careful consideration of the regional and temporal variability in biomass supply, as well as the ongoing requirement for a relatively homogenous feedstock. This is because the owner of the plant must maximize plant operating time in order to minimize the capital expenditure associated with the cost of producing biofuel. The biomass supply chain's logistical challenges are framed by these conditions. Momenitabar et al. (2022) viewed that the domains of energy are where the effects of machine learning technologies are most visible in terms of how fuels are delivered. This is because the supply curves are actually moved outward by more sophisticated energy supply systems. They take fewer resources and lower their cost. The different decisions that are made by Liao and Yao (2021) on an operational, tactical, and strategic level may have short-, medium-, and long-term implications, respectively. According to Ghaderi et al. (2016), optimization and simulation are two common techniques for modeling the supply chains for biofuels, and their use varies based on the intricacy, unpredictability, and uncertainty of the systems. According to recent research updates, machine learning holds great promise for reducing barriers to the expansion of biofuel production. The focus of today's analysis is on recent advancements in the use of basic machine learning techniques in bioenergy systems that use biofuels for sustainable development. This chapter could assist scientists in developing precise scientific hypotheses and creating machine learning-based trials for biofuel research in the future.

5.8 CONCLUSION

In conclusion, a move toward sustainable energy solutions is required due to the growing global population and energy demands, particularly in light of the depletion of fossil fuel supplies. Because they are environmentally friendly and consistent with the ideas of a circular economy, biofuels—especially those made from lignocellulosic biomass—are important. A multitude of process parameters impact productivity and quality in the complex biochemical and thermochemical processes involved in the generation of biofuels. The production of biofuel can be made more efficient and with better results by modeling and optimizing these processes using machine learning and artificial intelligence–based approaches. It will be essential to adopt these technical developments if we are to meet future energy demands in a sustainable and efficient manner. Artificial intelligence—especially machine learning and deep learning—has advanced significantly across a number of disciplines, including the study of biofuels. These technologies provide strong modeling tools for intricate systems such as CO_2 capture and biofuel production optimization. Response surface methodology, artificial neural networks (ANN), genetic algorithms, and other AI-based methods are among the many strategies that have been shown to be successful in raising biofuel yields and optimizing important process variables. Also, it may be possible to get past the challenges associated with designing bioprocesses and produce biofuel that is both more sustainable and productive. Multiple monitoring approaches combined with soft sensors can yield more accurate real-time data and enhance process management; nonetheless, the industry's primary hurdles still lie in the complexity of data processing and the availability of knowledge in remote places. Machine learning has enormous potential to break down obstacles and propel the sustainable growth of bioenergy systems as long as it keeps advancing. Finally, this chapter could help future researchers studying biofuels formulate accurate scientific hypotheses and design machine learning-based trials.

REFERENCES

Aghbashlo, M., Peng, W., Tabatabaei, M., Kalogirou, S. A., Soltanian, S., Hosseinzadeh-Bandbafha, H., Mahian, O., & Lam, S. S. (2021). Machine learning technology in biodiesel research: A review. *Progress in Energy and Combustion Science*, 85, 100904. https://doi.org/10.1016/j.pecs.2021.100904

Ambaye, T. G., Vaccari, M., Bonilla-Petriciolet, A., Prasad, S., Van Hullebusch, E., & Rtimi, S. (2021). Emerging technologies for biofuel production: A critical review on recent progress, challenges and perspectives. *Journal of Environmental Management,* 290, 112627. https://doi.org/10.1016/j.jenvman.2021.112627

Amini, Y., Gerdroodbary, M. B., Pishvaie, M. R., Moradi, R., & Monfared, S. M. (2016). Optimal control of batch cooling crystallizers by using genetic algorithm. *Case Studies in Thermal Engineering*, 8, 300–310. https://doi.org/10.1016/j.csite.2016.09.001

Asadullah, M. (2014). Barriers of commercial power generation using biomass gasification gas: A review. *Renewable and Sustainable Energy Reviews*, 29, 201–215. https://doi.org/10.1016/j.rser.2013.08.074

Awogbemi, O., & Von Kallon, D. V. (2023). Application of machine learning technologies in biodiesel production process—A review. *Frontiers in Energy Research*, 1122638. https://doi.org/10.3389/fenrg.2023.1122638

Berahas, A. S., Byrd, R. H., & Nocedal, J. (2019). Derivative-free optimization of noisy functions via Quasi-Newton Methods. *SIAM Journal on Optimization*, 29(2), 965–993. https://doi.org/10.1137/18m1177718

Biofuels. (n.d.). Google Books. https://books.google.com/books?hl=en&lr=&id=hy6RDwAAQBAJ&oi=fnd&pg=PA93&dq=+Nasreen+S,+Nafees+M,+Qureshi+LA,+Asad+MS,+Sadiq+A,+Ali+SD:++Biofuels-state+of+development:+review+of+catalytic+trans+esterification+methods+for+biodiesel+production.+Chapter+2013,++6.&ots=7162i19Hzw&sig=xDvphs3AEjMeUxLsn7_onYW9n6g

Boriah, S., Chandola, V., & Kumar, V. (2008). Similarity measures for categorical data: A comparative evaluation. In Proceedings of the 2008 SIAM International Conference on Data Mining (SDM 2008) (pp. 243–254). Society for Industrial and Applied Mathematics (SIAM).. https://doi.org/10.1137/1.9781611972788.22

Brodin, M., Vallejos, M., Opedal, M. T., Area, M. C., & Chinga-Carrasco, G. (2017b). Lignocellulosics as sustainable resources for production of bioplastics – A review. *Journal of Cleaner Production*, 162, 646–664. https://doi.org/10.1016/j.jclepro.2017.05.209

Caspeta, L., Caro-Bermúdez, M. A., Ponce-Noyola, T., & Martinez, A. (2014). Enzymatic hydrolysis at high-solids loadings for the conversion of agave bagasse to fuel ethanol. *Applied Energy*, 113, 277–286. https://doi.org/10.1016/j.apenergy.2013.07.036

Chen, J., Wang, Z., Wu, Y., Li, L., Li, B., Pan, D., & Zuo, T. (2019). Environmental benefits of secondary copper from primary copper based on life cycle assessment in China. *Resources Conservation and Recycling*, 146, 35–44. https://doi.org/10.1016/j.resconrec.2019.03.020

Cherad, R., Onwudili, J., Biller, P., Williams, P., & Ross, A. (2016). Hydrogen production from the catalytic supercritical water gasification of process water generated from hydrothermal liquefaction of microalgae. *Fuel*, 166, 24–28. https://doi.org/10.1016/j.fuel.2015.10.088

Cinar, S., Cinar, S. O., Wieczorek, N., Sohoo, I., & Kuchta, K. (2021). Integration of artificial intelligence into biogas plant operation. *Processes*, 9(1), 85. https://doi.org/10.3390/pr9010085

Del Rio-Chanona, E. A., Wagner, J. L., Ali, H., Fiorelli, F., Zhang, D., & Hellgardt, K. (2018). Deep learning-based surrogate modeling and optimization for microalgal biofuel production and photobioreactor design. *AIChE Journal*, 65(3), 915–923. https://doi.org/10.1002/aic.16473

Demirbas, A. H., & Demirbas, I. (2007). Importance of rural bioenergy for developing countries. *Energy Conversion and Management*, 48(8), 2386–2398. https://doi.org/10.1016/j.enconman.2007.03.005

Fischer, J., Lopes, V. S., Cardoso, S. L., Filho, U. C., & Cardoso, V. L. (2017). Machine learning techniques applied to lignocellulosic ethanol in simultaneous hydrolysis and fermentation. *Brazilian Journal of Chemical Engineering*, 34(1), 53–63. https://doi.org/10.1590/0104-6632.20170341s20150475

Garlapati, V. K., & Banerjee, R. (2010). Evolutionary and swarm intelligence-based approaches for optimization of lipase extraction from fermented broth. *Engineering in Life Sciences*, 10(3), 265–273. https://doi.org/10.1002/elsc.200900086

Ghaderi, H., Pishvaee, M. S., & Moini, A. (2016). Biomass supply chain network design: An optimization-oriented review and analysis. *Industrial Crops and Products*, 94, 972–1000. https://doi.org/10.1016/j.indcrop.2016.09.027

Ghaderi, M., Javadikia, H., Naderloo, L., Mostafaei, M., & Rabbani, H. (2019). An analysis of noise pollution emitted by moving MF285 Tractor using different mixtures of biodiesel, bioethanol and diesel through artificial intelligence. *Journal of Low Frequency Noise, Vibration and Active Control*, 38(2), 270–281. https://doi.org/10.1177/1461348418823572

Ghosh, D., Sobro, I. F., & Hallenbeck, P. C. (2012). Optimization of the hydrogen yield from single-stage photofermentation of glucose by *Rhodobacter capsulatus* JP91 using response surface methodology. *Bioresource Technology*, 123, 199–206. https://doi.org/10.1016/j.biortech.2012.07.061

Güleç, F., Şimşek, E. H., & Sarı, H. T. (2021). Prediction of biomass pyrolysis mechanisms and kinetics: Application of the Kalman Filter. *Chemical Engineering & Technology*, 45(1), 167–177. https://doi.org/10.1002/ceat.202100229

Guo, J., & Baghban, A. (2017). Application of ANFIS strategy for prediction of biodiesel production using supercritical methanol. *Energy Sources Part a Recovery Utilization and Environmental Effects*, 39(17), 1862–1868. https://doi.org/10.1080/15567036.2017.1380731

Guo, M., & Song, W. (2019). The growing U.S. bioeconomy: Drivers, development and constraints. *New Biotechnology*, 49, 48–57. https://doi.org/10.1016/j.nbt.2018.08.005

Gupta, K. K., Kalita, K., Ghadai, R. K., Ramachandran, M., & Gao, X. (2021). Machine learning-based predictive modelling of biodiesel production: A comparative perspective. *Energies*, 14(4), 1122. https://doi.org/10.3390/en14041122

Haenlein, M., & Kaplan, A. (2019). A Brief History of artificial intelligence: On the past, present, and future of artificial intelligence. *California Management Review*, 61(4), 5–14. https://doi.org/10.1177/0008125619864925

Haider, M. A., Pakshirajan, K., Singh, A., & Chaudhry, S. (2007). Artificial Neural Network-Genetic algorithm approach to optimize media constituents for enhancing lipase production by a soil microorganism. *Applied Biochemistry and Biotechnology*, 144(3), 225–235. https://doi.org/10.1007/s12010-007-8017-y

He, L., Bai, L., Dionysiou, D. D., Wei, Z., Spinney, R., Chu, C., Lin, Z., & Xiao, R. (2021). Applications of computational chemistry, artificial intelligence, and machine learning in aquatic chemistry research. *Chemical Engineering Journal,* 426, 131810. https://doi.org/10.1016/j.cej.2021.131810

Heidari, M., Norouzi, O., Salaudeen, S., Acharya, B., & Dutta, A. (2019). Prediction of hydrothermal carbonization with respect to the biomass components and severity factor. *Energy & Fuels*, 33(10), 9916–9924. https://doi.org/10.1021/acs.energyfuels.9b02291

Helei, L., Tantikhajorngosol, P., Chan, C., & Tontiwachwuthikul, P. (2021b). Technology development and applications of artificial intelligence for post-combustion carbon dioxide capture: Critical literature review and perspectives. *International Journal of Greenhouse Gas Control*, 108, 103307. https://doi.org/10.1016/j.ijggc.2021.103307

Huang, Y., Kangas, L. J., & Rasco, B. A. (2007). Applications of artificial neural networks (ANNs) in food science. *Critical Reviews in Food Science and Nutrition*, 47(2), 113–126. https://doi.org/10.1080/10408390600626453

Hudebine, D., & Verstraete, J. J. (2004). Molecular reconstruction of LCO gas oils from overall petroleum analyses. *Chemical Engineering Science*, 59(22–23), 4755–4763. https://doi.org/10.1016/j.ces.2004.09.019

Jha, S. K., Bilalovic, J., Jha, A., Patel, N., & Zhang, H. (2017). Renewable energy: Present research and future scope of Artificial Intelligence. *Renewable and Sustainable Energy Reviews*, 77, 297–317. https://doi.org/10.1016/j.rser.2017.04.018

Kadam, K., Rydholm, E., & McMillan, J. (2004). Development and validation of a kinetic model for enzymatic saccharification of lignocellulosic biomass. *Biotechnology Progress*, 20(3), 698–705. https://doi.org/10.1021/bp034316x

Kana, E. G., Oloke, J., Lateef, A., & Adesiyan, M. (2012). Modeling and optimization of biogas production on saw dust and other co-substrates using Artificial Neural network and Genetic Algorithm. *Renewable Energy*, 46, 276–281. https://doi.org/10.1016/j.renene.2012.03.027

Khishtandar, S. (2019). Simulation based evolutionary algorithms for fuzzy chance-constrained biogas supply chain design. *Applied Energy*, 236, 183–195. https://doi.org/10.1016/j.apenergy.2018.11.092

Kim, G., Hong, S., Lee, A., Lee, Y., & Im, S. (2017). Moisture content measurement of broadleaf litters using near-infrared spectroscopy technique. *Remote Sensing*, 9(12), 1212. https://doi.org/10.3390/rs9121212

Kiviharju, K., Salonen, K., Moilanen, U., & Eerikäinen, T. (2008). Biomass measurement online: The performance of in situ measurements and software sensors. *Journal of Industrial Microbiology & Biotechnology*, 35(7), 657–665. https://doi.org/10.1007/s10295-008-0346-5

Kraszkiewicz, A., Kachel, M., Parafiniuk, S., Zając, G., Niedziółka, I., & Sprawka, M. (2019). Assessment of the possibility of using hemp biomass (Cannabis sativa l.) for energy purposes: A case study. *Applied Sciences*, 9(20), 4437. https://doi.org/10.3390/app9204437

Leblon, B., Adedipe, O., Hans, G., Haddadi, A., Tsuchikawa, S., Burger, J., Stirling, R., Pirouz, Z., Groves, K., Nader, J., & LaRocque, A. (2013). A review of near-infrared spectroscopy for monitoring moisture content and density of solid wood. *The Forestry Chronicle*, 89(05), 595–606. https://doi.org/10.5558/tfc2013-111

Liew, W. H., Hassim, M. H., & Ng, D. K. (2014). Review of evolution, technology and sustainability assessments of biofuel production. *Journal of Cleaner Production*, 71, 11–29. https://doi.org/10.1016/j.jclepro.2014.01.006

Lu, L., Pecha, M. B., Wiggins, G. M., Xu, Y., Gao, X., Hughes, B., Shahnam, M., Rogers, W. A., Carpenter, D., & Parks, J. E. (2022). Multiscale CFD simulation of biomass fast pyrolysis with a machine learning derived intra-particle model and detailed pyrolysis kinetics. *Chemical Engineering Journal,* 431, 133853. https://doi.org/10.1016/j.cej.2021.133853

Macrelli, S., Mogensen, J., & Zacchi, G. (2012). Techno-economic evaluation of 2nd generation bioethanol production from sugar cane bagasse and leaves integrated with the sugar-based ethanol process. *Biotechnology for Biofuels*, 5(1), 22. https://doi.org/10.1186/1754-6834-5-22

Maltsoglou, I., Koizumi, T., & Felix, E. (2013). The status of bioenergy development in developing countries. *Global Food Security*, 2(2), 104–109. https://doi.org/10.1016/j.gfs.2013.04.002

Marison, I., Hennessy, S., Foley, R., Schuler, M., Sivaprakasam, S., & Freeland, B. (2012). The choice of suitable online analytical techniques and data processing for monitoring of bioprocesses. *Advances in Biochemical Engineering, Biotechnology*, 249–280. https://doi.org/10.1007/10_2012_175

Martínez, J. D. (2021). An overview of the end-of-life tires status in some Latin American countries: Proposing pyrolysis for a circular economy. *Renewable and Sustainable Energy Reviews*, 144, 111032. https://doi.org/10.1016/j.rser.2021.111032

Mashhadimoslem, H., & Ghaemi, A. (2022). Machine learning analysis and prediction of N_2, N_2O, and O_2 adsorption on activated carbon and carbon molecular sieve. *Environmental Science and Pollution Research*, 30(2), 4166–4186. https://doi.org/10.1007/s11356-022-22508-9

Mashhadimoslem, H., Vafaeinia, M., Safarzadeh, M., Ghaemi, A., Fathalian, F., & Maleki, A. (2021). Development of predictive models for activated carbon synthesis from different biomass for CO_2 adsorption using artificial neural networks. *Industrial & Engineering Chemistry Research*, 60(38), 13950–13966. https://doi.org/10.1021/acs.iecr.1c02754

Meena, M., Shubham, S., Paritosh, K., Pareek, N., & Vivekanand, V. (2021). Production of biofuels from biomass: Predicting the energy employing artificial intelligence modelling. *Bioresource Technology*, 340, 125642. https://doi.org/10.1016/j.biortech.2021.125642

Milano, J., Ong, H. C., Masjuki, H., Chong, W., Lam, M. K., Loh, P. K., & Vellayan, V. (2016). Microalgae biofuels as an alternative to fossil fuel for power generation. *Renewable and Sustainable Energy Reviews*, 58, 180–197. https://doi.org/10.1016/j.rser.2015.12.150

Mishra, P., & Lohumi, S. (2021). Improved prediction of protein content in wheat kernels with a fusion of scatter correction methods in NIR data modelling. *Biosystems Engineering*, 203, 93–97. https://doi.org/10.1016/j.biosystemseng.2021.01.003

Mohamed, M. S., Tan, J. S., Mohamad, R., Mokhtar, M. N., & Ariff, A. B. (2013). Comparative analyses of response surface methodology and artificial neural network on medium optimization for *Tetraselmis* sp. FTC209 grown under mixotrophic condition. *The Scientific World Journal*, 2013, 1–14. https://doi.org/10.1155/2013/948940

Momenitabar, M., Ebrahimi, Z. D., & Ghasemi, P. (2022). Designing a sustainable bioethanol supply chain network: A combination of machine learning and meta-heuristic algorithms. *Industrial Crops and Products*, 189, 115848. https://doi.org/10.1016/j.indcrop.2022.115848

Muller-Feuga, A., Guédes, R. L., Hervé, A., & Durand, P. (1998). Comparison of artificial light photobioreactors and other production systems using porphyridium cruentum. *Journal of Applied Phycology*, 10(1), 83–90. https://doi.org/10.1023/a:1008046814640

Mutlu, A. Y., & Yucel, O. (2018). An artificial intelligence-based approach to predicting syngas composition for downdraft biomass gasification. *Energy*, 165, 895–901. https://doi.org/10.1016/j.energy.2018.09.131

Nagata, Y., & Chu, K. H. (2003). Optimization of a fermentation medium using neural networks and genetic algorithms. *Biotechnology Letters*, 25(21), 1837–1842. https://doi.org/10.1023/a:1026225526558

Nguyen, D., Gadhamshetty, V., Nitayavardhana, S., & Khanal, S. K. (2015). Automatic process control in anaerobic digestion technology: A critical review. *Bioresource Technology*, 193, 513–522. https://doi.org/10.1016/j.biortech.2015.06.080

Nolan, A., Donnell, K. M., Devlin, G. J., Carroll, J. P., & Finnan, J. (2010). Economic Analysis of manufacturing costs of pellet production in the Republic of Ireland using Non-Woody Biomass~!2009-07-23~!2009-12-22~!2010-03-02~!. *The Open Renewable Energy Journal*, 3(1), 1–11. https://doi.org/10.2174/1876387101003010001

Okolie, J. A., Epelle, E. I., Tabat, M. E., Orivri, U., Amenaghawon, A. N., Okoye, P. U., & Gunes, B. (2022). Waste biomass valorization for the production of biofuels and value-added products: A comprehensive review of thermochemical, biological and integrated processes. *Process Safety and Environmental Protection*, 159, 323-344. https://doi.org/10.1016/J.PSEP.2021.12.049

Parsaeian, M., Rahimi, M., Rohani, A., & Lawson, S. S. (2022). Towards the modeling and prediction of the yield of oilseed crops: A Multi-machine learning approach. *Agriculture*, 12(10), 1739. https://doi.org/10.3390/agriculture12101739

Pasquini, C. (2003). Near Infrared Spectroscopy: Fundamentals, practical aspects and analytical applications. *Journal of the Brazilian Chemical Society*, 14(2), 198–219. https://doi.org/10.1590/s0103-50532003000200006

Qdais, H. A., Hani, K. B., & Shatnawi, N. (2010). Modeling and optimization of biogas production from a waste digester using artificial neural network and genetic algorithm. *Resources Conservation and Recycling, 54*(6), 359–363. https://doi.org/10.1016/j.resconrec.2009.08.012

Rahimi, M., Thanh, H. V., Ebrahimzade, I., Abbaspour-Fard, M. H., & Rohani, A. (2023). A multi-criteria decision-making (MCDM) approach to determine the synthesizing routes of biomass-based carbon electrode material in supercapacitors. *Journal of Cleaner Production*, 397, 136606. https://doi.org/10.1016/j.jclepro.2023.136606

Rathore, A. S., Nikita, S., & Jesubalan, N. G. (2022). Digitization in bioprocessing: The role of soft sensors in monitoring and control of downstream processing for production of biotherapeutic products. *Biosensors and Bioelectronics: X*, 12, 100263. https://doi.org/10.1016/j.biosx.2022.100263

Rentizelas, A. A., Tolis, A. J., & Tatsiopoulos, I. P. (2008). Logistics issues of biomass: The storage problem and the multi-biomass supply chain. *Renewable and Sustainable Energy Reviews*, 13(4), 887–894. https://doi.org/10.1016/j.rser.2008.01.003

Rinnan, Å., Van Den Berg, F., & Engelsen, S. B. (2009). Review of the most common pre-processing techniques for near-infrared spectra. *TrAC Trends in Analytical Chemistry*, 28(10), 1201–1222. https://doi.org/10.1016/j.trac.2009.07.007

Rivera, E. C., Rabelo, S. C., Garcia, D. D. R., Filho, R. M., & Da Costa, A. C. (2010). Enzymatic hydrolysis of sugarcane bagasse for bioethanol production: Determining optimal enzyme loading using neural networks. *Journal of Chemical Technology & Biotechnology,* 85(7), 983–992. https://doi.org/10.1002/jctb.2391

Ruiz, H. A., Rodríguez-Jasso, R. M., Fernandes, B. D., Vicente, A. A., & Teixeira, J. A. (2013). Hydrothermal processing, as an alternative for upgrading agriculture residues and marine biomass according to the biorefinery concept: A review. *Renewable and Sustainable Energy Reviews,* 21, 35–51. https://doi.org/10.1016/j.rser.2012.11.069

Safaei-Farouji, M., Thanh, H. V., Dai, Z., Mehbodniya, A., Rahimi, M., Ashraf, U., & Radwan, A. E. (2022). Exploring the power of machine learning to predict carbon dioxide trapping efficiency in saline aquifers for carbon geological storage project. *Journal of Cleaner Production*, 372, 133778. https://doi.org/10.1016/j.jclepro.2022.133778

Schmidhuber, J. (2014). Deep learning in neural networks: An overview. *Neural Networks*, 61, 85–117. https://doi.org/10.1016/j.neunet.2014.09.003

Schügerl, K. (2001). Progress in monitoring, modeling and control of bioprocesses during the last 20 years. *Journal of Biotechnology,* 85(2), 149-173. https://doi.org/10.1016/S0168-1656(00)00361-8

Sewsynker-Sukai, Y., Faloye, F., & Kana, E. B. G. (2016). Artificial neural networks: An efficient tool for modelling and optimization of biofuel production (a mini review). *Biotechnology & Biotechnological Equipment*, 31(2), 221–235. https://doi.org/10.1080/13102818.2016.1269616

Shafizadeh, A., Shahbeig, H., Nadian, M. H., Mobli, H., Dowlati, M., Gupta, V. K., Peng, W., Lam, S. S., Tabatabaei, M., & Aghbashlo, M. (2022). Machine learning predicts and optimizes hydrothermal liquefaction of biomass. *Chemical Engineering Journal,* 445, 136579. https://doi.org/10.1016/j.cej.2022.136579

Sharma, V., Tsai, M., Chen, C., Sun, P., Nargotra, P., & Dong, C. (2023). Advances in machine learning technology for sustainable biofuel production systems in lignocellulosic biorefineries. *The Science of the Total Environment*, 886, 163972. https://doi.org/10.1016/j.scitotenv.2023.163972

Shi, J., Thompson, V. S., Yancey, N. A., Stavila, V., Simmons, B. A., & Singh, S. (2012). Impact of mixed feedstocks and feedstock densification on ionic liquid pretreatment efficiency. *Biofuels*, 4(1), 63–72. https://doi.org/10.4155/bfs.12.82

Şimşek, E. H., Güleç, F., & Akçadağ, F. S. (2020). Understanding the liquefaction mechanism of Beypazarı lignite in tetralin with ultraviolet irradiation using discrete time models. *Fuel Processing Technology*, 198, 106227. https://doi.org/10.1016/j.fuproc.2019.106227

Sokhansanj, S., & Hess, J. R. (2009). Biomass supply logistics and infrastructure. *Methods in Molecular Biology*, 1–25. https://doi.org/10.1007/978-1-60761-214-8_1

Sonnleitner, B. (2012). Automated measurement and monitoring of bioprocesses: Key elements of the M3C strategy. *Advances in Biochemical Engineering, Biotechnology*, 132, 1–33. https://doi.org/10.1007/10_2012_173

Stephen, J. D., Mabee, W. E., & Saddler, J. N. (2011). Will second-generation ethanol be able to compete with first-generation ethanol? Opportunities for cost reduction. *Biofuels Bioproducts and Biorefining*, 6(2), 159–176. https://doi.org/10.1002/bbb.331

Sun, Y., Liu, L., Wang, Q., Yang, X., & Tu, X. (2016). Pyrolysis products from industrial waste biomass based on a neural network model. *Journal of Analytical and Applied Pyrolysis*, 120, 94–102. https://doi.org/10.1016/j.jaap.2016.04.013

Svendsen, C., Skov, T., & Van Den Berg, F. W. J. (2014). Monitoring fermentation processes using in-process measurements of different orders. *Journal of Chemical Technology & Biotechnology*, 90(2), 244–254. https://doi.org/10.1002/jctb.4483

Taleb, I., Dssouli, R., & Serhani, M. A. (2015). Big data pre-processing: A quality framework. *Proceedings - 2015 IEEE International Congress on Big Data, BigData Congress 2015*, 191–198. https://doi.org/10.1109/BIGDATACONGRESS.2015.35

Tan, C. K., Wilcox, S. J., & Ward, J. (2006). Use of artificial intelligence techniques for optimisation of co-combustion of coal with biomass. *Journal of the Energy Institute*, 79(1), 19–25. https://doi.org/10.1179/174602206x90913

Tyner, W. E. (2008). The US ethanol and biofuels boom: Its origins, current status, and future prospects. *Bioscience*, 58(7), 646–653. https://doi.org/10.1641/b580718

Ulber, R., Frerichs, J., & Beutel, S. (2003). Optical sensor systems for bioprocess monitoring. *Analytical and Bioanalytical Chemistry,* 376(3), 342–348. https://doi.org/10.1007/s00216-003-1930-1

Van Geem, K. M., Hudebine, D., Reyniers, M. F., Wahl, F., Verstraete, J. J., & Marin, G. B. (2007). Molecular reconstruction of naphtha steam cracking feedstocks based on commercial indices. *Computers & Chemical Engineering*, 31(9), 1020–1034. https://doi.org/10.1016/j.compchemeng.2006.09.001

Verstraete, J. J., Revellin, N., Dulot, H., & Hudebine, D. (2004). Molecular reconstruction of vacuum gas oils. *Prep. Am. Chem. Soc. Div. Fuel Chem*, 49(1), 20-21.

Wang, R., He, Z., Chen, H., Guo, S., Zhang, S., Wang, K., Wang, M., & Ho, S. (2024). Enhancing biomass conversion to bioenergy with machine learning: Gains and problems. *The Science of the Total Environment*, 927, 172310. https://doi.org/10.1016/j.scitotenv.2024.172310

Williams, C. L., Dahiya, A., & Porter, P. (2015). Introduction to bioenergy. In *Elsevier eBooks* (pp. 5–36). https://doi.org/10.1016/b978-0-12-407909-0.00001-8

Wiloso, E. I., & Heijungs, R. (2013). Key issues in conducting life cycle assessment of bio-based renewable energy sources. In *Green Energy and Technology* (pp. 13–36). https://doi.org/10.1007/978-1-4471-5364-1_2

Xing, Y., Zheng, Z., Sun, Y., & Alikhani, M. A. (2021). A review on Machine learning application in biodiesel production studies. International *Journal of Chemical Engineering*, 2021, 1–12. https://doi.org/10.1155/2021/2154258

Liao, M., & Yao, Y. (2021). Applications of artificial intelligence-based modeling for bioenergy systems: A review. *GCB Bioenergy*, *13*(5), 774–802. https://doi.org/10.1111/gcbb.12816

Zhang, Y., Teng, L., Quan, Y., Tian, H., Dong, Y., Meng, Q., Lu, J., Lin, F., & Zheng, X. (2010). Artificial Intelligence based optimization of fermentation medium for β-glucosidase production from newly isolated strain tolypocladium cylindrosporum. In *Lecture Notes in Computer Science* (pp. 325–332). https://doi.org/10.1007/978-3-642-15615-1_39

6 Machine Learning and Its Application in Biodiesel Production

Arpan Das

6.1 INTRODUCTION

Currently, more than 80% of the energy used worldwide comes from fossil fuels, which indicates a significant reliance of human civilization on these finite non-renewable sources of energy. For many decades, fossil fuels have been a widely used industrial energy source in both developed and developing countries due to their ideal combination of characteristics, such as easy transportability, versatility, and accessibility. Aside from resource constraints, the escalating use of fossil fuels for commercial and industrial purposes gives rise to several difficulties such as pollutants in the air, increasing temperatures, health impairments, global climate change concerns, and the emission of greenhouse gas (GHG) on a worldwide scale. Significantly, the majority of harmful petrol pollutants are emitted as a result of the extraction, chemical processing, and use of fossil fuels. In order to meet the projected 56% increase in global energy demand from 2010 to 2040, there is an urgent requirement for a sustainable alternative energy source (Sieminski, 2013; Ardabili et al., 2019). The energy crisis, together with substantial reliance on fossil fuels, rising resource volatility, and environmental issues, intensified the concern about the depletion of resources. As a result, the world is now shifting toward eco-friendly energy sources to ensure an adequate supply of energy and meet the growing energy demands from renewable sources. Of all the products obtained from crude oil, diesel is the predominant fuel, and its usage is steadily increasing globally because of the superior thermodynamic effectiveness and economic performance of diesel engines compared to petrol counterparts. Based on data published by OPEC (2019), the worldwide diesel and petrol (middle distillates) demand is projected to rise from 1660 billion liters (bl) in 2017 to 1834 bl in 2040. The rise may be attributed strictly to the increasing proliferation of vehicles powered by diesel in transport and industrial domains. Diesel-based compression ignition (CI) engines are widely used reciprocating engines in various applications. CI engines are commonly used in many on-road applications such as light vehicles, commercial buses, and trucks. On the other hand, there has been a rise in the utilization of diesel-fueled engines, namely in power generating, construction, and agricultural industries, resulting in an expansion of the market share for diesel engines. According to a report, the worldwide market

DOI: 10.1201/9781003571223-6

capitalization of diesel engines is projected to reach USD 266.3 billion by 2027, compared to the recorded value of USD 291 billion in 2018 (Awogbemi and Kallon, 2023). Although CI engines have made significant contributions to social, economic, and industrial advancements, their use leads to substantial harmful emissions that worsen human health and are responsible for global climate change. Despite the growing opportunities for utilizing CI engines and their projected market value, these engines have a negative impact on the environment. An effective approach to improve the situation is to create and implement cost-effective, ecologically friendly, and long-lasting renewable fuels for compression ignition engines. The utilization of consumable or non-consumable vegetable oils, alcohols, animal-based fats, or methane as a diesel substitute has encountered numerous disadvantages. The extensive use of clean vegetable oil-based fuels for compression ignition engines is limited as they have higher viscosity, lower calorific value, less volatility, can degrade during long-term storage, high production costs, and possible conflict with food distribution networks. Animal-based fats exhibit substandard calorific value, inadequate combustion, and insufficient vaporization. Additionally, they need preheating to function properly. Additional disadvantages, such as gum-like material formation, lubrication oil solidification, partial burning, blockage of the injector nozzle, problems with atomization, and harmful gas emissions, make it necessary to find alternative fuels that are superior (Tabatabaei et al., 2019). Among the several environmentally friendly alternatives, biodiesel is a viable option for diesel engines because of its reduced engine emissions (41% less greenhouse gas emissions), benefits in physical and chemical qualities, and the absence of substantial changes required (Sudhakar et al., 2012; Dharma et al., 2016; Jiaqiang et al., 2017; Knothe and Razon, 2017). Biodiesel is composed of long-chain fatty acid alkyl esters, produced via the process of transesterification of triglycerides or esterification of free fatty acids (FFAs) with an alcohol catalyst in acidic or basic environments or in the presence of an enzyme source. For most regulated pollutants emitted, including carbon monoxide, hydrocarbons, particulate matter, and smoke, biodiesel surpasses petrodiesel, with the exception of nitrogen oxides (NOx) (Aghbashlo et al., 2017). The benefits of this fuel, which is technically feasible, economically viable, and environmentally sustainable, in comparison to conventional diesel fuel, encompass its reproducibility, biodegradability, nontoxic nature, enhanced lubricity, lower flammability, and the lack of sulfur and benzene. Despite inherent disadvantages, including increased viscosity, reduced energy content, fuel pumping challenges, and increased copper-made strip corrosion, biodiesel can be effectively applied in diesel engines with minimal or no modifications. The ability of biodiesel and petrodiesel to mix in any proportion indicates that their combination is preferred over pure biodiesel, not only in developed countries but also in developing countries. Biodiesel is considered a solution to the limitations encountered when using diesel or vegetable oils as fuel for compression ignition (CI) engines. Research on the production, purification, and use of biodiesel has gained momentum nowadays. As a result, researchers and scholars have implemented several ways to improve and streamline the production process. An effective strategy involves utilizing cutting-edge information technology, such as machine learning, modeling, optimization, and statistical techniques. This is mostly due to the necessity of comprehensively grasping the highly intricate chemical process and several competing elements associated with biodiesel manufacturing. The intricacies associated with regulating the conflicting process parameters cannot be accomplished using trial-and-error experimental methods. Implementing suitable technology will optimize efficiency, save expenses, minimize material waste, eliminate redundant experiments, and address various contingencies. Extensive research has been conducted in recent years to explore the use of biodiesel as a cost-effective and environmentally friendly alternative in compression ignition engines. The results of these tests demonstrated the numerous advantages obtained from operating compression ignition (CI) engines using both blended and unblended biodiesel. Additional approaches to enhance the efficiency and emission properties of compression ignition (CI) engines using biodiesel involve mixing it with octanol and implementing combustion and emission techniques. Feedstocks for biodiesel consist of a diverse range of long-chain fatty acids, which significantly impact the esterification and subsequent procedures like purification and refining. The trans-esterification reaction can be complicated by errors and uncertainties that result from instrument selection, environmental conditions, calibration, and surroundings. These factors lead to significant intricacies in the (trans)esterification dynamics, thereby increasing the complexity and sophistication of accurately modeling the biodiesel production process. A suitable modeling strategy can assist in determining the most effective (trans) esterification techniques for feedstocks, optimizing design and operational parameters, regulating or modifying reaction conditions for improved quality of the product, lowering operational expenses, boosting production capacity, and enabling increased adaptability in feedstock choices. Nevertheless, the high computational requirements and numerical challenges associated with these methods have restricted their extensive use in (trans)esterification process modeling. Furthermore, the physicochemical characteristics of biodiesel can be accurately estimated by the use of physical models. Implementing such models on the (trans)esterification process is a challenging and computationally intensive endeavor. Nevertheless, the development of these idealized models has been predicated upon extensive assumptions and oversimplifications, posing significant challenges in implementing their findings. Reliable simulation of these computationally demanding numerical models necessitates the estimation or predefinition of several engine-specific parameters. Furthermore, statistical methodologies might be utilized in the field of biodiesel research. Nevertheless, these approaches frequently lack the necessary capacity to accurately represent the nonlinear and dynamic characteristics of biodiesel systems. The utility and precision of statistical approaches are consequently constrained by the availability of data. Hence, it is necessary to use efficient, rapid, precise, and resilient modeling methods to simulate and regulate intricate physical and chemical dynamics.

6.2 AN INTRODUCTION TO ARTIFICIAL INTELLIGENCE (AI) AND MACHINE LEARNING (ML)

Artificial Intelligence (AI) is the capacity of computers to emulate the functions of the human brain through various computerized methods, such as machine learning, fuzzy logic, and heuristic algorithms (Russell and Norvig, 2010; Ali et al., 2015; Liao and Yao, 2021). Predicting the characteristics of biomass and biofuels, the efficiency of the conversion process, supply chain modeling, and optimization are its main uses. Response surface methodology (RSM), genetic algorithms, and the Taguchi method are suggested approaches for optimization; in the meantime, trending modeling techniques in internal combustion engine research are artificial neural networks (ANN), regression, and analytical techniques (Traver et al., 1999; Shin et al., 2012; Betiku et al., 2015; Maran and Priya, 2015). ML algorithms can be developed and trained via supervised, semi-supervised, or unsupervised learning, deep learning, as well as reinforcement learning. The process entails using algorithms and statistical models to train a computer system to analyze provided data and make informed decisions, with the goal of continuously enhancing accuracy. This is used to forecast maintenance, diagnose errors, optimize, monitor, and regulate the systems, and alert workers to process attacks (Wong et al., 2013; Wu et al., 2013; Guang-Bin et al., 2004; Keliris et al., 2016; Lahdhiri et al., 2019). Undoubtedly, machine learning (ML) is experiencing rapid growth as a data science application, finding practical use in several industries such as healthcare, finance, travel, retail, social media, and scientific research (Anand et al., 2022). Due to the ongoing significance and practical uses of ML technologies in modern society, the market capital has consistently and gradually risen. The worldwide machine learning (ML) service, valued at USD 2.4 billion in 2019, increased to USD 15.47 billion in 2021. It is projected to reach USD 58.26 billion, USD 157.49 billion, and USD 305.62 billion in 2025, 2028, and 2030, respectively, according to Precedence Research (2022). The implementation of ML technology offers significant advantages such as dependability, ingenuity, precision, and versatile applications and has led to a decrease in effort and time, which in turn promotes more fast and dependable decision-making. Utilization of machine learning algorithms has resulted in enhanced precision and effectiveness in the provision of services without the need for human involvement. ML technologies effectively manage complex jobs in unpredictable and harsh environmental settings. Machine learning algorithms possess the ability to process vast amounts of data, detect patterns, and leverage this information to make decisions with high precision (Wazid et al., 2022). In order to achieve enhanced precision and dependability, machine learning necessitates vast, impartial datasets for training, which can be an arduous undertaking to acquire. Another limitation of using these technologies is the need for sufficient time and computer facilities to enable the algorithms to learn and run, comprehend, and build the capability to accurately evaluate the outcomes and make dependable decisions that align with the intended objective. Although machine learning algorithms rely on lengthy training to develop conclusions, they are not error-proof, especially in the absence of substantial and diverse training data (Aliramezani et al., 2022). ANN, genetic algorithm (GA), Linear Regression, Random Forest Regression (RF), and Support Vector Machines (SVM) are a few examples of algorithms for machine learning (ML) that have been widely applied in biodiesel investigation to track, anticipate, improve, supervise, and come to decisions in vital sectors. The ANN algorithm is widely considered to be the most commonly used machine learning algorithm due to its fundamental advantages, such as its ability to detect faults, its ease of adaptation, its accurate approximation of nonlinear and continuous functions, its multi-tasking ability, its ability to store large amounts of data, and its excellent fault-tolerance properties. Emerging from the field of mathematical neurobiology, the ANN methodology and its various hybrids present promising solutions for addressing the complexities and nonlinearities inherent in intricate biological processes, such as biodiesel systems, even in the presence of uncertain, partial, and missing data. ANN-based modeling techniques can effectively and rapidly represent biological processes based on a collection of examples, without necessitating an understanding of the underlying mechanisms and principles. In essence, ANN-based models possess the capability to simulate dynamic systems characterized by multivariate properties without requiring a comprehensive grasp of the detailed mechanisms involved. However, the limited transparency associated with many of these data-driven approaches is a primary factor contributing to the characterization of the entire range of models as "theory-free." In more straightforward terms, ANN-based modeling systems draw inspiration from the neurological processing capabilities of the human brain. Nevertheless, it is important to note that ANN's performance is influenced by the specific hardware it is implemented on. Additionally, the topology of the network cannot be preset and can only be discovered by practical experience, trial, and error. Recently, academics have conducted significant studies into the use of ML technologies for biodiesel research. However, only a small number of the previously stated initiatives have extensively explored the use of machine learning for the analysis, prediction, and improvement of process parameters in biodiesel synthesis. So far, attempts to address this clear research need have not produced the necessary outcomes, mainly due to insufficient workforce, lack of infrastructure, and other problems in the field of data science research. This provides another justification for the ongoing intervention. The use of ANN models typically follows a procedure that includes data gathering, data processing, network architecture construction, training, model testing, and outcome validation. The gathering and processing of data are undeniably crucial in determining the success of a modeling procedure. To guarantee the effectiveness and reliability of the model, it is important to gather a sufficient amount of data from reputable sources (Bali and Singla, 2022). An analysis of the gathered data is necessary to improve the training process and raise the reliability of the ANN model. The accuracy and reliability of the result must be verified to ensure that it fulfills the expectations. Several statistical metrics, such as standard deviation, standard error deviation, mean square error, root mean square error, and Pearson correlation coefficient squared, have been developed to assess the accuracy and dependability of the model's outcome.

6.3 BIODIESEL PRODUCTION

Biodiesel is a notable sustainable and renewable energy source that serves as a diesel substitute in compression ignition engines. Biodiesel has acquired widespread acceptance among consumers of renewable fuel as it is safe to handle, can be produced easily, and can be mixed with existing petroleum diesel without further modification. These variables have exerted a significant influence on the production and consumption of it. In order to satisfy the worldwide need for biodiesel, its amount of production is increasing globally. In 2021, the worldwide biodiesel market value was about USD 32.09 billion, and in 2030, it is expected to reach USD 189.7 billion, as forecasted by Precedence Research (2022). The anticipated expansion in market scope is driven by the rising demand for environmentally friendly biodiesel, particularly for the purpose of generating power and fueling automotive, farm, and marine engines on a global level. Investing more in biodiesel production and use is considered a viable approach to counter the concerning rise in greenhouse gas emissions.

6.4 FEEDSTOCK FOR BIODIESEL PRODUCTION

Choosing the appropriate feedstock for the manufacture of biodiesel is a critical factor that affects the expenses, facilities conversion process, efficiency of conversion, pump price, as well as the viability of the biodiesel technology. For instance, the raw material used for production, known as feedstock, represents around 75%–88% of the final production cost, and this should be reduced enough to minimize biodiesel selling price (Ahmed et al., 2021). The choice of feedstock significantly impacts the selection of catalysts, the parameters of the process, the strategies for intensification, and the methods of purification utilized. Examples of the biodiesel's performance attributes derived from several natural sources have been shown in Table 6.1. ANN, multiple regression, statistical regression, and multiple nonlinear regression models are among the most favored methods, according to research on machine learning implementations in the input phase. The temperature, speed, duration, and composition of the blend are the main input variables for blending. The viscosity, point of ignition, oxidation stability, density, methane proportion, heating values, and cetane number are among the output factors. In order to predict these parameters, Mairizal et al. (2020) investigated biodiesels produced from several sources, including oil extracted from walnut, sunflower seed, peanuts, rapeseed, hydrogenated oil from coconuts, copra oil, and beef tallow using multiple linear regressions. Experimental findings indicate that the inclusion of mono-unsaturated and polyunsaturated fatty acids as an independent parameter leads to an improvement in prediction accuracy. The model inputs consisted of the content of polyunsaturated fatty acids in the feedstock, the iodine value, and the saponification coefficient. Tchameni et al. (2019) proposed using multiple ANNs and nonlinear regression for the prediction of rheological characteristics of waste vegetable-based oil. The results demonstrated the higher performance of the ANN model compared to the MNLR approach. Utilizing non-edible feedstock, specifically residual cooking oil, greatly decreases the expenses associated with producing biodiesel. This approach also provides an effective means of its disposition and prevents environmental pollution. However, using edible vegetable oils has negative effects on food security, triggers the food versus fuel debate, and substantially contributes to higher food costs. Algae contain a high concentration of fatty acids, and they may be cultivated without the need for land, water, or fertilizer. Additionally, it can be quickly converted into biodiesel. While the process of converting algae into biodiesel necessitates sophisticated methods, it is not feasible to accomplish this at the individual household level. The scale of production

TABLE 6.1
The Biodiesel's Performance Attributes Derived from Several Natural Sources

Feedstock	Blend Content	Type of Engine	Test Situation	Brake-specific Fuel Consumption (BSFC)	References
Coconut	B5, B15	One cylinder, four stroke, DI diesel engine	100% load, variable speeds of 1500 to 2400	0.53% for B5 and 2.11% for B15	Liaquat et al. (2013)
Safflower	B100	One cylinder, four stroke, CI engine	Variable loads, constant speed of 1500 rpm	16.87% (full load)	Balasubramanian and Purushothaman (2019)
Peanut	B5, B20, B50, B100	Three-cylinder Yanmar 3009D diesel engine	Variable engine speeds (rated power = 14.2 kW)	Up to 9%, B100 (B100 > B50 > B20 > B5)	Santos et al. (2013)
Jatropha	B100	Double cylinder, DI, CI diesel engine	Variable loads, constant speed of 1500 rpm	56.55% at full load	Paul et al. (2014)
Soy	B20, B40, B100	Single cylinder, air cooled, vertical, DI diesel engine	Variable loads, constant speed of 1500 rpm	4.2% B20, 8.7% B40, 14.65% B100	Al-Dawody and Bhatti (2014)
Palm	B5	Mitsubishi 4D68 SCHC DI 4-stroke, 4-cylinder engine with EGR	Variable speeds (1000 to 3000 rpm)	3.80% (1000 rpm) 4.22% (3000 rpm)	Yasin et al. (2015)

and commercialization remains difficult, even with large-scale operations. Different researchers employed RSM in addition to artificial neural networks (ANNs) for the development of models, making predictions, and processing parameter optimization for biodiesel production from oils extracted from algae, spent cooking oil, and groundnut oil (Ayoola et al. 2019; Selvaraj et al. 2019; Garg and Jain 2020). The researchers determined that both the RSM and ANN models were highly effective in predicting and optimizing production process parameters, resulting in enhanced biodiesel yield.

6.5 TECHNIQUES FOR BIODIESEL PRODUCTION

Several techniques, including direct utilization, micro-emulsification, pyrolysis, transesterification, and supercritical, have been used to transform substrate into biodiesel. This can be classified into two methods: chemical and physical. Using vegetable oils directly or diluting them to substitute fossil-based diesel fuel in compression ignition (CI) engines is the most straightforward method of utilizing alternative fuels. Despite the method's inexpensive manufacturing and capital costs, the characteristics of vegetable oils pose challenges, rendering their direct use as fuel difficult, inappropriate, and impractical. High amounts of acidity are present in some vegetable oils. They additionally possess unsaturated hydrocarbon chain reactivity, a dense viscosity, as well as low volatility. Consequently, when vegetable oils are utilized in unaltered CI engines, they lead to inadequate combustion, engine degradation, excessive carbon buildup, and solidification of the engine lubricant. Additional drawbacks of directly using vegetable oils as a replacement fuel for unaltered CI engines encompass inadequate atomization, heightened emissions of pollutants, expedited engine deterioration, exorbitant engine upkeep expenses, and subpar engine performance (Dabi and Saha, 2019).

6.5.1 MICRO-EMULSIFICATION

The goal of the micro-emulsification method is to solve the problem of vegetable oils' viscosity being high when they are utilized as fuel in diesel engines. This process creates a transparent, homogeneous, thermodynamically stable liquid that can be used as fuel by mixing vegetable oils and animal fats with the appropriate solvents of choice, surfactants, or cetane improvers. These homogenized oils provide stable, uniformly distributed, isotropic in nature fluids with the required droplet size and microstructures. For micro-emulsion, 1-butanol, 2-octanol, ethanol, butanol, methanol, propanol, and hexanol are often used solvents. Vegetable oils are micro-emulsified using surfactants like rhamnolipid, sorbitan monooleate, and higher alcohols. Likewise, cetane improvers such as nitrocarbonates, nitroalkanes, and alkyl nitrates are frequently used for this purpose (Mishra and Goswami, 2018).

6.5.2 PYROLYSIS

Pyrolysis is a process that combines heat and chemical reactions to convert feedstocks into fuels and other valuable products. This can be done with or without the help of certain compounds called catalysts. It is a widely employed method for breaking down oils obtained from vegetable, animal-derived fats, and other raw materials by breaking their chemical bonds. This process produces sustainable fuels that have similar qualities and structures to petroleum diesel fuels. Biodiesel produced from the pyrolysis of vegetable oils exhibits favorable physicochemical characteristics and enhanced engine efficiency. Nevertheless, the technique faces limitations due to factors such as exorbitant production expenses, intricate and costly infrastructure needs, the insufficient presence of oxygen in the fuels, and the generation of chemically similar short-chain molecules to petrol (Su et al., 2022).

6.5.3 Transesterification

The most frequent reaction in the process of producing biodiesel is transesterification, which is enhanced under mild reaction conditions by the use of heterogeneous or homogeneous acid and base catalysts. The most common method of producing biodiesel is the transesterification, which uses methanol or ethanol to alkylate triglyceride with the help of a catalyst to produce glycerol as an intermediate and fatty acid ester as the main product. When compared to ethanol, methanol exhibits greater dispersion in homogeneous catalysts and demonstrates higher efficiency in reacting with triglycerides (Awogbemi et al., 2021). Nevertheless, due to the boiling point of methanol being 60°C, it is typically necessary to limit the maximum reaction temperature to 60°C in order to prevent the evaporation of methanol during a transesterification reaction. The transesterification process, in short, replaces the glycerol in triglycerides with primary alcohol by substitution. Triglycerides are changed into diglycerides (DGs), monoglycerides (MGs), and glycerol by a series of three consecutive reversible processes. The biodiesel produced is refined to satisfy the specified ASTM D6751 and EN 14214 criteria, ensuring that it is free from acetaldehyde, formaldehyde, and other harmful gases when burned. Additionally, to stop the production of dangerous gases during burning, the glycerol is removed (Zhang et al., 2022). In addition to being a raw material for the production of bioethanol, cosmetics, animal feed, and other chemicals, glycerol can be used as a fuel additive (Chilakamarry et al., 2022). Manufacturing process response features have a significant impact on the transesterification reaction, and productivity is affected by a number of transesterification-related parameters, including temperature, type of oil and catalyst, length of reaction, and oil to alcohol molar ratio (Meng et al., 2014; Guo et al., 2021). Due to optimizing transesterification, statistical techniques and numerous physical tests are required to anticipate reaction reactions and interactions between every parameter (Moayedi et al., 2020; Salam et al., 2020). Machine learning (ML) technology has the potential to improve estimation quality. Numerous studies exist about the use of machine learning technology in the modeling of combustion processes in biodiesel-fueled engines. The transesterification reaction can be readily accomplished using modest procedure parameters and can be performed easily without the need for advanced infrastructure or technical expertise. Nevertheless, the utilization of the transesterification process has been constrained due to inadequate mass transfer, variable product quality, the production of a substantial amount of effluent, and the requirement for numerous separation procedures.

6.6 CATALYSTS FOR BIODIESEL PRODUCTION

In order to optimize conversion effectiveness and accelerate biodiesel synthesis rate, catalysts are frequently employed. While it is possible to produce biodiesel in the absence of a catalyst, its inclusion in the production process is beneficial and widely preferred, despite the additional expense it incurs. Catalysts expedite reaction time, enhance conversion efficiency, and facilitate the reduction of reaction activation energy. The selection of a suitable catalyst relies on factors such as the nature of the feedstock, the concentration of free fatty acids, acid value, and the amount of moisture present in the feedstocks. Homogeneous catalysts generally remain in liquid form and are utilized for converting triglycerides that have a high amount of unbound fatty acids (FFA) and acid value. The limited solubility of the catalysts in glycerol and biodiesel presents difficulties in effectively isolating and recovering them from the final product (Tan et al., 2019). Acid and base catalysts are two categories for homogenous catalysts used in the transesterification procedure. Some noteworthy instances of acidic homogenous catalysts used in transesterification processes are H_2SO_4, H_3PO_4, HCl, $C_2HF_3O_2$, and sulfonated acids. NaOH, KOH, and $NaOCH_3$ are a few examples of basic homogeneous catalysts utilized in transesterification processes (Changmai et al., 2020). One issue that arises with homogeneous catalysts is the difficulty of separating them from glycerol and biodiesel, as they are typically in the same liquid phase. The purification procedure involves multiple washing phases to ensure thorough removal of catalysts, which necessitates a substantial amount of water and results in the generation of a significant volume of wastewater. Moreover, homogeneous catalysts are comparatively costly and their reuse is nearly impracticable (Mukhtar et al., 2022). In contrast, heterogeneous catalysts typically exist in a solid or gaseous state and are in a distinct phase from the substances undergoing the chemical reaction. The catalysts possess active sites containing reactive elements that facilitate rapid reactions. Heterogeneous catalysts such as oxides of metals or sulfated metal, and several cation exchange resins have shown effectiveness in the generation of biodiesel. These catalysts show exceptional efficacy under favorable reaction circumstances, remarkable recyclability, superior recoverability, effortless separability, cost-effectiveness in purification, and minimal wastewater formation. Nevertheless, these catalysts are only compatible with feedstocks that have a low level of free fatty acid (FFA) concentration and high saponication values (Mandari and Devarai, 2021). A variety of waste materials, such as crop residues, food wastes, crab shells, quail eggshells, and chicken eggshells, can be used to create heterogeneous base catalysts. This method lowers production costs, encourages resource recovery, and improves sustainability and cleanliness of the environment. Currently, research attention has been focused toward the use of enzyme catalysts or biocatalysts for the generation of biodiesel. Enzymatic catalytic biodiesel synthesis utilizes mainly microorganisms to induce chemical reactions. Lipase is the most frequent and widely utilized enzyme for biodiesel generation. Lipases are typically derived from microbes such as *B. cepacia*, *T. lanuginosus*, *C. antarctica*, *A. niger*, and *A. gonensins*. Under carefully controlled conditions, fermentation takes place (Altinok et al., 2023). Lipases can be classified as intracellular and extracellular lipases. Intracellular lipases use entire cells of bacteria or unicellular yeasts and are present inside their cell walls. Extracting intracellular lipases is a simple and cost-effective process that requires minimal purification. However, these lipases have the drawback of low biodiesel yield and low conversion rate. In contrast, extracellular lipases are obtained from a culture of microorganisms, then isolated and purified prior to use. Compared to intracellular lipase, extracellular lipase incurs a higher cost for downstream processing and is burdened by intricate separation methods (Chuengcharoenphanich et al., 2023).

6.7 PROCESS PARAMETERS AFFECTING BIODIESEL PRODUCTION

Once the selection of raw materials, production procedures, and reactor design has been determined, it is crucial to make a well-informed decision on the process parameters. The selection of production process parameters directly impacts the efficiency and overall yield of biodiesel. Variables such as the oil to alcohol ratio, catalyst dosage, production temperature, and reaction duration have a substantial impact on the outcome of biodiesel synthesis. Choosing the most appropriate course of action to achieve the highest possible conversion efficiency and product output, while considering these conflicting criteria, is a complex undertaking. The utilization of trial-and-error experimental methodologies for determining the best parameters of the production process is arduous, intricate, costly, time-consuming, and frequently leads to significant material inefficiency. Data-driven machine learning technology has been utilized to create well-informed judgments that ensure cost-effectiveness in the manufacture of biodiesel. In a research, Srivastava et al. (2018) examined the use of a genetic algorithm (GA) to forecast and optimize process parameters for biodiesel production. They successfully transformed microalgae oil into good-quality biodiesel. They suggested conducting additional research on the application of genetic algorithms (GA) to determine the most effective parameters for biodiesel production. This study focuses on examining different machine learning technologies used to forecast and enhance process parameters for the production of environmentally friendly biodiesel.

6.7.1 Alcohol to Oil Ratio

The production of biodiesel and the efficiency of conversion are significantly influenced by the alcohol to oil ratio. Alcohol dissolves and disperses the oil and catalyst during the transesterification cycle. Hence, in order to enhance the production of biodiesel, it is imperative to maintain a higher alcohol to oil ratio (Okolie et al., 2022). However, a system with an overly high alcohol to oil proportion may have a surplus of methanol, which would raise manufacturing expenses, complicate the steps of separation and purification, lead to lower conversion efficiency, and ultimately result in a lower yield of biodiesel through the transesterification reaction. Alcohol, particularly methanol, contains a polar hydroxyl group that promotes the emulsification of glycerol and biodiesel in the transesterification synthesis. The process of emulsifying the products enhances the occurrence of the reverse reaction, thereby diminishing the efficiency of conversion and the output of biodiesel (Gupta and Pal Singh, 2023). For optimal biodiesel production, a methanol to oil proportion of 6:1 and an ethanol to oil proportion of 9:1 are advised (Musa, 2016; Abusweireh et al., 2022). The best raw materials, catalyst types, alcohol to oil molar proportion, and other process parameters are found by simulating the process using ANN and other statistical techniques. The goal is to maximize the yield of biodiesel while maintaining the best efficiency of conversion.

6.7.2 Catalyst: Its Type and Concentration

The utilization of catalysts in the process of transesterification for biodiesel production aids in decreasing the amount of energy required to initiate the reaction, minimizing energy usage, reducing the time needed for the reaction to occur, enhancing the efficiency of conversion, and ultimately leading to a substantial increase in the yield of biodiesel. Several categories of catalysts, such as acid, alkali, biobased, homogenous type, heterogenous type, enzyme, and so on, have been employed for the transesterification process of converting glycerides into biodiesel. The choice of catalyst and its concentration has a significant impact on various aspects, including the excellence and purity of the final product, the cost of production, the convenience of separating and purifying the product, the rate of conversion, the efficiency of conversion, and the yield of biodiesel. For instance, it is possible to achieve biodiesel yields ranging from 92% to 98% even when using low-quality feedstock with high fatty acid concentration. Lower biodiesel yields are seen when base catalysts are employed with the same feedstock and production conditions. According to Sitepu et al. (2020), using heterogenous catalysts leads to greater conversion efficiency and biodiesel output compared to most homogenous catalysts. Generally, a greater concentration or dosage of catalysts typically leads to an increased biodiesel yield. Nevertheless, the uneven distribution of catalysts might have a negative impact by causing the formation of emulsions and the production of mixtures with high viscosity. This intensifies the process of product separation and purification, enhances the saponification reaction, and significantly decreases the amount of biodiesel produced (Nayab et al., 2022). Transesterification with insufficient catalyst quantity and concentration progresses at a sluggish rate, resulting in unreacted feedstock, producing biodiesel of poor quality, and yielding a low amount of biodiesel. However, an excessive concentration of catalysts in the transesterification reaction causes the reactants to clump together, reduces their mass and surface interaction, and has a considerable negative effect on the production of biodiesel (Xie and Li, 2023). However, it is important to use statistical tools to assess the reliability and validity of machine learning models in order to achieve cost-effective and optimal biodiesel yield.

6.7.3 Reaction Temperature

During the transesterification process, temperature is crucial for the efficient conversion of feedstock into biodiesel, as well as for achieving a high biodiesel yield and controlling the overall kinetics of the total reaction. Raising the temperature, decreases the thickness and evaporation tendency of the substances involved in the reaction, promotes proper mixing, and intensifies the molecular bonding between the reactants. Elevated temperature additionally facilitates the reduction in oil viscosity and promotes the separation of biodiesel (Bashir et al., 2022). Nevertheless, an excessively high reaction temperature can hinder the transesterification reaction. In addition to the high energy expenditure, raising the reaction temperature beyond the allowable limit can lead to saponification, as well as an increase in the reverse reaction. The alcohol should also be selected according to the reaction temperature. During the transesterification reaction, ethanol appears to exhibit a greater tolerance for elevated temperatures compared to methanol. Although ethanol can tolerate a temperature of approximately 78°F, for a methanol-based transesterification reaction, 65°F is the optimal temperature. Reaction temperature below 65°F will result in a decrease in reaction speed, lower effectiveness of conversion, and a decrease in biodiesel yield. On the other hand, raising the temperature of the reactions over 70°F will result in a drop in the methanol to oil ratio, more methanol evaporation, and a hindrance to the production of biodiesel (Mathew et al., 2021; Stevanato et al., 2023). The selection of the temperature should be made in conjunction with other parameters involved in the process, as these elements are interconnected and mutually reliant. In a distinct study, Ahmad et al. (2019), Gimbun et al. (2013), and Al-Saadi et al. (2020) documented biodiesel yields of 98.6%, 96.9%, and 95.1%, respectively, through the synthesis of biodiesel via transesterification at reaction temperatures of 59°F, 65°F, and 70°F. The researchers determined that many parameters, including the kind of feedstock, catalyst, grade of feedstock, methanol to oil ratio, and catalyst concentration, had an impact on the reaction temperature. An effective solution to this problem is to utilize machine learning models, simulation approaches, and statistical tools to determine the ideal reaction temperature that will maximize the output of biodiesel.

6.7.4 Reaction Time

The reaction time quantifies the complete duration of the transesterification reaction. The aforementioned factors, including the catalyst type, feedstock grade, reactor size, reactant quantity, and reaction temperature, are crucial considerations in any chemical reaction (Bashir et al., 2022). Selecting an appropriate reaction time guarantees the production of enough product and ensures that the entire reaction is finished within the allotted time. The reaction period should be sufficient to allow ample time for the various reacting substances to blend, interact, and undergo chemical reactions to generate the desired products. Extended reaction time has been found to enhance the production of biodiesel, as it allows for more extensive reaction completion. Elkelawy et al. (2020) and Narowska et al. (2019) found that increasing the reaction time resulted in a higher biodiesel output. This pattern was ascribed to the ample availability of time for the reactions to reach completion. As the reaction time increased, they observed a decrease in biodiesel production because the reaction reached equilibrium and the creation of soap was activated.

6.7.5 Mixing/Agitation Speed

The degree of agitation of the reactants has a substantial impact on the production of the product in all chemical reactions. During transesterification, the reactants are thoroughly combined and stirred to achieve their proper dispersion, homogeneous distribution, and uniform reaction throughout the reactor. If the triglycerides are fully soluble in alcohol, the reaction may only take place at the interface rather than throughout the entire reactor. This frequently results in delayed response, presence of unreacted substances in some areas of the reactor, and low productivity. The combination of the reacting components promotes enhanced solubility of the oil in methanol, facilitates more efficient reaction throughout the entire reactor, and results in increased product creation. Tabatabaei et al. (2019) conducted an experiment to investigate the impact of mixing speeds of 200 rpm, 400 rpm, and 600 rpm on the yield of biodiesel. In a separate study, Likozar and Levec (2014) examined mixing speeds ranging from 100 rpm to 600 rpm. According to their analysis, higher mixing speed results in improved product output, and an ideal speed of approximately 400 rpm is recommended for achieving the highest biodiesel yield. Increasing the mixing speed promotes the creation of soap and results in greater energy usage.

6.8 APPLICATION OF MACHINE LEARNING TECHNOLOGIES IN OPTIMIZATION OF PROCESS PARAMETERS

Currently, interest is growing in using ML methodologies such as ANN, RSM, and Adaptive Neuro-Fuzzy Inference System (ANFIS) to forecast and optimize the parameters of the biodiesel production process. This is because they have the ability to accurately forecast and optimize the process parameters. Tan Y. H. et al. (2019), Fangfang et al. (2021), and Farobie et al. (2015) conducted individual studies where they utilized ANN to create models and forecast the production of biodiesel. They used oil from jatropha, canola, and waste cooking oil as raw materials for their experiments. The presented findings indicated that the projected outcomes were consistent with the experimental results. Furthermore, the implementation of the ANN model effectively eliminated feedstock wastage and resulted in time savings. Previous research conducted by scholars in the field has demonstrated that ANN can effectively model, predict, and optimize the multiple variables that impact biodiesel production. This technique is known for its accuracy, ability to handle both linear and nonlinear scenarios, and information storing capacity within the network. Nevertheless, ANN has faced criticism due to its limited interpretability, lack of explainability, and high data requirements for training, validation, and testing. Additionally, ANN is reliant on certain hardware and necessitates a significant level of computational expertise and processing capability. These disadvantages have required the utilization of alternative tools. RSM, along with other modeling and optimization methods, has become prominent and widely accepted. This is a mathematical and statistical technique used to optimize a process that is influenced by multiple elements or variables, resulting in a specific response or output. The dependent variables refer to the responses or results of a study, while the independent variables are the input or predictor variables (Anggoro et al., 2022). Wahidin et al. (2018), Yesilyurt et al. (2019), and Anwar et al. (2018) examined the application of the RSM technique to model, predict, and optimize biodiesel yield. The researchers utilized RSM to simulate and enhance the process variables involved in the transesterification of *Nannochloropsis* sp. biomass, mustard seed oil, and stone fruit seed oil, with the aim of producing high-quality biodiesel. The reported values for R^2 and Adj. R^2 D are 0.96912 and 0.95059, 0.9818 and 0.9649, and 0.9781 and 0.9386, respectively, which indicate the accuracy and consistency of the model. These experiments have verified that RSM is a feasible technique for optimizing the reaction parameters in the transesterification process to get the highest possible biodiesel production. The study focus has shifted toward comparing ANN, RSM, and other comparable methodologies. The purpose of this study is to assist researchers in determining and choosing more efficient and economical methods for predicting, modeling, and optimizing biodiesel production. The ultimate goal is to contribute to the advancement of research in this field (Samuel et al., 2020). Examined and contrasted the utilization of RSM and Grey Wolf Optimizer (GWO) to simulate and enhance the production of biodiesel from waste sunflower oil through the transesterification process. The scientists concluded that both strategies were dependable in forecasting the output of waste sunflower biodiesel. However, they discovered that the GWO technique was more precise and easier to use. Fayyazi et al. (2015) and Kolakoti et al. (2020) conducted a comparison between RSM and genetic algorithm (GA). They found that the GA outperformed the RSM technique in terms of speed, efficiency, and improvement, showing a 4.96% advantage. The study conducted by Chizoo et al. (2022) examined the effectiveness of integrating GA and RSM with ANFIS. The authors found that ANFIS-GA exhibited superior predictive ability, a faster pace, and a more cost-effective approach compared to RSM. Jisieike et al. (2023) and Ishola et al. (2019) conducted separate research and found that while ANN and RSM can both model and predict biodiesel yield, ANFIS is more resilient, precise, and efficient compared to ANN and RSM. Researchers have increasingly adopted both ANN and RSM to anticipate and optimize the process of biodiesel production. This could be attributed to their precision, durability, and user-friendly interface. Multiple writers have used both strategies to create a model and make predictions about the yield of biodiesel. Additional methodologies, such as extreme learning machine (ELM), SVM, artificial bee colony (ABC), and others, have been utilized for the purpose of forecasting and simulating biodiesel production. An instance of ELM was utilized to simulate and enhance the production of biodiesel from *Ceiba pentandra* oil using a transesterification process aided by microwave irradiation. The ELM technique was determined to be efficacious and led to a substantial production of biodiesel (Silitonga et al., 2020). Faizollahzadeh Ardabili et al. (2019) also documented the effective application of SVM-RSM and ELM-RSM methods to optimize the process of producing biodiesel. The hybrid approaches performed well and demonstrated a strong ability to optimize the quality of biodiesel production. Rostami et al. (2016) demonstrated the utilization of the ABC algorithm to optimize the manufacture of biodiesel from Mountain Almond oil. The ABC algorithm technique exhibited exceptional precision in determining the ideal process parameters for biodiesel synthesis.

6.8.1 Consequences, Difficulties, and Future Directions

In future, the desire for cost-effective, readily available, and eco-friendly sources of energy is expected to rise. Hence, research efforts focused on enhancing the accessibility of renewable energy will persist as a top priority. Biodiesel, a significant constituent of renewable fuel, will persist in occupying a prominent position in the realm of renewable energy. Considering the ongoing implementation of cutting-edge technologies in several domains, including research and development, the use of machine learning (ML) and comparable technologies will streamline and generate technology-driven methods for advancing biodiesel research. Utilizing simulation, computational, mathematical, statistical, modeling, and optimization tools in the biodiesel industry can help make biodiesel manufacturing more accessible and contribute to achieving energy self-sufficiency. Researchers and biodiesel refiners need to develop rapid, cost-effective, and efficient procedures that minimize material wastage (Goswami et al., 2022). The demand for the utilization of machine learning and other innovative methods in the modeling, forecasting, estimating, and optimization of the production of biodiesel will persist. The continued utilization of these technologies in research on biodiesel will lead to the inventions and implementation of additional innovative and user-friendly technologies to advance the biodiesel industry. The actual implementation of biodiesel manufacturing poses significant challenges. Various nonlinear elements and unanticipated events can influence the reactions. The utilization of simulation, optimization, and modeling tools such as ANN, ML, numerical methods, mathematical techniques, simulations, and so on is not uncomplicated. It demands meticulous efforts and intense focus to navigate through the substantial levels of uncertainty and complexity associated with these tools. Key factors in advancing biodiesel production include the careful selection of feedstocks, the strategic choice of production procedures and methods, the optimization of reaction parameters, the improvement of procedures, and the scaling up of production and purification infrastructure. The various aspects related to the biodiesel value chain, such as economic, infrastructural, personnel, environmental issues, policy formation and implementation, biodiesel mandates, and standards, require a comprehensive analysis and enhancement. The necessary infrastructure for forecasting, predicting, and optimizing the biodiesel generating process must be established and enhanced to fulfill the requirements of the modern day. Undoubtedly, the implementation of machine learning technology and other prediction and optimization techniques has proven advantageous in biodiesel research. Nevertheless, it is not currently a perfect and comprehensive answer capable of resolving all the intricacies inherent in the research field. There are clear and fundamental limitations that need to be addressed directly. There are difficulties in understanding and interpreting the results of ANN, RSM, and other statistical analyses. Several of the outcomes are unfinished, incapable of existing independently, and challenging to comprehend. Certain aspects of these need to be adjusted, examined, and refined based on the training data. ML technologies are specifically developed to observe and analyze the many stages of biodiesel production, purification, characterization, and use in real-world scenarios. Further research is necessary to enhance and advance machine learning technologies and comparable tools for the purpose of monitoring and regulating the quality of biodiesel, while also ensuring adherence to established criteria.

6.9 CONCLUSION

Biodiesel is a type of liquid, eco-friendly, and renewable biofuel that can replace fossil-based diesel fuel. It is cost-effective and biodegradable. The production of biodiesel requires careful consideration of several nonlinear parameters for efficient transformation of different feedstocks into high-quality biodiesel. Various factors, including feedstock and their condition, concentration and size of the catalysts, proposed method, duration, pressure and temperature during reaction, agitation speed, molar ratio of oil to alcohol, and reactor type, are the major parameters being considered for the overall processes. It impacts not just the rate at which conversion occurs but also the amount of biodiesel produced, the characteristics of the biodiesel, and its potential uses. The use of machine learning technologies, statistical techniques, modeling, and optimization tools in biodiesel research has introduced a new perspective in the study of biodiesel production. Nowadays, different advanced technologies, including ANN, RSM, Adaptive Neuro-Fuzzy Inference Systems (ANFIS), Grey Wolf Optimizer (GWO), Genetic Algorithms, and SVM, are frequently employed in biodiesel research for modeling, forecasting, approximating, and enhancing the efficiency of biodiesel production through optimization of feedstock, production processes, and catalysts used in the biodiesel manufacturing process. Biodiesel is a highly favored and readily producible type of biofuel. Utilizing machine learning (ML) and other cutting-edge technology in the manufacturing of biodiesel will generate greater interest in the synthesis and utilization of this alternative fuel. It is recommended to initially adopt ML technologies for optimizing biodiesel production processes on a laboratory scale before scaling up to large industrial scales. By adopting this approach, it becomes more feasible to address and resolve challenges and problems with more efficiency and less expenditure. There is a need for the creation and application of straightforward, user-friendly, and strong cutting-edge technology to monitor and regulate the complete biodiesel manufacturing system. It is advisable to provide additional training and enhance capacity-building efforts to improve comprehension and expand the benefits of technologies across the whole biodiesel value chain. This includes areas such as selecting appropriate raw materials, extracting oil, pre-treating feedstock, selecting and configuring reactors, purifying and ensuring the quality of biodiesel, utilizing biodiesel, and reducing emissions in engines that use biodiesel as fuel. Legislation and enforcement should be put in place to support and regularly carry out policies and programs that promote the production and use of biodiesel for various purposes in order to reduce carbon emissions and make the environment more sustainable. The government should promote the democratization and adoption of biodiesel and other biofuels, especially at the home level, in order to alleviate the reliance on fossil-based diesel fuels and its associated consequences.

REFERENCES

Abusweireh, RS., Rajamohan, N., & Vasseghian, Y. (2022). Enhanced production of biodiesel using nanomaterials: A detailed review on the mechanism and influencing factors. *Fuel,* 319, 123862.

Aghbashlo, M., Tabatabaei, M., Khalife, E., Najafi, B., Mirsalim, S.M., Gharehghani, A., Mohammadi, P., Dadak, A., Shojaei, T.R., & Khounani, Z. (2017). A novel emulsion fuel containing aqueous nano cerium oxide additive in diesel–biodiesel blends to improve diesel engines performance and reduce exhaust emissions: Part II–exergetic analysis. *Fuel,* 205, 262–271.

Ahmad, T., Danish, M., Kale, P., Geremew, B., Adeloju, S. B., Nizami, M., et al. (2019). Optimization of process variables for biodiesel production by transesterification of flaxseed oil and produced biodiesel characterizations. *Renewable Energy,* 139, 1272–1280.

Ahmed, M., Abdullah, A., Patle, D.S., Shahadat, M., Ahmad, Z., Athar, M., et al. (2021). Feedstocks, catalysts, process variables and techniques for biodiesel production by one-pot extraction-transesterification: A review. *Environmental Chemistry Letters,* 20, 335–378.

Al-Dawody, M.F., & Bhatti, S.K. (2014). Experimental and computational investigations for combustion, performance and emission parameters of a diesel engine fueled with soybean biodiesel-diesel blends. *Energy Procedia,* 52, 421–430.

Ali, J.M., Hussain, M. A., Tade, M.O., Zhang, J. (2015). Artificial Intelligence techniques applied as estimator in chemical process systems-a literature survey. *Expert Systems with Applications*, 42(14), 5915–5931.

Aliramezani, M., Koch, C.R., & Shahbakhti, M. (2022). Modeling, diagnostics, optimization, and control of internal combustion engines via modern machine learning techniques: A review and future directions. *Progress in Energy and Combustion Science*, 88, 100967.

Al-Saadi, A., Mathan, B., & He, Y. (2020). Esterification and transesterification over SrO–ZnO/Al_2O_3 as a novel bifunctional catalyst for biodiesel production. *Renewable Energy,* 158, 388–399.

Altinok, F., Albayrak, S., Arslan, N.P., Taskin, M., Aygun, E., Sisecioglu, M., et al. (2023). Application of *Anoxybacillus gonensins* UF7 lipase as a catalyst for biodiesel production from waste frying oils. *Fuel,* 334, 126672.

Anand, A., Trivedi, N. K., Abdul Wassay, M., AlSaud, Y., & Maheshwari, S. (2022). Application and uses of big data analytics in different domain. In *Machine Intelligence and Data Science Applications.* Editors V. Skala, T. P. Singh, T. Choudhury, R. Tomar, and M. Abul Bashar (Singapore: Springer), 481–500.

Anggoro, P.W., Purharyono, Y., Anthony, A.A., Tauviqirrahman, M., Bayuseno, A.P., & Jamari, A. (2022). Optimisation of cutting parameters of new material orthotic insole using a Taguchi and response surface methodology approach. *Alexandria Engineering Journal*, 61(5), 3613–3632.

Anwar, M., Rasul, M.G., Ashwath, N., & Rahman, M.M. (2018). Optimisation of second-generation biodiesel production from Australian native stone fruit oil using response surface method. *Energies,* 11(10), 2566.

Ardabili, S., Mosavi, A., & V´arkonyi-K´oczy, A.R. (2019). Systematic review of deep learning and machine learning models in biofuels research. In *Proceedings of the International Conference on Global Research and Education*, Springer, Balatonf¨ured, Hungary.

Awogbemi, O. & Kallon, D.V.V. (2023). Application of machine learning technologies in biodiesel production process—A review. *Frontiers in Energy Research*, 11, 1122638.

Awogbemi, O., Kallon, D.V.V., & Aigbodion, V.S. (2021). Trends in the development and utilization of agricultural wastes as heterogeneous catalyst for biodiesel production. *Journal of the Energy Institute,* 98, 244–258.

Ayoola, A.A., Hymore, F.K., Omonhinmin, C.A., Olawole, O.C., Fayomi, O.S.I., Babatunde, D., et al. (2019). Analysis of waste groundnut oil biodiesel production using response surface methodology and artificial neural network. *Chemical Data Collections*, 22, 100238.

Balasubramanian, K., & Purushothaman, K. (2019). Performance, emission and combustion characteristics of safflower, neem and corn biodiesels fuelled in a CI engine. *Nature Environment & Pollution Technology,* 18, 1265–1273.

Bali, N., & Singla, A. (2022). Emerging trends in machine learning to predict crop yield & study its influential factors: A survey. *Archives of Computational Methods in Engineering,* 29(1), 95–112.

Bashir, M.A., Wu, S., Zhu, J., Krosuri, A., Khan, M.U., & Ndeddy Aka, R.J. (2022). Recent development of advanced processing technologies for biodiesel production: A critical review. *Fuel Processing Technology,* 227, 107120.

Betiku, E., Okunsolawo, S.S., Ajala, S.O., & Odedele, O.S. (2015). Performance evaluation of artificial neural network coupled with generic algorithm and response surface methodology in modeling and optimization of biodiesel production process parameters from shea tree (*Vitellaria paradoxa*) nut butter. *Renewable Energy*, 76, 408–417.

Changmai, B., Vanlalveni, C., Ingle, A. P., Bhagat, R., & Rokhum, S. L. (2020). Widely used catalysts in biodiesel production: A review. *RSC Advances,* 10(68), 41625–41679.

Chilakamarry, C.R., Sakinah, A., Zularisam, A., Sirohi, R., Khilji, I.A., Reddy, V.J., et al. (2022). Bioconversion of glycerol into biofuels—Opportunities and challenges. *BioEnergy Research,* 15(1), 46–61.

Chizoo, E., Augustine, S.C., Chukwu, E.G., & Gerald, U. (2022). Adaptive neuro fuzzy inference system-genetic algorithm versus response surface methodology desirability function algorithm modelling and optimization of biodiesel synthesis from waste chicken fat. *Journal of the Taiwan Institute of Chemical Engineers,* 136, 104389.

Chuengcharoenphanich, N., Watsuntorn, W., Qi, W., Wang, Z., Hu, Y., & Chulalaksananukul, W. (2023). The potential of biodiesel production from grasses in Thailand through consolidated bioprocessing using a cellulolytic oleaginous yeast, *Cyberlindnera rhodanensis* CU-CV7. *Energy,* 263, 125759.

Dabi, M., & Saha, U.K. (2019). Application potential of vegetable oils as alternative to diesel fuels in compression ignition engines: A review. *Journal of the Energy Institute,* 92(6), 1710–1726.

Dharma, S., Ong, H.C., Masjuki, H.H., Sebayang, A.H., & Silitonga, A.S. (2016). An overview of engine durability and compatibility using biodiesel-bioethanol-diesel blends in compression-ignition engines. *Energy Conversion and Management*, 128, 66–81.

Elkelawy, M., Bastawissi, A.E., Esmaeil, K.K., Radwan, A.M., Panchal, H., Sadasivuni, K.K., et al. (2020). Maximization of biodiesel production from sunflower and soybean oils and prediction of diesel engine performance and emission characteristics through response surface methodology. *Fuel,* 266, 117072.

Faizollahzadeh Ardabili, S., Najafi, B., & Shamshirband, S. (2019). Fuzzy logic method for the prediction of cetane number using carbon number, double bounds, iodic, and saponification values of biodiesel fuels. *Environmental Progress and Sustainable Energy,* 38, 584–599.

Fangfang, F., Alagumalai, A., & Mahian, O. (2021). Sustainable biodiesel production from waste cooking oil: ANN modeling and environmental factor assessment. *Sustainable Energy Technologies and Assessments,* 46, 101265.

Farobie, O., Hasanah, N., & Matsumura, Y. (2015). Artificial neural network modelling to predict biodiesel production in supercritical methanol and ethanol using spiral reactor. *Procedia Environmental Science,* 28, 214–223.

Fayyazi, E., Ghobadian, B., Najafi, G., Hosseinzadeh, B., Mamat, R., & Hosseinzadeh, J. (2015). An ultrasound-assisted system for the optimization of biodiesel production from chicken fat oil using a genetic algorithm and response surface methodology. *Ultrasonics Sonochemistry,* 26, 312–320.

Garg, A., & Jain, S. (2020). Process parameter optimization of biodiesel production from algal oil by response surface methodology and artificial neural networks. *Fuel,* 277, 118254.

Gimbun, J., Ali, S., Kanwal, C.C.S.C., Shah, L.A., Ghazali, N.H.M., Cheng, C.K., et al. (2013). Biodiesel production from rubber seed oil using activated cement clinker as catalyst. *Procedia Engineering,* 53, 13–19.

Goswami, L., Kayalvizhi, R., Dikshit, P.K., Sherpa, K.C., Roy, S., Kushwaha, A., et al. (2022). A critical review on prospects of bio-refinery products from second and third generation biomasses. *Chemical Engineering Journal,* 448, 137677.

Guang-Bin, H., Qin-Yu, Z., & Chee-Kheong, S. (2004). Extreme learning machine: A new learning scheme of feed forward neural networks. In *Proceedings of the 2004 IEEE International Joint Conference on Neural Networks (IEEE Cat. No. 04CH37541),* Budapest, Hungary.

Guo, M., Jiang, W., Chen C., et al. (2021). Process optimization of biodiesel production from waste cooking oil by esterification of free fatty acids using La3+/ZnO-TiO2 photocatalyst. *Energy Conversion and Management*, 229, 113745.

Gupta, V., & Pal Singh, K. (2023). The impact of heterogeneous catalyst on biodiesel production: A review. *Materials Today: Proceedings,* 78 (3), 364–371.

Ishola, N.B., Okeleye, A.A., Osunleke, A.S., & Betiku, E. (2019). Process modelling and optimization of sorrel biodiesel synthesis using barium hydroxide as a base heterogeneous catalyst: Appraisal of response surface methodology, neural network and neuro-fuzzy system. *Neural Computing and Application*, 31(9), 4929–4943.

Jiaqiang, E., Pham, M., Zhao, D., et al. (2017). Effect of different technologies on combustion and emissions of the diesel engine fueled with biodiesel: a review. *Renewable and Sustainable Energy Reviews*, 80, 620–647.

Jisieike, C.F., Ishola, N.B., Latinwo, L.M., & Betiku, E. (2023). Crude rubber seed oil esterification using a solid catalyst: Optimization by hybrid adaptive neuro-fuzzy inference esystem and response surface methodology. *Energy,* 263, 125734.

Keliris, A., Salehghaffari, H., Cairl, B., Krishnamurthy, P., Maniatakos, M., & Khorrami, F. (2016). Machine learning based defense against process-aware attacks on industrial control systems. In *Proceedings of the 2016 IEEE International Test Conference (ITC).*

Knothe, G., & Razon, L.F. (2017). Biodiesel fuels. *Progress in Energy and Combustion Science*, 58, 36–59.

Kolakoti, A., Jha, P., Mosa, P.R., Mahapatro, M., & Kotaru, T.G. (2020). Optimization and modelling of mahua oil biodiesel using RSM and genetic algorithm techniques. *Mathematical Models in Engineering*, 6(2), 134–146.

Lahdhiri, H., Said, M., Ben Abdellafou, K., Taouali, O., & Harkat, M.F. (2019). Supervised process monitoring and fault diagnosis based on machine learning methods.' *E-International Journal of Advanced Manufacturing Technology*, 102(5), 2321–2337.

Liao, M., & Yao, Y. (2021). Applications of artificial intelligence based modelling for bioenergy systems: a review. *GCB Bioenergy*, 13(5), 774–802.

Liaquat, A.M., Masjuki, H.H., Kalam, M.A., Fattah, I.M.R., Hazrat, M.A., Varman, M., Mofijur, M., & Shahabuddin, M. (2013). Effect of coconut biodiesel blended fuels on engine performance and emission characteristics. *Procedia Engineering,* 56, 583–590.

Likozar, B., & Levec, J. (2014). Effect of process conditions on equilibrium, reaction kinetics and mass transfer for triglyceride transesterification to biodiesel: Experimental and modeling based on fatty acid composition. *Fuel Processing Technology,* 22, 30–41.

Mairizal, A.Q., Awad, S., Priadi C.R., et al. (2020). Experimental study on the effects of feedstock on the properties of biodiesel using multiple linear regressions. *Renewable Energy*, 145, 375–381.

Mandari, V., & Devarai, S. K. (2021). Biodiesel production using homogeneous, heterogeneous, and enzyme catalysts via transesterification and esterification reactions: Acritical review. *Bio Energy Research*, 15, 935–961.

Maran, J.P., & Priya, B. (2015). Modeling of ultrasound assisted intensification of biodiesel production from neem (*Azadirachta indica*) oil using response surface methodology and artificial neural network. *Fuel*, 143, 262–267.

Mathew, G.M., Raina, D., Narisetty, V., Kumar, V., Saran, S., Pugazhendi, A., et al. (2021). Recent advances in biodiesel production: Challenges and solutions. *Science of The Total Environment,* 794, 148751.

Meng, X., Jia, M., & Wang, T. (2014). Neural network prediction of biodiesel kinematic viscosity at 313K. *Fuel*, 121,133–140.

Mishra, V.K., & Goswami, R. (2018). A review of production, properties and advantages of biodiesel. *Biofuels,* 9(2), 273–289.

Moayedi, H., Aghel, B., Foong, L.K., & Bui, D.T. (2020). Feature validity during machine learning paradigms for predicting biodiesel purity. *Fuel*, 262, 116498.

Mukhtar, A., Saqib, S., Lin, H., Hassan Shah, M.U., Ullah, S., Younas, M., et al. (2022). Current status and challenges in the heterogeneous catalysis for biodiesel production. *Renewable and Sustainable Energy Reviews,* 157, 112012.

Musa, I.A. (2016). The effects of alcohol to oil molar ratios and the type of alcohol on biodiesel production using transesterification process. *Egyptian Journal of Petroleum,* 25(1), 21–31.

Narowska, B., Kułażyński, M., Łukaszewicz, M., & Burchacka, E. (2019). Use of activated carbons as catalyst supports for biodiesel production. Renew. *Energy,* 135, 176–185.

Nayab, R., Imran, M., Ramzan, M., Tariq, M., Taj, M.B., Akhtar, M.N., et al. (2022). Sustainable biodiesel production via catalytic and non-catalytic transesterification of feedstock materials: A review. *Fuel,* 328, 125254.

Okolie, J.A., Ivan Escobar, J., Umenweke, G., Khanday, W., & Okoye, P.U. (2022). Continuous biodiesel production: A review of advances in catalysis, microfluidic and cavitation reactors. *Fuel,* 307, 121821.

Paul, G., Datta, A., & Mandal, B.K. (2014). An experimental and numerical investigation of the performance, combustion and emission characteristics of a diesel engine fueled with jatropha biodiesel. *Energy Procedia,* 54, 455–467.

Precedence Research (2022). Biodiesel market size. Available at: https://www.precedenceresearch.com (Accessed November 4, 2022).

Rostami, S., Hosseinzadeh, S.B., & Saeidi, K. (2016). Optimization of biodiesel production from prunus scoparia using artificial bee colony algorithm. *Journal of Renewable Energy and Environment,* 3(1), 52–58.

Russell, S., & Norvig, P. (2010). *Artificial Intelligence-A Modern Approach.* M. Hirsch, Ed., Pearson Education, Inc, Hoboken, NJ, USA, 3rd edition.

Salam, S., Choudhary, T., Pugazhendhi, A., Verma, T.N., & Sharma, A. (2020). A review on recent progress in computational and empirical studies of compression ignition internal combustion engine. *Fuel*, 279, 118469.

Samuel, O.D., Okwu, M.O., Oyejide, O.J., Taghinezhad, E., Afzal, A., & Kaveh, M. (2020). Optimizing biodiesel production from abundant waste oils through empirical method and grey wolf optimizer. *Fuel*, 281, 118701.

Santos, B.S., Capareda, S.C., & Capunitan, J.A. (2013). Engine performance and exhaust emissions of peanut oil biodiesel. *Journal of Sustainable Bioenergy Systems,* 3, 272–286.

Selvaraj, R., Moorthy, I.G., Kumar, R.V., & Sivasubramanian, V. (2019). Microwave mediated production of FAME from waste cooking oil: modelling and optimization of process parameters by RSM and ANN approach. *Fuel*, 237, 40–49.

Shin, H.-Y.,Lim, S.-M., Kang, S. C., & Bae, S.-Y. (2012). Statistical optimization for biodiesel production from rapeseed oil via transesterificaion in supercritical methanol. *Fuel Processing Technology*, 98, 1–5.

Sieminski, A. (2013). International energy outlook 2013, Report Number: DOE/EIA-0484, 2013, US Energy Information Administration (EIA), Washington, DC, USA.

Silitonga, A.S., Shamsuddin, A.H., Mahlia, T.M.I., Milano, J., Kusumo, F., Siswantoro, J.,et al. (2020). Biodiesel synthesis from Ceiba pentandra oil by microwave irradiation-assisted transesterification: ELM modeling and optimization. *Renewable Energy*, 146, 1278–1291 .

Sitepu, E.K., Heimann, K., Raston, C.L., & Zhang, W. (2020). Critical evaluation of process parameters for direct biodiesel production from diverse feedstock. *Renewable and Sustainable Energy Reviews*, 123, 109762.

Srivastava, G., Paul, A.K., & Goud, V.V. (2018). Optimization of non-catalytic transesterification of microalgae oil to biodiesel under supercritical methanol condition. *Energy Conversion and Management*, 156, 269–278.

Stevanato, N., de Mello, B.T.F., Saldaña, M.D.A., Cardozo-Filho, L., & da Silva, C. (2023). Production of ethyl esters from forage radish seed: An integrated sequential route using pressurized ethanol and ethyl acetate. *Fuel*, 332, 126075.

Su, G., Ong, H.C., Mofijur, M., Mahlia, T.M.I., & Ok, Y.S. (2022). Pyrolysis of waste oils for the production of biofuels: A critical review. *Journal of Hazardous Materials*, 424, 127396.

Sudhakar, K.,Rajesh, M., & Premalatha, M. (2012).Carbon mitigation potential of Jatropha biodiesel in Indian context. *Energy Procedia*, 14, 1421–1426.

Tabatabaei, M., Aghbashlo, M., Dehhaghi, M., Panahi, H.K.S., Mollahosseini, A., Hosseini, M., et al. (2019). Reactor technologies for biodiesel production and processing: A review. *Progress in Energy and Combustion Science*, 74, 239–303.

Tan, S.X., Lim, S., Ong, H.C., Pang, Y.L., Fitranto, K., Goh, B.H.H., et al. (2019). Two step catalytic reactive extraction and transesterification process via ultrasonic irradiation or biodiesel production from solid Jatropha oil seeds. *Chemical Engineering and Processing - Process Intensification*, 146, 107687.

Tchameni, A.P., Zhao, L., Ribeiro, J.X.F. & Li. T. (2019). Predicting the rheological properties of waste vegetable oil biodiesel modified water-based mud using artificial neural network. *Geosystem Engineering*, 22(2), 101–111.

Traver, M.L., Atkinson, R.J., & Atkinson, C.M. (1999). Neural network-based diesel engine emissions prediction using in cylinder combustion pressure. *SAE Transactions*, 108,1166–1180.

Wahidin, S., Idris, A., Yusof, N.M., Kamis, N.H.H., & Shaleh, S.R.M. (2018).Optimization of the ionic liquid-microwave assisted one-step biodiesel production process from wet microalgal biomass. *Energy Conversion and Management*, 171, 1397–1404.

Wazid, M., Das, A.K., Chamola, V., & Park, Y. (2022). Uniting cyber security and machine learning: Advantages, challenges and future research. *ICT Express*, 8(3), 313–321.

Wong, K.I., Wong, P.K., Cheung, C.S., & Vong, C.M. (2013). Modeling and optimization of biodiesel engine performance using advanced machine learning methods. *Energy*, 55, 519–528.

Wu, S., Wang, Y., & Cheng, S. (2013). Extreme learning machine based wind speed estimation and sensor less control for wind turbine power generation system. *Neurocomputing*, 102,163–175.

Xie, W., & Li, J. (2023). Magnetic solid catalysts for sustainable and cleaner biodiesel production: A comprehensive review. *Renewable and Sustainable Energy Reviews*,171, 113017.

Yasin, M.H.M., Paruka, P., Mamat, R., Yusop, A.F., Najafi, G., & Alias, A. (2015). Effect of low proportion palm biodiesel blend on performance, combustion and emission characteristics of a diesel engine. *Energy Procedia,* 75, 92–98.

Yesilyurt, M.K., Arslan, M., & Eryilmaz, T. (2019). Application of response surface methodology for the optimization of biodiesel production from yellow mustard (*Sinapis alba* L.) seed oil. *International Journal of Green Energy,* 16(1), 60–71.

Zhang, J., Wang, Y., Muldoon, V. L., & Deng, S. (2022). Crude glycerol and glycerol as fuels and fuel additives in combustion applications. *Renewable and Sustainable Energy Reviews*, 159, 112206.

7 Machine Learning and Its Application in Bioethanol Production

Gboyega Ebenezer Adebami, Silas Olukunle Fagbenro, Benjamin Ayodipupo Babalola, Khushboo Negi, Samakshi Verma, and Bukola Christianah Adebayo-Tayo

7.1 INTRODUCTION

Bioethanol, a renewable biofuel produced from biomass such as sugar crops, grains, and lignocellulosic materials, serves as a sustainable alternative to fossil fuels by reducing greenhouse gas emissions and enhancing energy independence (Broda et al. 2022). Bioethanol can be blended with gasoline or used in flexible fuel vehicles, making it a key component of modern energy strategies (Bušić et al. 2018). The production of bioethanol involves complex biochemical processes, including feedstock pretreatment, enzymatic hydrolysis, fermentation, and distillation (Adebami and Adebayo-Tayo, 2020; Shukla et al. 2023). Firstly, feedstock pretreatment prepares raw materials for fermentation by breaking down complex structures. This is followed by enzymatic hydrolysis, where enzymes convert polysaccharides into fermentable sugars, a crucial step for lignocellulosic biomass. Next, the fermentation process, influenced by factors like temperature, pH, and nutrient availability, uses yeast or other microorganisms to convert sugars into ethanol and carbon dioxide under anaerobic conditions. Finally, the distillation stage separates ethanol from other components, concentrating it to ensure high purity and quality for commercial use (Shukla et al. 2023). Each stage presents unique challenges that can impact the overall efficiency and yield of bioethanol production (Adebami et al., 2021). Traditional methods of optimizing these processes often fall short due to the inherent variability and complexity of biological systems. Machine learning (ML), a subset of artificial intelligence (AI), offers powerful tools for addressing these challenges (Soori et al. 2022).

Machine learning (ML) is a transformative technology enabling computers to learn from data and make decisions without explicit programming (Javaid et al. 2021). It encompasses algorithms like classification, regression, clustering, and feature engineering, enhancing applications across diverse fields (Taye, 2023). ML's ability to analyze vast datasets allows it to identify patterns and insights, making it

 DOI: 10.1201/9781003571223-7

invaluable in fraud detection, recommendation systems, image recognition, and autonomous vehicles (Pattnaik et al. 2024). The effectiveness of ML models relies heavily on the quality of training data; poor-quality data can lead to inaccurate predictions (Gong et al. 2023). As industries increasingly adopt ML solutions, its potential to drive innovation and efficiency expands, paving the way for advancements in healthcare, finance, and smart technologies. Overall, ML not only improves operational capabilities but also opens new avenues for research and development in artificial intelligence (Pattnaik et al. 2024). In biological sciences, ML aids in disease diagnosis, drug discovery, personalized medicine, gene prediction, and the analysis of genetic variations, revolutionizing healthcare and genomics. It also optimizes biofuel production processes, enhancing efficiency, and reducing costs (Bhardwaj et al. 2022; Su et al. 2024).

The application of ML in bioethanol production is not just theoretical; it has shown practical potential to transform the industry. For example, artificial neural networks (ANNs) have been used to model the enzymatic hydrolysis of lignocellulosic biomass, accurately predicting the yield of fermentable sugars (Coşgun et al. 2023). Similarly, support vector machines (SVMs) have optimized fermentation conditions, leading to higher ethanol yields and lower production costs (Zhang et al. 2020a; Coşgun et al. 2023). Additionally, decision trees and random forests have identified critical factors influencing bioethanol production, enabling more targeted and efficient process improvements (Long & Liu, 2023). These advancements highlight the significant role of ML in enhancing bioethanol production processes.

Despite these advancements, integrating ML into bioethanol production is still in its early stages. Challenges such as data availability, model interpretability, and the need for interdisciplinary collaboration persist (Wang et al. 2024). Additionally, the complexity of bioethanol production processes requires robust and adaptable ML models to handle diverse data types and sources (Owusu & Marfo, 2022). However, the potential benefits of ML—such as increased efficiency, reduced environmental impact, and enhanced sustainability—make it a promising area for future research and development (Raman et al. 2024). By leveraging ML, researchers can optimize various stages of bioethanol production, from feedstock selection to fermentation and distillation, ultimately leading to more cost-effective and environmentally friendly biofuel solutions (Coşgun et al. 2023). Addressing these challenges could unlock significant improvements in bioethanol production processes, paving the way for a more sustainable energy future (Bušić et al. 2018; Long & Liu, 2023).

This chapter aims to provide a comprehensive review of the application of ML in bioethanol production, drawing on recent studies and case examples. It discusses the various ML techniques used, their applications in different stages of the production process, and the outcomes achieved. Additionally, the chapter addresses the challenges and limitations of using ML in this context and propose future directions for research.

7.2 ROLES OF MACHINE LEARNING IN BIOETHANOL PRODUCTION

Machine learning (ML) plays a significant role in enhancing bioethanol production, particularly from lignocellulosic biomass. By employing advanced algorithms, ML can optimize various stages of the bioethanol production process, including pretreatment, fermentation, and conversion. For instance, artificial neural networks (ANNs) have demonstrated high accuracy in predicting ethanol yields by analyzing complex data patterns related to biomass composition and processing conditions.

7.2.1 Machine Learning Models and Algorithms in Bioethanol Production

Machine learning encompasses a wide range of algorithmic frameworks, including supervised, unsupervised, and reinforcement learning (Sarker, 2021). Each of these approaches offers distinct advantages when applied to bioethanol production, which involves a multifaceted series of processes such as feedstock preprocessing, enzymatic hydrolysis, microbial fermentation, and product recovery. Supervised learning algorithms, such as support vector machines (SVM), random forests, and deep neural networks (DNN), have shown significant promise in predictive modeling for bioethanol production (Velidandi et al. 2023). These algorithms rely on labeled datasets, where input parameters (e.g., feedstock characteristics, fermentation temperature, pH levels) are paired with known output responses (e.g., ethanol yield, production rate) (Lawson et al. 2021). A trained supervised model can predict future outcomes and optimize process conditions by correlating input parameters with output performance metrics. For instance, a neural network could predict ethanol concentration as a function of complex interactions between temperature, substrate concentration, enzyme activity, and microbial growth rates (Sipos et al. 2020). Sipos et al. (2020) highlighted the adaptability of their application, which effectively incorporates numerous variables and presents a comparative analysis of the results obtained from different neural networks (feedback vs. feed-forward) and learning algorithms (Back-Propagation versus Levenberg–Marquardt). Their simulation results demonstrated that the feedback neural network outperformed the feed-forward one. While the neural network configuration is quite flexible, allowing for varying hidden layers and nodes, the number of input and output nodes is dependent on the fermentation process parameters. After extensive simulations, they found that using pH and CO_2 as input variables significantly reduced the prediction errors of the neural network. ML can rapidly identify optimal operating conditions that would be time-consuming or costly to determine experimentally.

In unsupervised learning, the goal is to uncover hidden patterns or relationships in data without explicit labels (Jafari-Marandi 2021). Algorithms such as k-means clustering, principal component analysis (PCA), and t-SNE can segment or reduce large datasets obtained from omics technologies (e.g., genomics, proteomics, metabolomics) used in bioethanol production. For example, unsupervised learning can be employed to group microbial strains based on their metabolic profiles or cluster fermentation batches with similar performance traits (Patra et al. 2023). This clustering allows for the identification of high-performing strains or process conditions that share common biological or operational characteristics. Furthermore, unsupervised learning techniques can be deployed for anomaly detection, identifying outliers in fermentation data that may indicate process faults or contamination events, thereby improving quality control.

Additionally, Reinforcement Learning (RL) is particularly well-suited for process control and optimization in bioethanol production, as it allows systems to learn optimal actions through trial and error in an adaptive environment (Yoo et al. 2021). In an RL framework, an agent interacts with the bioethanol production system (the environment) and learns to maximize long-term rewards (e.g., ethanol yield, energy efficiency) by choosing actions (e.g., adjusting feed rates, controlling temperature or pH). Over time, the RL model improves through feedback from the system, adjusting its strategies to achieve optimal performance under fluctuating conditions. RL offers a more flexible approach to process optimization, particularly in bioethanol plants where conditions such as feedstock variability, microbial activity, or enzyme efficacy may change unpredictably.

7.2.2 Multidimensional Data Integration in Bioethanol Production via ML

Bioethanol production is inherently a multidimensional process involving various biological, chemical, and physical phenomena. The integration of data from multiple sources (e.g., feedstock properties, process conditions, microbial genetics, real-time fermentation monitoring) is crucial for understanding and optimizing the production process (Fisher et al. 2020). ML algorithms, particularly deep learning models, excel at integrating such high-dimensional data. For example, convolutional neural networks (CNNs) and recurrent neural networks (RNNs) can process time-series data from sensors used in fermentation bioreactors, identifying temporal correlations between real-time monitoring parameters and downstream ethanol yields (Goshisht, 2024). These models can even incorporate real-time data on substrate consumption, by-product formation, and microbial viability to predict optimal feeding strategies, adjust oxygen supply, or dynamically control pH (Müller et al. 2023).

Moreover, in multi-omics studies, ML-based integrative analyses can identify key biological markers (e.g., specific genes, proteins, or metabolites) that influence the efficiency of microbial strains in converting lignocellulosic biomass to ethanol (Zhang et al. 2023; Ahlawat et al. 2024). For instance, ensemble learning methods like random forests and gradient boosting can be used to rank features based on their importance, pinpointing the most critical biological pathways or genetic determinants that govern high ethanol productivity under different environmental conditions. This not only accelerates the discovery of robust microbial strains but also supports metabolic engineering efforts aimed at optimizing strains for higher ethanol production and resistance to fermentation inhibitors (Adegboye et al. 2021).

7.2.3 Bioethanol Feedstock Optimization Using ML

Another critical area where ML proves invaluable is in the selection and optimization of feedstocks for bioethanol production. Lignocellulosic biomass, including agricultural residues (e.g., corn stover, sugarcane bagasse) and energy crops (e.g., switchgrass, Miscanthus), exhibits considerable variability in composition, such as cellulose, hemicellulose, and lignin content, which directly impacts hydrolysis efficiency and ethanol yield (Deshavath et al. 2019). ML models can be developed to predict the suitability of various feedstocks by correlating compositional and structural features of biomass with ethanol conversion efficiency (Jeon et al. 2023). Random forests or decision trees, for example, can model the complex relationship between feedstock variability and enzymatic saccharification performance, allowing producers to select the most cost-effective and efficient feedstock based on real-time data from preprocessing operations (Sharma et al. 2023). Furthermore, ML has the potential to optimize pretreatment processes such as acid hydrolysis or steam explosion by modeling the interplay between feedstock characteristics, pretreatment conditions, and enzymatic digestibility (Zhang et al. 2021). Such predictive models can significantly reduce experimental time, facilitating the fine-tuning of process parameters to achieve maximal sugar release while minimizing inhibitor formation (e.g., furfural, hydroxymethylfurfural) that can impair microbial fermentation.

7.2.4 ML-driven Microbial Strain Engineering for Enhanced Bioethanol Production

Microbial fermentation lies at the heart of bioethanol production, and engineering microbial strains with enhanced ethanol-producing capabilities is a key area where ML can make transformative contributions. The integration of ML with synthetic biology enables the rational design of microbial strains by predicting the effects of genetic modifications on cellular metabolism (Wu et al. 2016). Metabolic pathway optimization can be achieved using constraint-based metabolic modeling combined with ML, where the algorithms predict flux distributions through different metabolic networks and identify potential bottlenecks limiting ethanol production (Khaleghi et al. 2021).

For example, ML can be used in combination with genome-scale metabolic models (GEMs) to simulate and predict how genetic alterations will affect cellular processes (Kim et al. 2021; Kundu et al. 2024). Techniques such as reinforcement learning or evolutionary algorithms can guide the selection of gene knockouts, over-expressions, or pathway modifications that are most likely to enhance ethanol yield. In addition, ML can optimize the balance between cellular growth and ethanol production, minimizing the trade-offs that often arise between biomass formation and product synthesis in engineered strains.

7.2.5 Process Control and Real-time Monitoring via ML

In an industrial bioethanol production plant, process control is paramount to maintaining optimal conditions for fermentation, and this is where ML excels in providing predictive and prescriptive insights. Through the application of ML models, such as Bayesian networks or long short-term memory (LSTM) networks, it is possible to predict deviations in real-time fermentation parameters (e.g., substrate concentration, ethanol content, microbial growth) and automatically adjust operational parameters (e.g., feed rate, temperature, agitation) to bring the system back to optimal conditions (Patra et al. 2023). This type of real-time process monitoring and adaptive control offers significant advantages over traditional static control systems, enabling continuous optimization and reducing downtime due to process inefficiencies.

Moreover, ML techniques, such as Gaussian processes or Markov decision processes, can predict long-term trends in bioethanol production systems by modeling the stochastic nature of microbial activity and environmental variables (Patra et al. 2023). This facilitates proactive decision-making, enabling plant operators to anticipate and mitigate risks such as microbial contamination or feedstock variability before they adversely impact production. Therefore, the integration of machine learning into bioethanol production is not merely a trend but a critical advancement that holds the potential to revolutionize the industry. ML offers unprecedented capabilities to model the inherent complexity of bioethanol production processes, from feedstock variability and pretreatment optimization to microbial strain engineering and real-time process control. By leveraging large-scale data and sophisticated algorithmic models, ML enhances productivity, reduces operational costs, and improves the overall efficiency of bioethanol production systems. As the bioethanol industry continues to evolve, the role of ML in bioenergy optimization will likely expand, leading to more sustainable and economically viable production processes.

7.3 VARIOUS MACHINE LEARNING ALGORITHMS IN BIOETHANOL PRODUCTION

The optimization of bioethanol production processes presents unique challenges due to the complex interplay of biological, chemical, and physical factors (Broda et al. 2022). Traditional optimization methods are often inadequate in dealing with the nonlinearity, high-dimensionality, and variability inherent to bioethanol systems. In response to these challenges, machine learning (ML) algorithms have emerged as powerful tools for modeling, predicting, and optimizing key processes in bioethanol production. By leveraging vast datasets and computational techniques, ML enables bioethanol producers to improve yields, optimize resource use, and reduce operational costs. In this section, we explore several commonly used ML algorithms—Artificial Neural Networks (ANNs), Random Forest, XGBoost, Support Vector Machines (SVM), and Decision Trees and their relevance to bioethanol production.

7.3.1 Artificial Neural Networks

Artificial Neural Networks (ANNs) are inspired by the structure and function of biological neurons. Biological neurons are specialized cells in the nervous system that transmit information through electrical impulses known as action potentials. These neurons communicate at synapses, where neurotransmitters are released to influence the activity of adjacent neurons. ANNs consist of multiple layers of interconnected "neurons" that can model complex, nonlinear relationships between input features and output variables (Abdolrasol et al. 2021).

In bioethanol production, ANNs have been employed to model various processes, including enzymatic hydrolysis, microbial fermentation, and feedstock pretreatment. Their capability to capture intricate interactions among process parameters—such as temperature, substrate concentration, and microbial activity—makes them particularly effective for predicting bioethanol yield (Voisin et al. 2024). Studies have shown that ANNs surpass traditional modeling techniques such as Response Surface Methodology (RSM) and Adaptive Neuro-Fuzzy Inference System (ANFIS) in predicting ethanol yields. ANNs achieve impressive correlation coefficients, with R^2 values nearing 0.90 (Onu et al. 2022). Through the analysis of data from diverse feedstocks and fermentation conditions, ANNs can determine the optimal parameters that enhance ethanol yield while reducing costs. This predictive ability is especially advantageous for scaling up production in industrial environments, positioning ANNs as essential tools in the shift toward more efficient and sustainable biofuel production (Taiwo & Musonge, 2023).

7.3.2 Random Forest

Random Forest is an ensemble method that builds multiple decision trees and aggregates their predictions to improve accuracy and reduce overfitting (Zhou et al. 2020). Random Forest excels at handling both categorical and continuous data, making it particularly useful for feature selection and predictive modeling in bioethanol production. By analyzing historical data, Random Forest can identify the key variables influencing ethanol yield and optimize the production process accordingly (Fischer et al. 2017; Bosu et al. 2022; Momenitabar et al. 2023).

In bioethanol production, Random Forest can be used to predict the efficiency of different feedstocks, such as lignocellulosic biomass, based on their composition (Long and Liu, 2023). By ranking the importance of features like cellulose, hemicellulose, and lignin content, the algorithm aids in selecting the best feedstocks for maximum ethanol yield. Moreover, Random Forest can optimize fermentation conditions by analyzing process data, helping operators adjust parameters like temperature and pH to achieve optimal yields. Its robust nature allows it to be employed in both lab-scale research and large-scale industrial settings. Long and Liu (2023) illustrates the performance of Random Forest, along with XGBoost and ANN, in predicting ethanol yield, showing that Random Forest demonstrates competitive accuracy and R^2 values across 100 replicative tests.

7.3.3 XGBoost

Extreme Gradient Boosting (XGBoost) is a gradient boosting algorithm that builds sequential decision trees, where each tree tries to correct the errors of its predecessor (Sagi and Rokach, 2021). XGBoost is known for its speed, accuracy, and ability to handle large, high-dimensional datasets (Sagi and Rokach 2021). This makes it a powerful tool in bioethanol production, where the complexity of the process and the size of datasets from sensors and omics studies are often prohibitive for traditional algorithms.

XGBoost can optimize enzymatic hydrolysis by modeling the effects of enzyme dosage, substrate composition, and pretreatment conditions on sugar release rates. Its ability to handle missing data and noise makes it ideal for real-time monitoring and control of fermentation processes in industrial bioreactors. Additionally, XGBoost has been applied to predictive maintenance by analyzing time-series data from equipment sensors to forecast potential system failures, allowing for proactive interventions that could minimize downtime and ensure continuous bioethanol production (Gardas & Narwane 2024)

7.3.4 Support Vector Machines

Support Vector Machines (SVM) are supervised learning models that can perform both classification and regression tasks. SVM works by finding the hyperplane that maximizes the margin between different classes in the data, making it effective for high-dimensional datasets (Maldonado et al. 2014). In bioethanol production, SVM is particularly useful for classifying microbial strains or identifying optimal process conditions based on a large number of variables (Maldonado et al. 2014; Coşgun et al. 2023).

SVM has been applied in strain classification, where it analyzes genomic and proteomic data to predict which microbial strains are most likely to produce high ethanol yields under specific fermentation conditions (Naveed et al. 2024; Khaleghi et al. 2021; Xu et al. 2024). It is also useful in outlier detection, helping to identify faulty batches or suboptimal conditions in real-time fermentation monitoring (Wang et al. 2023). By quickly flagging anomalies, SVM enables operators to correct process deviations before they negatively impact ethanol production efficiency.

7.3.5 Decision Trees

Decision Trees are intuitive and interpretable models that split data into branches based on the value of input variables. Each internal node represents a decision based on one feature, and the branches lead to possible outcomes. Decision Trees are widely used in bioethanol production due to their simplicity and ease of interpretation (Aldrich & Auret 2013). They provide clear decision-making paths, making them suitable for process control and optimization tasks where quick decisions are required (Chen & Xia 2024). Decision Trees are used to rank the importance of process variables, helping operators focus on the factors that have the greatest influence on ethanol yield. This feature selection capability is crucial in reducing the complexity of bioethanol production models and improving computational efficiency. Decision Trees help in feature selection and ranking the most influential variables affecting ethanol yield (Sharma et al. 2023).

Each machine learning algorithm has its own strengths, making it suitable for different aspects of bioethanol production. Artificial Neural Networks are best suited for capturing nonlinear relationships in fermentation and strain improvement. Random Forest and Decision Trees excel at feature selection and process control, while XGBoost's efficiency and accuracy make it ideal for real-time monitoring and predictive maintenance. Support Vector Machines, with their robustness in high-dimensional data, are invaluable for strain classification and outlier detection. As the bioethanol industry continues to evolve, the integration of these ML algorithms will be crucial in advancing process optimization, improving yields, and driving the industry towards greater efficiency and sustainability (İşçen et al. 2023).

7.3.6 Hybrid Models

Some studies have explored hybrid models that combine different machine learning techniques to enhance prediction accuracy and optimize production processes (Kenny et al. 2024). Hybrid models in machine learning are increasingly utilized in bioethanol production to enhance efficiency and yield. These models combine mechanistic approaches, such as flux balance analysis (FBA), with data-driven techniques like artificial neural networks (ANNs). This integration allows for a more comprehensive understanding of the metabolic pathways involved in ethanol biosynthesis (Lawson et al. 2021). For instance, studies have shown that hybrid models can accurately predict the impact of genetic modifications on ethanol yield by analyzing complex datasets derived from metabolic fluxes and fermentation conditions (Wu et al. 2024). In another instance, the technique of hybrid model was proven beneficial for designing sustainable bioethanol supply chains by integrating ML with quantitative optimization methods. The flexibility of these models facilitates optimization across various stages of bioethanol production, ultimately contributing to sustainable energy solutions (Arias et al. 2023).

Hybrid models offer several advantages in bioethanol production, including enhanced predictive accuracy. By leveraging both mechanistic and data-driven insights, these models provide more precise predictions of fermentation outcomes compared to traditional methods (Wu et al. 2024). Additionally, hybrid models can be utilized to optimize bioethanol production conditions by adjusting various parameters, such as temperature and substrate concentration, which are critical for maximizing ethanol yield (Desmal et al. 2022). Finally, the flexibility of ML techniques allows these models to adapt to different scales of production, ranging from laboratory settings to industrial applications.

7.4 APPLICATIONS OF ML IN BIOETHANOL PRODUCTION

Several machine learning techniques can significantly enhance bioethanol production. These various ML models facilitate the prediction of bioethanol production quality and yield estimation. Additionally, machine learning is utilized in the design, management, control, optimization, and monitoring of systems, resulting in highly accurate predictions (Xing et al. 2021). The application of these machine learning models is invaluable for examining and predicting the numerous biological processes that occur during bioethanol production. This understanding is crucial for improving process execution and increasing bioethanol yields, particularly in predicting the final concentration of ethanol through ionic liquids and enzymatic hydrolysis (Smuga-Kogut et al. 2021). Different models and algorithms can be developed using these methods, allowing for process optimization.

The production of bioethanol using ML can be effectively integrated with life cycle assessment (LCA). ML enhances the precision and comprehensiveness of

LCA results by predicting data gaps and addressing the absence of information. Additionally, machine learning's ability to rapidly and securely analyze vast amounts of data makes it feasible to evaluate increasingly complex models. Specifically, it is being applied to predict net positive energy output for various feedstocks under specific conditions, as well as to forecast bioethanol consumption in supply chain networks (Long & Liu 2023). Surrogate models developed with the assistance of machine learning have improved process performance. This research demonstrates that machine learning models can effectively forecast mass and energy balances, serving as valuable sources for traditional life cycle assessments, even though they have not yet provided precise guidelines for predicting LCA outcomes. Furthermore, a program was developed to automate downstream separation, product recovery, purification strategy design, and cost analysis to complement the surrogate modeling approach (Huntington et al. 2023). The manufacturing, use, and ecological implications of biodiesel can also be modeled, optimized, and managed through the application of machine learning and deep learning techniques. These approaches take into account how various parameters affect biofuel production rates, necessitating the use of predictive models to achieve desired outcomes. Such techniques can accurately represent the production process and offer an alternative modeling strategy to traditional mathematical models or the fundamental essence of the procedure itself (Ardabili et al. 2019).

7.4.1 Case Studies of ML Optimizing Operational Parameters to Enhance Efficiency and Yield

With the help of machine learning (ML) and its algorithms, the yield and efficiency of bioethanol production can be significantly enhanced. One notable technique within ML is Artificial Neural Networks (ANN), which are part of computational models that can be employed to achieve high yields in bioethanol production. Various computational and mathematical approaches can be utilized to improve bioethanol production. Significant models for optimizing bioethanol production include Response Surface Methodology (RSM), Genetic Algorithms, Fuzzy Logic, and ANN (Burugari et al., 2020). In recent years, the combination of ANN with Particle Swarm Optimization (PSO) has been used to refine the input variables of ANN and identify optimal sets of independent variables to enhance bioethanol yield. This optimization of input variables has revealed several operational solutions that can boost industrial bioethanol production, resulting in a 10% increase in output (Pereira et al. 2020). The ANN-PSO model has demonstrated remarkable results in optimizing process variables for bioethanol production. As a part of machine learning algorithms, ANN can also predict the concentration of bioethanol produced through fermentation methods. ANN models possess the capability to understand relationships between data that exhibit nonlinear dependencies. Consequently, once trained, these models become highly accurate in their predictions (Niaze et al. 2023). Key operational factors that significantly influence bioethanol production include cell concentration, water fermentation pH (WFMpH), pre-fermentation (PFM), and water fermentation hardness WFH).

An increase of 8.45% in bioethanol production was achieved at a sugar mill in India (Niaze et al. 2023). Artificial neural networks (ANNs) can be characterized by their use of mathematical and computational models for nonlinear function approximation, data classification, clustering, and non-parametric regression. These networks have demonstrated remarkable capabilities in solving complex nonlinear system identification and control problems. ANNs are employed across various domains to analyze input-output data, allowing for the description and forecasting of system behaviors. Such equations enable the estimation of ethanol production in relation to several parameters, including temperature, pH, and partial oxygen levels. By evaluating these parameters, the efficiency and yield of bioethanol can be enhanced (Ahmadian-Moghadam et al. 2013).

Bioethanol can be produced from lignocellulosic biomass (LCB). To generate ethanol from LCB, it must first undergo pretreatment and hydrolysis before the fermentation process. The optimization of fermentation can be accomplished using Central Composite Design (CCD), a major Response Surface Methodology (RSM) approach. Various parameters, such as incubation time, pH, and temperature, are optimized as independent variables, while the concentration of ethanol (measured in gram per liter) serves as the dependent variable for regression analysis. This statistical approach leads to a variety of calculations that provide data for optimizing operational parameters, resulting in enhanced bioethanol yield and efficiency (Sharma & Sharma 2024).

The hydrolysis process can also be optimized through ANN and Genetic Algorithm approaches to determine optimal enzyme loading and biomass digestibility. These models take multiple inputs and perform nonlinear transfer function operations on them, producing results in the form of weights. Factors such as biomass loading and substrate particle size play a crucial role in optimizing the hydrolysis process and predicting fermentable sugar concentrations like glucose (Vani et al. 2015). Additionally, mathematical modeling can be utilized to optimize the fermentation process by developing kinetic models based on glucose consumption rates, glucose concentration, *S. cerevisiae* growth, and ethanol concentration during batch fermentation (Damayanti et al. 2022).

The sugar yields from biomass content have been predicted as a function of pretreatment parameters using an artificial neural network (ANN) model. However, most research studies have relied on either randomly selected network configurations or small, homogeneous datasets that limit the range of pretreatment conditions. The reliability of the programming approach also depends on optimizing the high-level parameters of the ANN design. When a predictive model is developed using specific biological feedstock data—such as limited pretreatment conditions or minimal variation in anatomical traits—the model's generalization potential may be inadequate (Haldar et al. 2023).

7.4.2 Real-world Applications Showing Improvements in Process Control and Cost Reduction

ML can effectively control the production process of bioethanol. Through ML, machines can independently acquire information from data, sometimes with human assistance. This field involves creating algorithms and frameworks to perform complex operations. Proportional-Integral-Derivative (PID) controllers are straightforward yet reliable tools for production control, typically adjusted manually. While they may incorporate technology to assist in selecting control parameters, they primarily fulfill the fundamental control requirements for most Single Input Single Output (SISO) operations. However, they struggle with complex dynamics such as inverse response and long dead times. The concept of self-adapting SISO PID controllers through reinforcement learning has been expanded to accommodate Multiple-Input Multiple-Output (MIMO) systems or Model Predictive Control (MPC) methods (Raven et al. 2024). Both Brazilian sugarcane-based ethanol and U.S. corn-based ethanol have reported significant cost reductions primarily due to the implementation of advanced learning techniques that optimize manufacturing expenses and operational performance through machine learning (Hansen and Mirkouei 2018). Understanding the relationship between input parameters and productivity facilitates effective system optimization, scalability, cost reduction, long-term viability, and enhanced productivity (Okolie 2024). Machine learning has the capacity to forecast and impact critical process parameters (CPPs) as well as product critical quality attributes (CQAs). It governs processes in response to parameter variations and enhances the overall information processing related to the production cycle (Mondal et al. 2023). The application of machine learning in bioethanol production has not only reduced costs but also improved the control and optimization of the entire production process. This version maintains your original ideas while enhancing clarity and coherence.

7.4.3 Integration of ML with Other Technologies

The increasing demand for renewable and sustainable energy sources has put bioethanol, a leading biofuel, at the forefront of research and industrial production (Maity & Mallick 2022). Bioethanol production presents an alternative solution as concerns about fossil fuel depletion and environmental pollution increase (Adewuyi 2020). Nevertheless, the complexity of bioethanol production processes which range from feedstock pretreatment to fermentation and distillation poses great difficulties. Among those technological innovations that have shown much promise in increasing bioethanol production are metabolic engineering and process automation (Adewuyi 2020). These two technologies have been crucial in enhancing efficiency as well as optimizing biochemical pathways involved in bioethanol synthesis. This is further elevated by incorporating ML, which provides unique opportunities for optimizing production pipelines leading to desired output.

In terms of bioethanol production, there is a synergistic technological framework incorporating metabolic engineering, process automation, and ML that improves productivity and promotes sustainability. Important functions are performed by metabolic engineering where modifications are done on microorganisms such as bacteria and yeasts in order to optimize their metabolic pathways so as to increase the amount of ethanol produced and extend feedstock choices such as lignocellulosic biomass (Bajpai 2016). Despite challenges faced in conserving microbial efficiency through genetic modifications, metabolic engineers have been able to reduce expensive enzyme usage and enhance the growth of these microorganisms under industrial conditions. Such foundational work lays the groundwork for process automation thereby streamlining bioethanol production via regulated automatic systems controlling various stages right from raw material preparation to fermentation (Yousuf et al. 2020). These systems foster uniformity by maintaining an optimal environment that maximizes labor expenses and human errors resulting to the achievement of a reliable and effective bioethanol production process.

The integration of ML makes both metabolic engineering and process automation significantly better (Lawson et al. 2021). In the case of bioethanol generation, ML algorithms may help analyze extensive databases to represent intricate biochemical links, foretell results induced by genetic modifications and adjust fermentation conditions (Bosu et al. 2022). Consequently, with improved predictive knowledge, ML has made it possible to smartly adjust production processes and more dynamically so that the microorganisms are constantly operating at their optimal levels. Thereby, real-time optimization backed by data from automatic sensors and apparatus enhances yields of bioethanol and reduces waste as well as energy costs. Therefore, integrating ML with metabolic engineering and process automation in an intelligent system has the potential to revolutionize the production of bioethanol through constant adaptation, predictive decision-making and increased efficiency (Wang et al. 2022). These three technologies will serve as key catalysts for the further progress of renewable energy since bioethanol production is an ongoing process.

7.5 SYNERGY BETWEEN ML AND METABOLIC ENGINEERING

The blend of metabolic engineering and ML can considerably speed up progress in bioethanol generation through the optimization of microbial pathways. For instance, metabolic engineering genetic and regulatory systems in microorganisms can be altered to produce more of certain metabolites bioethanol (Wu et al. 2016). Nonetheless, it is difficult for researchers to foresee the results of these genetic changes due to the complexity of biological systems combined with a myriad of factors influencing metabolic activities. One possible avenue to addressing this problem lies in machine learning whereby one may model, predict and optimize such pathways with the aid of extensive experimental data sets that have been produced through laboratory work (Wu et al. 2016; Zhang et al. 2020b). The utilization of machine learning algorithms, notably Artificial Neural Networks (ANNs) and Support Vector Machines (SVMs), has been productive in metabolic engineering. The ANNs designed to emulate the brain's anatomy, can learn complex and nonlinear connections among input and output variables (Samanta et al. 2003). Therefore, ANNs are most suitable for processing intricate metabolic networks with highly nonlinear interactions between genes, proteins and metabolites that cannot easily be captured through traditional techniques (Samanta et al. 2003). In metabolic engineering, ANNs may be employed to predict how certain genetic changes will affect microbial performance (e.g., ethanol yield, growth rate, stress tolerance). Through experiential data training of ANNs researchers can make accurate models that forecast optimal genetic configurations to improve bioethanol production.

Support Vector Machines (SVMs) have become yet another powerful ML tool utilized in metabolic engineering. SVMs are especially helpful for tasks about classification, which entail identifying genetic modifications that would result in positive metabolic traits (Samanta et al. 2003). By analyzing large datasets of microbial genetic and phenotypic data, SVMs can classify different genetic modifications based on their impact on bioethanol production. For instance, SVMs can distinguish between genetic modifications that produce high ethanol yield and those which are incapable of doing so (Coşgun et al. 2023). This ability to make differentiations between genetic interventions enables researchers to dedicate their attention to promising ones, hence saving time and other resources necessary for experimental purposes.

The optimization of microbial fermentation processes is a case in point of how ML and metabolic engineering can work together seamlessly. In bioethanol production, fermentation is an important stage where engineered microorganisms such as yeast (*Saccharomyces cerevisiae*) are used to convert sugars into ethanol. For maximum yield, it is necessary to spit out metabolic pathways controlling sugar uptake, glycolysis, and ethanol synthesis processes. Conventional approaches to fermentation optimization rely on laborious experiments which include trial and error. However, ML algorithms can facilitate this approach by predicting which genetic alterations promote fermentation efficiency (Parekh et al. 2000). For instance, researchers can make use of ANNs applied to experimental data in order to model the effects of overexpressing or knocking out specific genes involved in fermentation. These models then serve to identify optimal genetic configurations for achieving the highest ethanol yields under given fermentation circumstances. Alongside fermentation, ML algorithms can also be employed to optimize metabolic pathways involved in exploiting alternative feedstocks including lignocellulosic biomass (Cheng et al. 2023). Bioethanol production from lignocellulose, formed from the walls of plant cells, provides a large and renewable source of sugars; however, its intricate structure prevents microorganisms from breaking it down into ethanol (Chen & Qiu 2010). Metabolic engineering has been utilized in an attempt to redesign microorganisms so that they can degrade lignocellulosic materials effectively and ferment the resulting sugars. Prediction by ML about which genetic modifications will increase the capacity of these microorganisms to metabolize lignocellulosic sugars can enhance this effort (Chen & Qiu 2010). For instance, authors could classify genetic modifications that boost up enzymes involved in lignocellulose degradation using SVMs and hence hasten the development of microbial strains able to make good use of this kind of feedstock.

In addition to genetic modifications, ML can assist in optimizing microbial growth conditions. Some of these environmental factors include temperature, pH, and nutrient availability all of which are very essential for good performance by microbes during bioethanol production (Chen and Fu 2016). Using data from fermentation experiments, ML algorithms can then be used to predict the best environmental conditions that would maximize ethanol yield (Chen & Fu 2016). This process may involve predicting, for example, the best combinations of temperature and pH levels for particular microbial strains or establishing appropriate concentrations of nutrients that will promote growth of such microbes as well as the production of ethanol. However, by integrating information on both genetic and environmental variables, ML covers all aspects of improving bioethanol production. One of the key advantages of ML in metabolic engineering is its ability to manage high-dimensional data. Biological systems are inherently complex, characterized by a multitude of variables that interact in dynamic and often unpredictable ways (Kim et al. 2020). ML algorithms, particularly deep learning models like ANNs, are proficient at analyzing high-dimensional datasets comprising several variables; thus, making them apt for modelling metabolic pathways. With the increasing prevalence of biological data due to advancements in omics technologies (such as genomics, transcriptomics and proteomics), ML can process and examine these vast collections of data in order to discover hidden patterns and relationships that were previously unnoticed (Goshisht, 2024). This capacity to extract valuable insights from extensive databases is significant for metabolic engineers who want to optimize microbial performance for bioethanol production.

7.5.1 Synergy between ML and Process Automation

In the context of bioethanol production, process automation means advanced control systems and technologies are being used to monitor and regulate various stages of the production process—such as feedstock preparation, fermentation, and distillation—with minimal human intervention (Vinitha et al. 2022; Long & Liu 2023). The aim is to enhance efficiency while minimizing operational costs and maintaining consistent quality by ensuring optimal conditions in real time. At present, traditional automation approaches like rule-based programming and static control algorithms used in bioethanol processing plants become less effective as these become more complex (Vyas et al. 2015). This is where the integration of ML with process automation requires a major improvement. Machine learning is able to handle large quantities of real-time data generated by sensors, controllers, and production equipment; hence it allows for dynamic adjustments on operational parameters due to changes in conditions (Khamwachirapithak et al. 2023). In contrast to conventional control systems relying upon fixed set points, ML algorithms can learn continuously from what they see through their eyes all around them progressively adapting so that there will always be an appropriate response for any new piece of information coming up at that time. ML-driven process automation in bioethanol production guarantees optimal operation of the entire production pipeline, enhancing yield, consistency, and overall efficiency (Khamwachirapithak et al. 2023).

One of the most significant areas of applying machine learning in process automation is real-time monitoring and control of fermentation, where microorganisms convert sugars into ethanol (Khamwachirapithak et al. 2023). The process of fermentation is very sensitive with multiple variables like temperature, pH, nutrient concentration, and oxygen levels influencing the microbial action that determines ethanol yield (Tse et al. 2021). Traditional automation systems can regulate these variables partially but they are limited by their static control algorithms which are unable to respond to minute changes in microbial metabolism or environmental fluctuations.

On the other hand, ML models can continuously process sensor data from fermentation tanks and predict when changes should be made in operational parameters. For example, an ML algorithm may observe data about bacterial growth rates, ethanol levels, and sugar consumption patterns to determine when to change the temperature or pH. By dynamically altering these aspects through real-time feedback, ML-guided system guarantees that microorganisms always operate under their best circumstances hence producing maximum ethanol quantity with fewer unwanted by-products (Behera et al. 2019). Such real-time optimization minimizes manual intervention and allows more uniform production results, particularly in large-scale industrial bioethanol plants.

Similarly, ML can improve the distillation process which involves separating and purifying ethanol from fermentation broth (Vane 2008; Khalid et al. 2019). Distillation requires much energy as well as accurate control over temperature and pressure for efficient recovery of ethanol (Khalid et al. 2019). Such models can be integrated with automated distillation systems so that real-time data such as vapor composition, power consumption, and temperature profiles are continually being monitored. In this case, ML algorithms are able to utilize these data types to estimate when temperature or pressure changes must take place in order for optimal separation efficiency to remain intact. It is not only an aid to the final ethanol product purity but also a means of reducing energy consumption, hence it makes the process cheaper and environmentally friendly. Another major benefit of process automation through machine learning (ML) is predictive and preventive maintenance. A typical bioethanol production plant has continuous operations with harsh conditions on pumps, fermenters, and distillation columns (Galbe et al. 2013). However, if these machines fail mechanically or suffer from any malfunction this will cause delays in production, lower efficiency rates, and affect product quality. The trainers can analyze historical data about the performance of different equipment such as sensors and maintenance logs to develop algorithms that predict their eventual failure or malfunction. The models may be used for alerting the operator when specific components are likely to fail or when any machine needs more handling, thus enabling the manager to take preventive measures so that expensive interruptions do not occur. On top of that, ML based predictive maintenance improves equipment life cycle while minimizing unplanned shut downs enabling more reliable production of bioethanol (Sharmila et al. 2024).

The optimization of raw materials and energy resources can also be achieved through ML-driven process automation (Jawad & Balázs 2024). For instance, when it comes to feedstock preparation, ML models can make predictions about the most effective processing methods by analyzing data on the composition and characteristics of raw materials such as lignocellulosic biomass. Maximum amounts of fermentable sugars are extracted from the feedstock by tuning parameters like feedstock input rate, pretreatment intensities (for example, steam explosion, or chemical hydrolysis) and enzyme dosages, as dictated by ML algorithms. Then lesser resource use will help decrease waste, thereby reducing total cost of bioethanol production. An extra layer of optimization is added as machine learning can combine data from all parts of the production chain. The traditional bioethanol production process has each stage (such as feedstock processing, fermentation, distillation etc.) controlled separately, with limited interactions among them. On the other hand, ML models can analyze information from different stages at the same time and find out how changes occurring at one point affect others. For instance, an ML model may recognize that certain adjustments in feedstock processing conditions could lead to an increase in fermentation efficiency although these two stages are separate by space (Wang et al. 2024). Machine learning enables cross-stage optimization making it possible for bioethanol production process to work like a completely closed system wherein every stage is optimized relative to the overall production pipeline.

Moreover, reinforcement learning (RL), being a subset of machine learning (ML) has the potential for automating the processes associated with bioethanol production (Saha et al. 2024). In RL, algorithms learn about the best control measures by interacting with production environment and getting feedback concerning the correlations of different control actions (Saha et al. 2024). Over time, RL algorithms may build very effective control policies that harmonize several objectives like maximizing ethanol yield, reducing energy consumption, and maintaining process stability. Unlike traditional control systems which require heavy tuning by human operators, RL-based systems can always discover optimal strategies automatically through continuous real-time data learning (Saha et al. 2024). This could improve both process efficiency and flexibility since RL algorithms can quickly react to any surprises or disturbances in the course of production.

The combination of ML and process automation signifies a notable breakthrough in bioethanol production. To achieve this, ML algorithms have been designed to monitor and optimally adjust important operational parameters like temperature, pH, or feedstock flow rates. With such dynamic optimization, there will be little human interference, constant production, less wastage, and energy consumption. Furthermore, predictive maintenance is possible through machine learning (ML), which optimizes resource utilization as well as pooling data from all stages of production thereby resulting to highly efficient and sustainable processes for bioethanol manufacture. To satisfy the increasing global demand for more biofuels and their respective applications refreshing this direction is essential in process automation development efforts.

7.6 CHALLENGES AND LIMITATIONS OF MACHINE LEARNING IN BIOETHANOL PRODUCTION

Machine learning (ML) has the potential to enhance bioethanol production processes by optimizing various parameters, improving yield, and facilitating resource management (Owusu and Marfo, 2022). However, several challenges and limitations hinder its effective application in this field such as data availability and quality, issues related to model interpretability, and the need for interdisciplinary collaboration (Ali et al. 2023).

7.6.1 Data Availability and Quality

One of the primary challenges in applying machine learning to bioethanol production is the availability and quality of data. Effective ML models require large datasets that accurately represent the variables influencing bioethanol yield (Wang et al. 2024). However, data on agricultural residues, fermentation conditions, and biochemical processes can be sparse or inconsistent. For instance, studies have shown that the availability of agricultural residues varies significantly by region, impacting the data collected for ML applications (Javaid et al. 2023). Moreover, differences in the ways data are collected, variations in experimental conditions and noise present in biological systems can make datasets poor quality thus hindering high levels of predictive accuracy by ML models. For process automation to be achieved, sensor data of production facilities is vital for real-time monitoring and controlling purposes (Liu et al. 2020).

Scarcity of High-Quality Data coupled with inconsistencies in data collection methods, incomplete dataset, and environmental factors can lead to variability in the dataset quality, directly affects model performance and leading to unreliable model predictions (Munappy et al. 2022). Furthermore, many existing datasets may not capture the full complexity of biological processes involved in bioethanol production. The lack of comprehensive datasets that include diverse feedstock types, fermentation conditions, and microbial interactions limits the ability of ML algorithms to generalize findings across different contexts (Wang et al. 2024). Thus, improving data collection methodologies and ensuring high-quality datasets are crucial for advancing ML applications in bioethanol production (Lawson et al. 2021).

7.6.2 Issues Related to Model Interpretability

Another significant limitation of machine learning in bioethanol production is model interpretability. Many ML algorithms, particularly deep learning models, operate as "black boxes," making it challenging for researchers to understand how decisions are made based on input data (Patnaik et al. 2024). For example, deep learning and artificial neural networks (ANN) are complex algorithms which are hard to interpret even though these models can predict outcomes and optimize processes with high accuracies due to the elusive rationale behind the decision-making process of the ANNs (Abdolrasol et al. 2021). This lack of transparency can hinder trust in ML predictions and complicate the validation of models against experimental results where regulatory compliance and process optimization are critical (Barbierato & Gatti, 2023). In metabolic engineering, researchers find it difficult to determine why certain models suggest some metabolic pathway optimizations or genetic modifications. If there are no clear explanations then biotechnologists will rely on the model's predictions less in guiding experiments (Cheng et al. 2023).

In the context of bioethanol production, stakeholders need to comprehend the rationale behind ML-driven recommendations for process optimization. The inability to explain how certain inputs affect outputs limits trust in ML applications among researchers and industry professionals (Gardas & Narwane 2024). For example, understanding which specific parameters (e.g., temperature, pH levels) most significantly impact ethanol yield is essential for practical implementation. Therefore, developing interpretable models or employing techniques that enhance model transparency is vital for fostering confidence among practitioners and facilitating the adoption of ML technologies in bioethanol production (Khamwachirapithak et al. 2023).

7.6.3 Need for Interdisciplinary Collaboration

Addressing the challenges associated with machine learning in bioethanol production necessitates interdisciplinary collaboration among experts from various fields (Wang et al. 2024). Successful integration of ML into bioprocessing requires knowledge from domains such as microbiology, chemical engineering, data science, and environmental science (Helleckes et al. 2023). For instance, insights from microbiologists can inform data collection on microbial behavior during fermentation processes, while chemical engineers can provide expertise on optimizing bioreactor designs (Jiang et al. 2022). Therefore, each discipline brings unique insights that are essential for developing comprehensive models that account for biological variability and process dynamics. Collaborative efforts can also enhance the development of robust datasets that reflect real-world conditions encountered during bioethanol production. This collaborative approach will not only improve the efficacy of machine learning applications but also promote innovation in sustainable bioethanol production practices (Wang et al. 2024).

7.6.4 Scalability of Models

Many machine learning models developed in laboratory settings may not scale effectively to industrial applications due to differences in operational conditions and constraints encountered at larger scales (Lwakatare et al. 2020; Khuat et al. 2024). The scalability of machine learning models in bioethanol production presents several challenges and limitations. One primary issue is the data heterogeneity involved in bioethanol production processes, which include diverse biomass types, varying fermentation conditions, and different enzymatic activities (Coşgun et al. 2023). This diversity makes it difficult to create a one-size-fits-all model that can be effectively scaled across different production setups. While ML algorithms can effectively analyze small datasets and optimize specific processes, their performance may deteriorate when applied to larger, more complex datasets typical in bioethanol production (Long & Liu, 2023). As the volume of data increases, issues such as computational resource limitations and longer processing times can hinder model training and deployment (Singh 2023). Additionally, the diversity of feedstock types and variations in production conditions complicate the generalization of ML models, making it difficult to create universally applicable solutions. Ensuring that models remain accurate and efficient across different scales requires continuous retraining and fine-tuning, which can be resource-intensive (Lawson et al. 2021; Akinpelu et al. 2023). Furthermore, integrating ML solutions into existing production systems poses logistical challenges, as it necessitates alignment with current operational practices and infrastructure. Addressing these scalability issues is crucial for realizing the full potential of ML in enhancing bioethanol production processes.

7.6.5 Regulatory and Ethical Considerations

The use of machine learning in bioethanol production must navigate regulatory frameworks that govern data usage, especially when involving sensitive information or proprietary technologies (Lawson et al. 2021). Ethical considerations regarding data privacy, ownership, security, and the implications of automated decision-making also pose challenges (Peng et al. 2022). For instance, production facilities collect sensitive information often belonging to owners' proprietary processes making it necessary for the use of strict data governance protocols (Osasona et al. 2024). On top of that, if an ML based system fails—for example when fermentation parameters are adjusted inaccurately—this could lead to expensive losses or safety hazards during production operations (Khuat et al. 2024). Therefore, there is a need for stringent regulations and guidelines which aim at ensuring safe and responsible deployment of machine learning in bioethanol production.

7.6.6 Integration with Existing Systems

Many bioethanol production plants have already conventional automation systems that operate based on predefined control algorithms which makes the integration of ML-driven systems within such existing infrastructures a technical challenge (Bušić et al. 2018; Peng et al. 2022). To retrofit an established plant for advanced algorithm-based ML, substantial amounts of money must be invested into new sensors, control systems and data storage capabilities (Soori et al. 2022). Also, specialized interfaces and software may be needed to make traditional automation systems compatible with ML models which complicates the integration process (Mumuni & Mumuni 2024).

7.7 Future Prospects of Machine Learning and Its Application in Bioethanol Production

7.7.1 Potential Advancements in ML Applications for Bioethanol Production

Machine learning (ML) has the potential to revolutionize bioethanol production by optimizing various stages of the process, from biomass selection to fermentation (Sharma et al. 2023). Here are some key advancements:

Enhanced Biomass Selection and Pretreatment: ML algorithms can analyze large datasets to identify the most efficient biomass sources for bioethanol production. This includes evaluating the cellulose, hemicellulose, and lignin content of different biomass types. Advanced ML models can predict the optimal pretreatment methods to maximize the yield of fermentable sugars. This involves optimizing parameters such as temperature, pH, and enzyme concentrations (Pushpa et al. 2024).

Feedstock Optimization: ML algorithms can analyze vast amounts of data on soil health, weather patterns, and nutrient levels to predict optimal conditions for feedstock growth. This can lead to increased yields and more efficient use of resources (Elbasi et al. 2022).

Process Modeling and Fermentation Optimization: ML models can simulate the complex chemical reactions and heat/mass transfer mechanisms involved in bioethanol production. ML can be used to model and control the fermentation process, ensuring optimal conditions for microbial activity (Caro et al. 2024). This includes monitoring and adjusting factors like pH, temperature, and nutrient levels in real time. Predictive models can forecast fermentation outcomes based on initial conditions, allowing for adjustments to be made proactively to improve yield and efficiency (Khuat et al. 2024).

Process Integration and Automation: Integrating ML with process control systems can lead to fully automated bioethanol production facilities. This reduces human error and increases consistency in production (Bušić et al. 2018; Arias et al. 2023). ML algorithms can continuously learn and adapt to changing conditions, improving the overall efficiency and sustainability of the production process (Mana et al. 2024).

Waste Minimization and By-product Utilization: ML can help in identifying and minimizing waste streams in bioethanol production. This includes optimizing the use of by-products such as lignin for other industrial applications (Ge et al. 2023). Advanced data analytics can uncover new ways to utilize waste materials, turning them into valuable co-products and reducing the environmental impact of bioethanol production (Broda et al. 2022).

Supply Chain Optimization: ML can be used to optimize the bioethanol supply chain, from feedstock procurement to distribution. By analyzing data on transportation costs, storage requirements, and market demand, ML algorithms can help producers make more informed decisions and reduce waste (Odimarha et al. 2024).

Sustainability Improvements: ML can contribute to the sustainability of bioethanol production by identifying opportunities for energy efficiency, waste reduction, and emissions control (Long & Liu 2023). For example, ML models can analyze data on energy consumption and greenhouse gas emissions to pinpoint areas for improvement.

Quality Control: ML can be used to monitor and control the quality of bioethanol throughout the production process. By analyzing real-time data on parameters like purity, density, and octane rating, ML algorithms can quickly identify and correct any deviations, ensuring consistent product quality (Aghbashlo et al. 2021).

7.7.2 Importance of Continued Research and Development

Continued research and development (R&D) in ML applications for bioethanol production are crucial for several reasons.

Addressing Technical Challenges: Bioethanol production involves complex biochemical processes that are influenced by numerous variables. Ongoing

R&D is needed to refine ML models and improve their accuracy in predicting and optimizing these processes (Aghbashlo et al. 2021). Research can help in developing new ML algorithms that are better suited to handle the specific challenges of bioethanol production, such as nonlinear interactions between variables (Wang et al. 2024).

Enhancing Economic Viability: Continued R&D can lead to cost reductions in bioethanol production by improving process efficiency and reducing waste. This makes bioethanol a more competitive alternative to fossil fuels (Bušić et al. 2018). Investment in R&D can also lead to the discovery of new, cost-effective biomass sources, and pretreatment methods, further lowering production costs (Rajagopal et al. 2009).

Promoting Environmental Sustainability: Research can help in developing ML models that optimize bioethanol production processes to minimize environmental impact. This includes reducing greenhouse gas emissions and improving the overall sustainability of the production process (Momenitabar et al. 2023). Continued R&D can also lead to the development of new technologies for utilizing waste materials and by-products, further enhancing the environmental benefits of bioethanol production (Bušić et al. 2018).

Driving Innovation: Ongoing research fosters innovation by encouraging the development of new technologies and approaches. This can lead to breakthroughs that significantly improve the efficiency and sustainability of bioethanol production (Carrillo-Nieves et al. 2019). Collaboration between researchers, industry, and policymakers can accelerate the adoption of ML technologies in bioethanol production, driving further innovation and progress (Arias et al. 2023).

Feedstock Diversification: While corn and sugarcane are currently the most common feedstocks for bioethanol production, there is a growing interest in using alternative feedstocks like lignocellulosic biomass and agricultural waste (Bušić et al. 2018). Continued R&D is needed to develop efficient conversion processes for these feedstocks and optimize their use in conjunction with ML technologies.

Regulatory Compliance: As governments around the world implement stricter regulations on greenhouse gas emissions and biofuel mandates, the bioethanol industry must adapt to remain compliant (Saini et al. 2021). R&D efforts focused on improving the sustainability and efficiency of bioethanol production can help producers stay ahead of regulatory changes and maintain their competitive edge.

7.7.3 The Role of ML in Driving Innovation and Sustainability in the Biofuel Industry

Machine learning plays a pivotal role in driving innovation and sustainability in the biofuel industry, particularly in bioethanol production (Tse et al. 2021):

Innovation in Process Optimization: ML enables the development of advanced process optimization techniques that can significantly improve the efficiency of bioethanol production (Momenitabar et al. 2023). This includes optimizing fermentation conditions, enzyme usage, and biomass pretreatment methods. Predictive analytics powered by ML can help in identifying potential issues before they arise, allowing for proactive adjustments and continuous improvement of the production process (Soori et al. 2022).

Sustainability through Resource Efficiency: ML can optimize the use of resources in bioethanol production, reducing waste and improving the overall sustainability of the process (Okolie 2024). This includes optimizing water and energy usage, as well as minimizing the environmental impact of waste streams. By improving the efficiency of bioethanol production, ML can help in reducing the reliance on fossil fuels and promoting the use of renewable energy sources (Tse et al. 2021).

Data-Driven Decision-Making: ML provides valuable insights through data analysis, enabling informed decision-making in bioethanol production (Akinpelu et al. 2023). This includes identifying trends, predicting outcomes, and optimizing processes based on data-driven insights. Data-driven decision-making can lead to more efficient and sustainable production practices, as well as the development of new technologies and approaches (Pantović et al. 2023).

Collaboration and Knowledge Sharing: ML facilitates collaboration and knowledge sharing among researchers, industry professionals, and policymakers. This can lead to the development of best practices and the adoption of new technologies in bioethanol production (Bušić et al. 2018; Arias et al. 2023). Collaborative efforts can accelerate the pace of innovation and drive the adoption of sustainable practices in the biofuel industry.

Competitiveness: As the biofuel industry continues to evolve, producers who embrace ML technologies will be better positioned to compete in the global market (Javaid et al. 2023). By optimizing their production processes and supply chains, ML-enabled producers can reduce costs, improve product quality, and respond more quickly to changing market demands (Adenekan et al. 2024).

7.8 CONCLUSION

In conclusion, the integration of machine learning (ML) into bioethanol production represents a significant advancement in the quest for sustainable energy solutions. The role of ML is pivotal, as it enhances the efficiency of various stages in bioethanol production, from biomass pretreatment to fermentation and conversion processes. By employing sophisticated algorithms, such as artificial neural networks and regression trees, researchers and industry practitioners can predict outcomes with remarkable accuracy, thereby optimizing operational parameters and maximizing ethanol yields.

The diverse applications of ML in bioethanol production extend beyond mere predictive modeling. These technologies facilitate real-time monitoring and control of production processes, allowing for adaptive management that responds to changing conditions. Furthermore, the integration of ML with other technologies—such as enzymatic hydrolysis systems and life cycle assessments—creates a holistic approach that not only improves efficiency but also promotes sustainability by minimizing resource consumption and environmental impact.

However, despite these promising advancements, several challenges remain. Issues related to data acquisition, model validation, and the scalability of ML applications must be addressed to unlock their full potential. As the field evolves, ongoing research is crucial to refine existing algorithms and expand their applicability across various biomass sources. This will ensure that ML techniques can be effectively utilized in diverse contexts, enhancing their relevance in the bioethanol sector.

The transformative potential of machine learning in bioethanol production underscores the growing importance of sustainable energy solutions in today's world. As global energy demands continue to rise, leveraging advanced technologies like ML will be essential for developing efficient and environmentally friendly alternatives to fossil fuels. Therefore, it is imperative to advocate for further exploration and integration of ML techniques within bioethanol production processes. By doing so, we can pave the way for a more sustainable future that balances energy needs with ecological preservation.

REFERENCES

Abdolrasol, M. G. M., Hussain, S. M. S., Ustun, T. S., Sarker, M. R., Hannan, M. A., Mohamed, R., Ali, J. A., Mekhilef, S., & Milad, A. (2021). Artificial neural networks based optimization techniques: A review. *Electronics*, *10*(21), 2689. https://doi.org/10.3390/electronics10212689

Adebami G. E., Kuila A., Ajunwa O. M., Fasiku S. A., & Asemoloye M. D. (2021). Genetics and metabolic engineering of yeast strain development for efficient ethanol production. *Journal of Food Process Engineering,* e13798, 1–30. https://doi.org/10.1111/jfpe.13798.

Adebami, G. E., & Adebayo-Tayo, B. C. (2020). Development of cellulolytic strain by genetic engineering approach for enhanced cellulase production. In *Genetic and Metabolic Engineering for Improved Biofuel Production from Lignocellulosic Biomass* (pp. 103–136). Amsterdam, Netherland: Elsevier. https://doi.org/10.1016/B978-0-12-817953-6.00008-7.

Adegboye, M.F., Ojuederie, O.B., Talia, P.M. and Babalola, O.O., 2021. Bioprospecting of microbial strains for biofuel production: Metabolic engineering, applications, and challenges. *Biotechnology for Biofuels*, *14*(1), 5. https://doi.org/10.1186/s13068-020-01853-2

Adenekan O.A., Solomon N.O., Simpa P., & Obasi S.C. (2024). Enhancing manufacturing productivity: A review of AI-driven supply chain management optimization and ERP systems integration. *International Journal of Management & Entrepreneurship Research*, 6(5), 1607–1624, https://doi.org/10.51594/ijmer.v6i5.1126.

Adewuyi, A. (2020). Challenges and prospects of renewable energy in Nigeria: A case of bioethanol and biodiesel production. *Energy Reports*, *6*(4), 77–88. https://doi.org/10.1016/j.egyr.2019.12.002

Aghbashlo, M., Peng, W., Tabatabaei, M., Kalogirou, S. A., Soltanian, S., Hosseinzadeh-Bandbafha, H., Mahian, O., & Lam, S. S. (2021). Machine learning technology in biodiesel research: A review. *Progress in Energy and Combustion Science*, *85*, 100904. https://doi.org/10.1016/j.pecs.2021.100904.

Ahlawat, Y.K., Srivastava, V., Samani, M., Harun, S., Rajuloo, V., & Chaudhary, D. (2024). Computational strategies for maximizing biomass and metabolite yields for bioproduction. In: *Medicinal and Aromatic Plants: Current Research Status, Value-Addition to Their Waste, and Agro-Industrial Potential* (Vol I, pp. 143–161). Cham: Springer Nature Switzerland. https://doi.org/10.1007/978-3-031-60117-0_11

Ahmadian-Moghadam, H., Elegado, F. B., & Nayve, R. (2013). Prediction of ethanol concentration in biofuel production using artificial neural networks. *American Journal of Modeling and Optimization*, *1*(3), 31–35. https://doi.org/10.12691/ajmo-1-3-2

Akinpelu, D. A., Adekoya, O. A., Oladoye, P. O., Ogbaga, C. C., & Okolie, J. A. (2023). Machine learning applications in biomass pyrolysis: From biorefinery to end-of-life product management. *Digital Chemical Engineering*, *8*, 100103. https://doi.org/10.1016/j.dche.2023.100103.

Aldrich, C., & Auret, L. (2013). *Unsupervised Process Monitoring and Fault Diagnosis with Machine Learning Methods*. *16*(3), 593–606. https://doi.org/10.1007/978-1-4471-5185-2

Ali, Y., Hussain, F., & Haque, M. M. (2023). Advances, challenges, and future research needs in machine learning-based crash prediction models: A systematic review. *Accident Analysis & Prevention*, *194*, 107378. https://doi.org/10.1016/j.aap.2023.107378

Ardabili, S., Mosavi, A., & Várkonyi-Kóczy, A. R. (2019, September). Systematic review of deep learning and machine learning models in biofuels research. In *International Conference on Global Research and Education* (pp. 19–32). Cham: Springer International Publishing. https://doi.org/10.1007/978-3-030-36841-8_2

Arias, A., Feijoo, G., & Moreira, M. T. (2023). How could Artificial Intelligence be used to increase the potential of biorefineries in the near future? A review. *Environmental Technology & Innovation*, *32*, 103277. https://doi.org/10.1016/j.eti.2023.103277.

Bajpai, P. (2016). Structure of lignocellulosic biomass. *SpringerBriefs in Molecular Science*, 7–12. https://doi.org/10.1007/978-981-10-0687-6_2

Barbierato, E., & Gatti, A. (2023). The challenges of machine learning: A critical review. *Electronics*, *13*(2), 416. https://doi.org/10.3390/electronics13020416.

Behera, S. S., Ray, R. C., Das, U., Panda, S. K., & Saranraj, P. (2019). Microorganisms in fermentation. *Essentials in Fermentation Technology*, 1–39. https://doi.org/10.1007/978-3-030-16230-6_1

Bhardwaj, A., Kishore, S., & Pandey, D. K. (2022). Artificial intelligence in biological sciences. *Life*, *12*(9), 1–20. https://doi.org/10.3390/life12091430.

Bosu, S., Pooja, R. P., & Rajasimman, M. (2022). Role of nanomaterials in enhanced ethanol production through biological methods – Review on operating factors and machine learning applications. *Fuel*, *320*, 123905. https://doi.org/10.1016/j.fuel.2022.123905

Broda, M., Yelle, D. J., & Serwańska, K. (2022). Bioethanol production from lignocellulosic biomass-challenges and solutions. *Molecules, 27*(24), 8717. https://doi.org/10.3390/molecules27248717

Burugari, V. K., Selvaraj, P., Praneel, V., Kondaveeti, H. K., & Kumar, M. P. (2020). The application of computational modeling for the optimization of bio fuel production processes. International *Journal of Advanced Trends in Computer Science and Engineering, 9*(5) 7883–7893. https://doi.org/10.30534/ijatcse/2020/140952020

Bušić, A., Marđetko, N., Kundas, S., Morzak, G., Belskaya, H., Ivančić Šantek, M., Komes, D., Novak, S., & Šantek, B. (2018). Bioethanol production from renewable raw materials and its separation and purification: A review. *Food Technology and Biotechnology*, *56*(3), 289–311. https://doi.org/10.17113/ftb.56.03.18.5546

Caro, I., Marzo-Gago, C., Díaz, A. B., & Blandino, A. (2024). Modelling and optimization of simultaneous saccharification and fermentation of agro-food residues. *Journal of Environmental Chemical Engineering*, *12*(1), 111862. https://doi.org/10.1016/j.jece.2023.111862.

Carrillo-Nieves, D., Rostro Alanís, M. J., De la Cruz Quiroz, R., Ruiz, H. A., Iqbal, H. M., & Parra-Saldívar, R. (2019). Current status and future trends of bioethanol production from agro-industrial wastes in Mexico. *Renewable and Sustainable Energy Reviews*, *102*, 63–74. https://doi.org/10.1016/j.rser.2018.11.031.

Chen, H., & Fu, X. (2016). Industrial technologies for bioethanol production from lignocellulosic biomass. *Renewable and Sustainable Energy Reviews*, *57*, 468–478. https://doi.org/10.1016/j.rser.2015.12.069

Chen, H., & Qiu, W. (2010). Key technologies for bioethanol production from lignocellulose. *Biotechnology Advances*, *28*(5), 556–562. https://doi.org/10.1016/j.biotechadv.2010.05.005

Chen, M., & Xia, J. (2024). Optimization parameters for efficient scale-up of fermentation process. In *Scale-Up and Chemical Process for Microbial Production of Plant-Derived Bioactive Compounds* (pp. 29–42). Academic Press. https://doi.org/10.1016/B978-0-443-15584-0.00002-1

Cheng, Y., Bi, X., Xu, Y., Liu, Y., Li, J., Du, G., Xueqin, L., & Liu, L. (2023). Machine learning for metabolic pathway optimization: A review. *Computational and Structural Biotechnology Journal*, *21*, 2381–2393. https://doi.org/10.1016/j.csbj.2023.03.045

Coşgun, A., Günay, M. E., & Yıldırım, R. (2023). A critical review of machine learning for lignocellulosic ethanol production via fermentation route. *Biofuel Research Journal*, *10*(2), 1859–1875. https://doi.org/10.18331/BRJ2023.10.2.5

Damayanti, A., Bahlawan, Z. A. S., & Kumoro, A. C. (2022). Modeling of bioethanol production through glucose fermentation using Saccharomyces cerevisiae immobilized on sodium alginate beads. *Cogent Engineering*, *9*(1), 2049438. https://doi.org/10.1080/23311916.2022.2049438

Deshavath, N. N., Veeranki, V. D., & Goud, V. V. (2019). Lignocellulosic feedstocks for the production of bioethanol: Availability, structure, and composition. In *Sustainable Bioenergy* (pp. 1–19). Elsevier. https://doi.org/10.1016/B978-0-12-817654-2.00001-0

Desmal, H. A., Abdallah, M., Inayat, A., & Shanableh, A. (2022). Optimization of ethanol fermentation using a hybrid model of support vector machine and genetic algorithm. *Advances in Science and Engineering Technology International Conferences (ASET)*, 1–5. https://doi.org/10.1109/ASET53988.2022.9734876.

Elbasi, E., Zaki, C., Topcu, A. E., Abdelbaki, W., Zreikat, A. I., Cina, E., Shdefat, A., & Saker, L. (2022). Crop prediction model using machine learning algorithms. *Applied Sciences*, *13*(16), 9288. https://doi.org/10.3390/app13169288.

Fischer, J., Lopes, V. S., Cardoso, S. L., Coutinho, U. and Cardoso, V. L. (2017). Machine learning techniques applied to lignocellulosic ethanol in simultaneous hydrolysis and fermentation. *Brazilian Journal of Chemical Engineering*, *34*, 53–63. https://doi.org/10.1590/0104-6632.20170341s20150475

Fisher, O. J., Watson, N. J., Escrig, J. E., Witt, R., Porcu, L., Bacon, D., Rigley, M., & Gomes, R. L. (2020). Considerations, challenges and opportunities when developing data-driven models for process manufacturing systems. *Computers & Chemical Engineering*, *140*, 106881. https://doi.org/10.1016/j.compchemeng.2020.106881

Galbe, M., Wallberg, O., & Zacchi, G. (2013). Cellulosic bioethanol production. *Separation and Purification Technologies in Biorefineries*, 487–501. https://doi.org/10.1002/9781118493441.ch18

Gardas, R., & Narwane, S. (2024). An analysis of critical factors for adopting machine learning in manufacturing supply chains. *Decision Analytics Journal*, *10*, 100377. https://doi.org/10.1016/j.dajour.2023.100377.

Ge, H., Zheng, J., & Xu, H. (2023). Advances in machine learning for high value-added applications of lignocellulosic biomass. *Bioresource Technology*, *369*, 128481. https://doi.org/10.1016/j.biortech.2022.128481

Gong, Y., Liu, G., Xue, Y., Li, R., & Meng, L. (2023). A survey on dataset quality in machine learning. *Information and Software Technology*, *162*, 107268. https://doi.org/10.1016/j.infsof.2023.107268.

Goshisht, M. K. (2024). Machine learning and deep learning in synthetic biology: Key architectures, applications, and challenges. ACS omega, 9(9), 9921–9945. https://doi.org/10.1021/acsomega.3c05913

Haldar, D., Shabbirahmed, A. M., & Mahanty, B. (2023). Multivariate regression and artificial neural network modelling of sugar yields from acid pretreatment and enzymatic hydrolysis of lignocellulosic biomass. *Bioresource Technology*, *370*, 128519. https://doi.org/10.1016/j.biortech.2022.128519

Hansen, S., & Mirkouei, A. (2018). Past infrastructures and future machine intelligence (MI) for biofuel production: a review and MI-based framework. In *International Design Engineering Technical Conferences and Computers and Information in Engineering Conference* (Vol. 51791, p. V004T05A022). American Society of Mechanical Engineers. https://doi.org/10.1115/DETC2018-86150

Helleckes, L. M., Hemmerich, J., Wiechert, W., Von Lieres, E., & Grünberger, A. (2023). Machine learning in bioprocess development: From promise to practice. *Trends in Biotechnology*, *41*(6), 817–835. https://doi.org/10.1016/j.tibtech.2022.10.010.

Huntington, T., Baral, N. R., Yang, M., Sundstrom, E., & Scown, C. D. (2023). Machine learning for surrogate process models of bioproduction pathways. *Bioresource Technology*, *370*, 128528. https://doi.org/10.1016/j.biortech.2022.128528

İşçen, A., Öznacar, K., Tunç, K. M., & Günay, M.E., (2023). Exploring the critical factors of biomass pyrolysis for sustainable fuel production by machine learning. *Sustainability*, *15*(20), 14884. https://doi.org/10.3390/su152014884

Jafari-Marandi, R., (2021). Supervised or unsupervised learning? Investigating the role of pattern recognition assumptions in the success of binary predictive prescriptions. *Neurocomputing*, *434*, 165–193. https://doi.org/10.1016/j.neucom.2020.12.063

Javaid, M., Haleem, A., Khan, I. H., & Suman, R. (2023). Understanding the potential applications of Artificial Intelligence in Agriculture Sector. *Advanced Agrochem*, *2*(1), 15–30. https://doi.org/10.1016/j.aac.2022.10.001

Javaid, M., Haleem, A., Pratap Singh, R., Suman, R., & Rab, S. (2021). Significance of machine learning in healthcare: Features, pillars and applications. *International Journal of Intelligent Networks*, *3*, 58–73. https://doi.org/10.1016/j.ijin.2022.05.002.

Jawad, Z. N., & Balázs, V. (2024). Machine learning-driven optimization of enterprise resource planning (ERP) systems: a comprehensive review. *Beni-Suef University Journal of Basic and Applied Sciences*, *13*(1). https://doi.org/10.1186/s43088-023-00460-y

Jeon, P. R., Moon, J. H., Ogunsola, N. O., Lee, S. H., Ling, J. L. J., You, S., & Park, Y. K., (2023). Recent advances and future prospects of thermochemical biofuel conversion processes with machine learning. *Chemical Engineering Journal*, *471*, 144503. https://doi.org/10.1016/j.cej.2023.144503

Jiang, Y., Luo, J., Huang, D., Liu, Y., & Li, D. (2022). Machine learning advances in microbiology: A review of methods and applications. *Frontiers in Microbiology*, *13*, 925454. https://doi.org/10.3389/fmicb.2022.925454.

Kenny, E. M., Ruelle, E., Keane, M. T., & Shalloo, L. (2024). A hybrid model that combines machine learning and mechanistic models for useful grass growth prediction. *Computers and Electronics in Agriculture*, *219*, 108805. https://doi.org/10.1016/j.compag.2024.108805

Khaleghi, M. K., Savizi, I. S. P., Lewis, N. E., & Shojaosadati, S. A. (2021). Synergisms of machine learning and constraint-based modeling of metabolism for analysis and optimization of fermentation parameters. *Biotechnology Journal*, *16*(11), 2100212. https://doi.org/10.1002/biot.202100212

Khalid, A., Aslam, M., Qyyum, M. A., Faisal, A., Khan, A. L., Ahmed, F., et al. (2019). Membrane separation processes for dehydration of bioethanol from fermentation broths: Recent developments, challenges, and prospects. *Renewable and Sustainable Energy Reviews*, *105*, 427–443. https://doi.org/10.1016/j.rser.2019.02.002

Khamwachirapithak, P., Sae-Tang, K., Mhuantong, W., Tanapongpipat, S., Zhao, X., Liu, C., et al. (2023). Optimizing ethanol production in *Saccharomyces cerevisiae* at Ambient and Elevated temperatures through machine learning-guided combinatorial promoter modifications. *ACS Synthetic Biology*, *12*(10), 2897–2908. https://doi.org/10.1021/acssynbio.3c00199

Khuat, T. T., Bassett, R., Otte, E., Grevis-James, A., & Gabrys, B. (2024). Applications of machine learning in antibody discovery, process development, manufacturing and formulation: Current trends, challenges, and opportunities. *Computers & Chemical Engineering*, *182*, 108585. https://doi.org/10.1016/j.compchemeng.2024.108585.

Kim, Y., Kim, G. B., & Lee, S.Y. (2021). Machine learning applications in genome-scale metabolic modeling. *Current Opinion in Systems Biology*, *25*, 42–49. https://doi.org/10.1016/j.coisb.2021.03.001.

Kim, G. B., Kim, W. J., Kim, H. U., & Lee, S. Y. (2020). Machine learning applications in systems metabolic engineering. *Current Opinion in Biotechnology*, *64*, 1–9. https://doi.org/10.1016/j.copbio.2019.08.010

Kundu, P., Beura, S., Mondal, S., Das, A.K. and Ghosh, A., (2024). Machine learning for the advancement of genome-scale metabolic modeling. *Biotechnology Advances*, 108400. https://doi.org/10.1016/j.biotechadv.2024.108400

Lawson, C. E., Martí, J. M., Radivojevic, T., Jonnalagadda, S. V. R., Gentz, R., Hillson, N. J., et al. (2021). Machine learning for metabolic engineering: A review. *Metabolic Engineering*, *63*, 34–60. https://doi.org/10.1016/j.ymben.2020.10.005

Liu, Y., Cao, Y., Du, G., & Liu, L. (2020). Systems and synthetic metabolic engineering: Challenges and prospects. *Systems and Synthetic Metabolic Engineering*, 237–264.

Long, F., & Liu, H. (2023). An integration of machine learning models and life cycle assessment for lignocellulosic bioethanol platforms. *Energy Conversion and Management*, *292*, 117379–117379. https://doi.org/10.1016/j.enconman.2023.117379

Lwakatare, L. E., Raj, A., Crnkovic, I., Bosch, J., & Olsson, H. H. (2020). Large-scale machine learning systems in real-world industrial settings: A review of challenges and solutions. *Information and Software Technology*, *127*, 106368. https://doi.org/10.1016/j.infsof.2020.106368.

Maity, S., & Mallick, N. (2022). Trends and advances in sustainable bioethanol production by marine microalgae: A critical review. *Journal of Cleaner Production*, *345*, 131153. https://doi.org/10.1016/j.jclepro.2022.131153

Maldonado, S., Weber, R., & Famili, F. (2014). Feature selection for high-dimensional class-imbalanced data sets using support vector machines. *Information Sciences*, *286*, 228–246. https://doi.org/10.1016/j.ins.2014.07.015

Mana, A., Allouhi, A., Hamrani, A., Rehman, S., El Jamaoui, I., & Jayachandran, K. (2024). Sustainable AI-based production agriculture: Exploring AI applications and implications in agricultural practices. *Smart Agricultural Technology*, *7*, 100416. https://doi.org/10.1016/j.atech.2024.100416.

Momenitabar, M., Dehdari Ebrahimi, Z., Abdollahi, A., Helmi, W., Bengtson, K., & Ghasemi, P. (2023). An integrated machine learning and quantitative optimization method for designing sustainable bioethanol supply chain networks. *Decision Analytics Journal*, *7*, 100236. https://doi.org/10.1016/j.dajour.2023.100236.

Mondal, P. P., Galodha, A., Verma, V. K., Singh, V., Show, P. L., Awasthi, M. K., ... & Jain, R. (2023). Review on machine learning-based bioprocess optimization, monitoring, and control systems. *Bioresource Technology, 370*, 128523. https://doi.org/10.1016/j.biortech.2022.128523

Müller, D. F., Wibbing, D., Herwig, C., & Kager, J., (2023). Simultaneous real-time estimation of maximum substrate uptake capacity and yield coefficient in induced microbial cultures. *Computers & Chemical Engineering, 173*, 108203. https://doi.org/10.1016/j.compchemeng.2023.108203.

Mumuni, A., & Mumuni, F. (2024). Automated data processing and feature engineering for deep learning and big data applications: A survey. *Journal of Information and Intelligence, 3*(2), 113–153. https://doi.org/10.1016/j.jiixd.2024.01.002.

Munappy, A. R., Bosch, J., Olsson, H. H., Arpteg, A., & Brinne, B. (2022). Data management for production quality deep learning models: Challenges and solutions. *Journal of Systems and Software, 191*, 111359. https://doi.org/10.1016/j.jss.2022.111359

Naveed, M.H., Khan, M.N.A., Mukarram, M., Naqvi, S.R., Abdullah, A., Haq, Z.U. et al. (2024). Cellulosic biomass fermentation for biofuel production: Review of artificial intelligence approaches. *Renewable and Sustainable Energy Reviews, 189*, 113906. https://doi.org/10.1016/j.rser.2023.113906

Niaze, A. A., Sahu, R., Sunkara, M. K., & Upadhyayula, S. (2023). Model construction and optimization for raising the concentration of industrial bioethanol production by using a data-driven ANN model. *Renewable Energy, 216*, 119031. https://doi.org/10.1016/j.renene.2023.119031

Odimarha, A. C., Ayodeji, S. A., & Abaku, E. A. (2024). Machine learning's influence on supply chain and logistics optimization in the oil and gas sector: A comprehensive analysis. *Computer Science & IT Research Journal, 5*(3), 725–740. https://doi.org/10.51594/csitrj.v5i3.976.

Okolie, J. A. (2024). Introduction of machine learning and artificial intelligence in biofuel technology. *Current Opinion in Green and Sustainable Chemistry, 47*, 100928. https://doi.org/10.1016/j.cogsc.2024.100928

Onu, C. E., Igbokwe, P. K., Nwabanne, J. T., & Ohale, P. E. (2022). ANFIS, ANN, and RSM modeling of moisture content reduction of cocoyam slices. *Journal of Food Processing and Preservation*, 46, e16032. https://doi.org/10.1111/jfpp.16032.

Osasona, F., Amoo, O. O., Atadoga, A., Abrahams, T. O., Farayola, O. A., & Ayinla, B. S. (2024). Reviewing the ethical implications of AI in decision making processes. *International Journal of Management & Entrepreneurship Research, 6*(2), 322–335. https://doi.org/10.51594/ijmer.v6i2.773

Owusu, W. A., & Marfo, S. A. (2022). Artificial intelligence application in bioethanol production. *International Journal of Energy Research, 2023*(1), 7844835. https://doi.org/10.1155/2023/7844835

Pantović, V., Vidojević, D., Vujičić, S., & Sofijanić, S. (2023). Data-driven decision making for sustainable IT project management excellence. *Sustainability, 16*(7), 3014. https://doi.org/10.3390/su16073014.

Parekh, S., Vinci, V. A., & Strobel, R. J. (2000). Improvement of microbial strains and fermentation processes. *Applied Microbiology and Biotechnology, 54*(3), 287–301. https://doi.org/10.1007/s002530000403

Patnaik, S., Khatri, N., & Rene, E. R. (2024). Fueling the future: Exploring the synergy of artificial intelligence-based algorithms and the use of biofuels in engine development. *Journal of the Taiwan Institute of Chemical Engineers*, 105729. https://doi.org/10.1016/j.jtice.2024.105729

Patra, P., Disha, B.R., Kundu, P., Das, M., & Ghosh, A., (2023). Recent advances in machine learning applications in metabolic engineering. *Biotechnology Advances, 62*, 108069. https://doi.org/10.1016/j.biotechadv.2022.108069

Pattnaik, D., Ray, S., & Raman, R. (2024). Applications of artificial intelligence and machine learning in the financial services industry: A bibliometric review. *Heliyon, 10*(1), e23492. https://doi.org/10.1016/j.heliyon.2023.e23492

Peng, J., Schwalbe-Koda, D., Akkiraju, K., Xie, T., Giordano, L., Yu, Y., et al. (2022). Human- and machine-centred designs of molecules and materials for sustainability and decarbonization. *Academic and Research Output, 7*(12), 991–1009. https://doi.org/10.1038/s41578-022-00466-5

Pereira, R. D., Badino, A. C., & Cruz, A. J. (2020). Framework based on artificial intelligence to increase industrial bioethanol production. *Energy & Fuels, 34*(4), 4670–4677. ttps://doi.org/10.1021/acs.energyfuels.0c00033

Pushpa, S. R., Awoyale, A. A., Lokhat, D., Sukumaran, R. K., & Savithri, S. (2024). Infrared-based machine learning models for the rapid quantification of lignocellulosic multi-feedstock composition. *Bioresource Technology Reports, 25*, 101747. https://doi.org/10.1016/j.biteb.2023.101747

Rajagopal, D., Sexton, S., Hochman, G., & Zilberman, D. (2009). Recent developments in renewable technologies: R&D investment in advanced biofuels. *Annual Review of Resource Economics, 1*, 621–644. https://doi.org/10.1146/annurev.resource.050708.144259.

Raman, R., Pattnaik, D., Lathabai, H. H., Kumar, C., Govindan, K., & Nedungadi, P. (2024). Green and sustainable AI research: An integrated thematic and topic modeling analysis. *Journal of Big Data, 11*(1), 1–28. https://doi.org/10.1186/s40537-024-00920-x

Raven, D. B., Chikkula, Y., Patel, K. M., Al Ghazal, A. H., Salloum, H. S., Bakhurji, A. S., & Patwardhan, R. S. (2024). Machine learning & conventional approaches to process control & optimization: Industrial applications & perspectives. *Computers & Chemical Engineering, 189*, 108789. https://doi.org/10.1016/j.compchemeng.2024.108789

Sagi, O., & Rokach, L., (2021). Approximating XGBoost with an interpretable decision tree. *Information Sciences, 572*, 522–542. https://doi.org/10.1016/j.ins.2021.05.055

Saha, R., Chauhan, A., & Smita Rastogi Verma. (2024). Machine learning: an advancement in biochemical engineering. *Biotechnology Letters, 46*(4), 497–519. https://doi.org/10.1007/s10529-024-03499-8

Saini, R., Osorio-Gonzalez, C. S., Brar, S. K., & Kwong, R. (2021). A critical insight into the development, regulation and future prospects of biofuels in Canada. *Bioengineered, 12*(2), 9847–9859. https://doi.org/10.1080/21655979.2021.1996017.

Samanta, B., Al-Balushi, K. R., & Al-Araimi, S. A. (2003). Artificial neural networks and support vector machines with genetic algorithm for bearing fault detection. *Engineering Applications of Artificial Intelligence, 16*(7–8), 657–665. https://doi.org/10.1016/j.engappai.2003.09.006

Sarker, I.H., (2021). Machine learning: Algorithms, real-world applications and research directions. *SN computer science, 2*(3), 160. https://doi.org/10.1007/s42979-021-00592-x

Sharma, P., & Sharma, N. (2024). RSM approach to pre-treatment of lignocellulosic waste and a statistical methodology for optimizing bioethanol production. *Waste Management Bulletin, 2*(1), 49–66. https://doi.org/10.1016/j.wmb.2023.12.004

Sharma, V., Tsai, M., Chen, C., Sun, P., Nargotra, P., & Dong, C. (2023). Advances in machine learning technology for sustainable biofuel production systems in lignocellulosic biorefineries. *Science of The Total Environment, 886*, 163972. https://doi.org/10.1016/j.scitotenv.2023.163972

Sharmila, V. G., Shanmugavel, S. P., & Banu, J. R. (2024). A review of emerging technologies and machine learning approaches for sustainable production of biofuel from biomass waste. *Biomass and Bioenergy, 180*, 106997. https://doi.org/10.1016/j.biombioe.2023.106997

Shukla, A., Kumar, D., Girdhar, M., Kumar, A., & Goyal, A. (2023). Strategies of pretreatment of feedstocks for optimized bioethanol production: Distinct and integrated approaches. *Biotechnol Biofuels* 16, 44 https://doi.org/10.1186/s13068-023-02295-2.

Singh, P. (2023). Systematic review of data-centric approaches in artificial intelligence and machine learning. *Data Science and Management*, *6*(3), 144–157. https://doi.org/10.1016/j.dsm.2023.06.001.

Sipos, A., Florea, A., Arsin, M., & Fiore, U. (2020). Using neural networks to obtain indirect information about the state variables in an alcoholic fermentation process. *Processes*, *9*(1), 74. https://doi.org/10.3390/pr9010074

Smuga-Kogut, M., Kogut, T., Markiewicz, R., & Słowik, A. (2021). Use of machine learning methods for predicting amount of bioethanol obtained from lignocellulosic biomass with the use of ionic liquids for pretreatment. *Energies*, *14*(1), 243. https://doi.org/10.3390/en14010243

Soori, M., Arezoo, B., & Dastres, R. (2022). Artificial intelligence, machine learning and deep learning in advanced robotics: A review. *Cognitive Robotics*, *3*, 54–70. https://doi.org/10.1016/j.cogr.2023.04.001

Su, J., Yang, L., Sun, Z., & Zhan, X. (2024). Personalized drug therapy: Innovative concept guided with proteoformics. *Molecular & Cellular Proteomics: MCP*, *23*(3). https://doi.org/10.1016/j.mcpro.2024.100737

Taiwo, A. E., & Musonge, P. (2023). Comparative evaluation of bioethanol fermentation process parameters using RSM, ANN and ANFIS. *Biofuels, Bioproducts and Biorefining*, *17*(4), 961–975. https://doi.org/10.1002/bbb.2490

Taye, M. M. (2023). Understanding of machine learning with deep learning: architectures, workflow, applications and future directions. *Computers*, *12*(5), 91. https://doi.org/10.3390/computers12050091.

Tse, T. J., Wiens, D. J., & Reaney, M. J. (2021). Production of bioethanol: A review of factors affecting ethanol yield. *Fermentation*, *7*(4), 268. https://doi.org/10.3390/fermentation7040268.

Vane, L. M. (2008). Separation technologies for the recovery and dehydration of alcohols from fermentation broths. *Biofuels, Bioproducts and Biorefining*, *2*(6), 553–588. https://doi.org/10.1002/bbb.108

Vani, S., Sukumaran, R. K., & Savithri, S. (2015). Prediction of sugar yields during hydrolysis of lignocellulosic biomass using artificial neural network modeling. *Bioresource Technology*, *188*, 128–135. https://doi.org/10.1016/j.biortech.2015.01.083

Velidandi, A., Gandam, P.K., Chinta, M.L., Konakanchi, S., reddy Bhavanam, A., Baadhe, R.R., et al. (2023). State-of-the-art and future directions of machine learning for biomass characterization and for sustainable biorefinery. *Journal of Energy Chemistry*, *81*, 42–63. https://doi.org/10.1016/j.jechem.2023.02.020

Vinitha, N., Vasudevan, J., & Gopinath, K. P. (2022). Bioethanol production optimization through machine learning algorithm approach: biomass characteristics, saccharification, and fermentation conditions for enzymatic hydrolysis. *Biomass Conversion and Biorefinery*, *13*(8), 7287–7299. https://doi.org/10.1007/s13399-022-03163-z

Voisin, E., Thakur, S., Nayak, J., Chakrabortty, S., & Pal, P. (2024). Application of neural network in prediction of acetic acid yield by Acetobacters. *South African Journal of Chemical Engineering*, *50*, 427–436. https://doi.org/10.1016/j.sajce.2024.10.001.

Vyas, R., Goel, P., & Tambe, S. S. (2015). Genetic programming applications in chemical sciences and engineering. *Springer EBooks*, 99–140. https://doi.org/10.1007/978-3-319-20883-1_5

Wang, K., Chen, J., Martiniuk, J., Ma, X., Li, Q., Measday, V., & Lu, X., (2023). Species identification and strain discrimination of fermentation yeasts Saccharomyces cerevisiae and Saccharomyces uvarum using Raman spectroscopy and convolutional neural networks. *Applied and Environmental Microbiology*, *89*(12), e01673–23. https://doi.org/10.1128/aem.01673-23

Wang, R., He, Z., Chen, H., Guo, S., Zhang, S., Wang, K., Wang, M., & Ho, S. (2024). Enhancing biomass conversion to bioenergy with machine learning: Gains and problems. *Science of The Total Environment*, *927*, 172310. https://doi.org/10.1016/j.scitotenv.2024.172310.

Wang, Z., Peng, X., Xia, A., Shah, A. A., Huang, Y., Zhu, X., Zhu, X., & Liao, Q. (2022). The role of machine learning to boost the bioenergy and biofuels conversion. *Bioresource Technology*, *343*, 126099. https://doi.org/10.1016/j.biortech.2021.126099

Wu, D., Xu, F., Xu, Y., Huang, M., Li, Z., & Chu, J. (2024). Towards a hybrid model-driven platform based on flux balance analysis and a machine learning pipeline for biosystem design. *Synthetic and Systems Biotechnology*, *9*(1), 33–42. https://doi.org/10.1016/j.synbio.2023.12.004.

Wu, S. G., Shimizu, K., Tang, J. K.-H., & Tang, Y. J. (2016). Facilitate collaboration among synthetic biology, metabolic engineering and machine learning. *ChemBioEng Reviews*, *3*(2), 45–54. https://doi.org/10.1002/cben.201500024

Xing, Y., Zheng, Z., Sun, Y., & Agha Alikhani, M. (2021). A review on machine learning application in biodiesel production studies. *International Journal of Chemical Engineering*, *2021*(1), 2154258. https://doi.org/10.1155/2021/2154258

Xu, L., Yang, T., Zhang, X.J., Chai, L.J., Li, X., Shi, J.S., et al. (2024). Predicting the multispecies solid-state vinegar fermentation process using single-cell Raman spectroscopy combined with machine learning. *LWT*, *192*, p.115708. https://doi.org/10.1016/j.lwt.2023.115708

Yoo, H., Byun, H.E., Han, D., & Lee, J.H., 2021. Reinforcement learning for batch process control: Review and perspectives. *Annual Reviews in Control*, *52*, 108–119. https://doi.org/10.1016/j.arcontrol.2021.10.006

Yousuf, A., Pirozzi, D., & Sannino, F. (eds.) (2020, January 1). *Chapter 1 - Fundamentals of Lignocellulosic Biomass*. ScienceDirect; Academic Press. https://www.sciencedirect.com/science/article/pii/B9780128159361000010

Zhang, H., Han, L. & Dong, H., (2021). An insight to pretreatment, enzyme adsorption and enzymatic hydrolysis of lignocellulosic biomass: Experimental and modeling studies. *Renewable and Sustainable Energy Reviews*, *140*, 110758. https://doi.org/10.1016/j.rser.2021.110758

Zhang, J., Petersen, S. D., Radivojevic, T., Ramirez, A., Perez-Manríquez, A., Abeliuk, E., et al. (2020b). Combining mechanistic and machine learning models for predictive engineering and optimization of tryptophan metabolism. *Nature Communications*, *11*(1), 4880. https://doi.org/10.1038/s41467-020-17910-1

Zhang, L., Chao, B., & Zhang, X. (2020a). Modeling and optimization of microbial lipid fermentation from cellulosic ethanol wastewater by Rhodotorula glutinis based on the support vector machine. *Bioresource Technology*, *301*, 122781. https://doi.org/10.1016/j.biortech.2020.122781.

Zhang, X., Wang, Y., Jiao, P., Zhang, M., Deng, Y., Jiang, C., et al. (2023). Microbiome-functionality in anaerobic digesters: A critical review. *Water Research*, 120891. https://doi.org/10.1016/j.watres.2023.120891.

Zhou, X., Lu, P., Zheng, Z., Tolliver, D., & Keramati, A., 2020. Accident prediction accuracy assessment for highway-rail grade crossings using random forest algorithm compared with decision tree. *Reliability Engineering & System Safety*, *200*, 106931. https://doi.org/10.1016/j.ress.2020.106931.

8 Artificial Intelligence in Enhancement of Bioethanol Production from Lignocellulosic Biomass

Arpan Das

8.1 INTRODUCTION

The annual demand for energy is on the rise, with a significant dependence on non-renewable energy sources, particularly natural gas and crude oil, for the synthesis of several kinds of compounds and market products. However, these fossil resources are acknowledged to be finite in terms of their production and processing capabilities. The limitations of these resources, in addition to the detrimental effects on the environment that come with their usage and extraction, underscore the urgent need to explore sustainable, greener alternatives such as biofuels. Biofuels, which include bioethanol, biodiesel, along with biogas, are fuels obtained from renewable resources (Cheah et al., 2020). While the applications of biofuels offer substantial advantages similar to those of fossil fuels, they are increasingly favored for their potential to mitigate global warming and address energy and environmental challenges. Research in this domain is reportedly expanding, driven by growing awareness of the sustainability, availability, and ecological benefits of these alternative fuels. The production of biofuels from biomass presents a sustainable solution to the increasing demand for fuel, characterized by minimal adverse environmental effects due to the rapid bioenergy cycle (Velazquez et al., 2018). Among these alternatives, bioethanol is particularly noted for its low emissions in automotive applications and its ability to achieve high compression ratios, which enhance the performance of spark-ignition engines. Its high-octane rating positions it as a valuable green additive for gasoline, contributing to improved engine efficiency. The production of bioethanol can be achieved through the fermentation of sugars, such as glucose, followed by distillation and additional purification processes, or through the hydrolysis of starch, which necessitates an extra step prior to fermentation. While the fermentation of simple sugars or starch-based biomass to produce ethanol is relatively uncomplicated, the use of these first-generation feedstocks has become less favorable due to the significant land

 DOI: 10.1201/9781003571223-8

requirements for agricultural production, which leads to competition with food resources. A highly promising approach for generating renewable energy in an environmentally sustainable manner involves the utilization of inexpensive and plentiful biomass sources, particularly lignocellulosic biomass (LCB) (Gupta and Chundawat, 2020). They are not directly associated with food production and are abundantly available globally as agricultural byproducts or forest residues. Furthermore, LCB represents the most abundant raw material on the planet suitable for biofuel production. The appeal of lignocellulosic waste is rooted not only in its plentiful availability but also in its ability to transform waste into a valuable commodity. This shift embodies the principles of a circular economy, wherein waste materials are repurposed for bioethanol production, thereby lessening landfill pressures and reducing dependence on limited fossil resources. The potential for bioethanol production from agricultural waste and crop residues is significant, with estimates suggesting a possible global output exceeding 491 gigaliters per year. Various technological approaches are available for converting lignocellulosic biomass (LCB) into ethanol; among these, gasification is done in confined air at high temperatures, producing primary products such as charcoal, syngas, and liquefied hydrocarbons; the syngas can then be fermented through anaerobic processes in order to make ethanol (Griffin and Schultz, 2012). Furthermore, liquefaction and pyrolysis are processes used to break down biomass into bio-oil-like products. It is also possible to take out and ferment the amount of sugar from bio-oil to produce bioethanol (Li et al., 2018). However, the most popular technological approach for producing bioethanol is still direct fermentation. In this context, lignocellulose-based ethanol production emerges as a viable technological solution, with various methods proposed to mitigate the resistance exhibited by the lignocellulosic structure, lower enzyme production costs, and enhance both enzymatic hydrolysis and fermentation processes. Despite substantial advancements made by both industry and academia, enzymatic hydrolysis continues to pose a significant challenge due to the substantial quantities of enzymes required, the intricacies of mass transfer, and the multitude of chemical reactions that lead to the formation of inhibitory byproducts. While both historical and contemporary research recognize the potential of sustainable fuels derived from various biomass sources, the processes of pretreatment and conversion are influenced by multiple factors that ultimately impact yield outcomes. These processes typically involve hydrolysis and fermentation methods, which are significantly influenced by variables such as feedstock type, duration, or temperature. The modeling of bioprocesses, which relies on these parameters and methodologies, presents ongoing challenges that cannot be addressed solely through experimental means; rather, they necessitate the implementation of straightforward and effective predictive techniques, including artificial intelligence (AI) methods. Given the intricate nature of the process variables involved in converting LCB to ethanol, machine learning (ML) offers a promising avenue for identifying optimal experimental conditions that maximize conversion efficiency through the most effective combination of variables. ML, a subset of artificial intelligence, aims to develop and refine computer programs capable of autonomously learning from historical data by employing various algorithms. These programs facilitate the analysis of datasets, uncover previously unrecognized trends, patterns, generalize data, construct models, or formulate heuristic-based rules. They excel in identifying nonlinear relationships between input as well as output variables (Bannor and Acheampong, 2019). While a multitude of ML algorithms exists and continues to expand, they primarily serve specific functions such as predicting or classifying outcomes based on new descriptor sets, clustering data according to descriptor similarities, or establishing associations between descriptors and output variables (Larose and Larose, 2014). Various algorithms are typically employed for distinct tasks; however, certain algorithms may serve multiple purposes, and conversely, a single task can often be addressed by several algorithms. There is considerable enthusiasm surrounding the application of machine learning (ML) techniques, such as artificial neural networks (ANNs), classification trees (CTs), and random forests (RFs), in facilitating the efficient production of ethanol from lignocellulosic biomass via SSF. While particular attention is required for specific tasks and algorithms, the implementation of machine learning (ML) can be distilled into three primary stages: the construction of the dataset (such as collecting information from literature-based experimental investigations), the selection and application of ML algorithms (for example, choosing ANNs for yield prediction, adjusting hyperparameters, and confirming the effectiveness of the model), and the interpretation of results (including the analysis of the effects and significance of various descriptors) (Alpaydin, 2020). Each of these stages is crucial for the effective execution of ML. Machine learning (ML) has experienced substantial advancements in recent years across multiple domains, particularly in energy and fuels. Notably, there are numerous successful implementations of ML within biofuel research, as highlighted in comprehensive reviews that encompass the entire discipline (Wang et al., 2022) and in specific investigations such as those focusing on ANN applications. Additionally, various research teams have examined the applications of ML in lignocellulosic biomass (LCB). Research has been done to assess the impacts of various biomass sources, solid loading, inhibitions, and degradation parameters on the feasibility of producing ethanol by means of concomitant fermentation with saccharification. There is considerable enthusiasm surrounding the application of machine learning (ML) techniques, such as ANNs, classification trees (CTs), and RFs, in facilitating the efficient production of ethanol from lignocellulosic biomass.

8.2 BIOMASS COMPLEXITY IN LIGNOCELLULOSIC MATERIALS

According to a 2005 USDA and DOE joint study, plants from conserved reserve areas combined with forest and agricultural leftovers might produce up to 1.3 billion tons of biomass annually for use in production (Perlack et al., 2005). With regard to a potential conversion process, it is assumed that this biomass can be used to produce enough biofuel annually to replace between 30% and 50% of the petrol used in the US. In contrast to sugar and starch-based feedstocks, lignocellulosic biomass (LCB) presents a more intricate composition, necessitating a thorough understanding of its molecular structure (Liu et al., 2019). Lignocellulosic biomass represents a multifaceted source of organic material, primarily consisting of three key components: cellulose, hemicellulose, and lignin, which are interwoven to form lignin-carbohydrate complexes. Table 8.1 shows the composition of different types of lignocellulosic

TABLE 8.1
Typical Compositions of Lignocellulosic Biomass

Feedstock	Cellulose	Hemicellulose	Lignin
Rice straw	38	32	12
Rice husk	37	29	24
Corn stover	38	26	19
Corn cob	45	35	15
Sugarcane bagasse	42	25	20
Switchgrass	40–45	30–35	12
Wheat straw	33–40	20–25	15–20
Agriculture residues	37–50	25–50	5–15
Industrial waste from chemical pulp	50–70	12–20	6–10
Waste paper	65	13	1

Source: Cheah et al. (2020).

biomass. Cellulose, the predominant sugar polymer (approximately 40–60% of the biomass's weight) in LCB, consists of a linear arrangement of glucose units connected primarily through β-1,4-glycosidic bonds and is the most structurally resilient element within lignocellulosic biomass. The structure of cellulose is characterized by a crystalline content of 60–80% and an amorphous fraction of 20–40%, which together form the essential structural and functional components of plant material (Bonechi et al., 2017; Tursi, 2019). Dimos et al. (2019) conducted a study utilizing cotton fiber as a primary material to investigate the structural characteristics of cellulose within plant cell walls. Cotton fiber is recognized as the sole natural source of pure cellulose, with its cellulose content ranging from 95% to 97% and exhibiting a crystallinity of approximately 70%. Through scanning electron microscopy, it was determined that the diameter of the fibrils measures around 500 nm, identifying it as the largest structural component of cellulose. Each fibril consists of intertwined microfibrils, which confer a tensile strength surpassing that of a steel wire of equivalent thickness. Microfibrils aggregate to form a network that serves as the fundamental framework of the cell wall, with diameters ranging from 10 to 25 nm. Furthermore, microfibers are constructed from elementary fibrils that are aligned in parallel, with the diameter of these elementary fibrils falling between 2 and 4 nm. In certain areas of the microfibrils, cellulose molecules are organized in a systematic manner, imparting crystalline properties to the cellulose.

Hemicelluloses, accounting for about 20–35% of the weight, represent the second most prevalent carbohydrate in LCB, where they act as binding agents within the biomass. Hemicelluloses are a branched heteropolymer composed of various sugars, including hexoses such as D-glucose, D-galactose, and D-mannose, as well as pentoses like D-xylose and L-arabinose, along with sugar acids such as D-glucuronic acid (Brandt et al., 2013). They exhibit a less crystalline structure that facilitates easier hydrolysis. The composition of hemicellulose varies among different types of lignocellulosic biomass, which can influence the efficiency of conversion processes. The diverse sugar composition necessitates the use of multiple enzymes to effectively hydrolyze these polymers into their monomeric forms.

Lignin, from a botanical perspective, serves to encase bundle cells, including wood fibers and sclerenchyma cells. Chemically, it is primarily composed of phenylpropanoid derivatives, which link together to form high molecular weight compounds through ether or carbon-carbon bonds. In terms of physical properties, lignin is characterized by its hardness, which contributes to the overall rigidity of the cell wall. Typically, plant cell walls that provide structural support and mechanical strength exhibit a significant lignin concentration. In woody plants, lignin content ranges from approximately 27% to 32%, while in herbaceous plants, it varies between 14% and 25% (Alvira et al., 2010; Jorgensen et al., 2007). Lignin is a high molecular weight, aromatic, water-insoluble, complex, and irregular polymer that endows plants with waterproofing properties and resilience (Zoghlami and Paes, 2019). The structural integrity of biomass is enhanced by the strong covalent and hydrogen bonds that link cellulose and hemicellulose to lignin, providing significant resistance to biological and physical degradation and serving as a protective structural element. However, it can also hinder the effective production of bioethanol.

8.2.1 Pretreatment

The primary goal of the pretreatment process is to optimize the effectiveness of bioethanol production by promoting the release of fermentable sugars from biomass while minimizing the generation of fermentation inhibitors. The biotransformation of lignocellulosic biomass (LCB) to ethanol presents significant challenges due to its intricate composition. The structural integrity of LCB is maintained by robust covalent bonds, Van der Waals forces, and several types of intermolecular interactions, resulting in a highly stable configuration that resists hydrolysis (Agrawal et al., 2021). Notably, lignin exhibits a higher resistance to biological degradation in contrast to cellulose and hemicellulose, serving as a defensive structural element. Lignocellulosic biomass (LCB) transformation to bioethanol presents significant challenges due to its intricate composition. The pretreatment phase encompasses a variety of techniques aimed at removing lignin, thereby enhancing the accessibility of carbohydrates, specifically cellulose and hemicellulose, for effective enzymatic hydrolysis. In the final phase, these fermentable sugars are transformed into ethanol. Notably, the phase of pretreatment represents the most significant financial investment in the overall process. Additionally, the substrate hydrolysis and subsequent fermentation stages are complicated by the presence of a diverse array of sugars, which adversely affects the economic viability of the entire operation. Various pretreatment techniques are available, including physical/physicochemical methods, chemical methods, and biological treatments. Among the physical methods, chopping serves to enhance the surface area of lignocellulosic biomass (LCB) for enzymatic action. The ultra-fine milling process is another option that can diminish cellulose polymerization and crystallinity, albeit at a higher cost. Microwave-based heating presents a promising substitute for pretreating lignocellulosic substances, since it eliminates the need for additional chemicals such as separating agents and solvents. When compared to conventional heating systems, this technology produces little to no smoke or waste end products and drastically cuts down on operating time and energy usage. Furthermore, hot-water-based pretreatment and steam explosion represent additional illustrations of physicochemical and physical preparation techniques. While concentrated acid can effectively decompose cellulose at lower temperatures, its practicality is limited due to the substantial environmental waste it generates. Consequently, dilute acid pretreatment is preferred over alternative techniques because it is more economical and efficient at breaking down hemicellulose into units of monomer and its ability to induce structural changes that enhance the accessibility of enzymes for cellulose bioconversion. Sulfuric acid is the most utilized acid for this pretreatment, although alternatives such as nitric acid, hydrochloric acid, and phosphoric acid are also applicable (Xu and Huang, 2014). Nevertheless, this approach is not without drawbacks, including the requirement for costly corrosion-resistant equipment and the necessity to neutralize acidic hydrolyzates prior to sugar fermentation (Zheng et al., 2009). The hydrolysate derived from dilute acid pretreatment often contains inhibitors such as acetate, furfural, and hydroxymethylfurfural, which can impede the growth of microorganisms. The detrimental effects of these inhibitory compounds may be mitigated through the application of detoxifying agents, such as *Amorphotheca resinae* ZN1, and neutralizing agents like calcium hydroxide [$Ca(OH)_2$], which can enhance the production cost (Zhang et al., 2010). Alkaline pretreatment offers a cost-effective alternative by reducing the necessity for expensive substances and customized designs aimed at resisting corrosion during prolonged reaction conditions. This process is typically conducted under moderate conditions, occasionally at room temperature, by immersing the biomass in solutions of sodium hydroxide or ammonium hydroxide. These treatments serve to dismantle lignin and eliminate acetyl groups from hemicellulose, thereby improving the digestibility of the material by enzymes. Certain alkaline pretreatment methods also facilitate retrieving and reusing chemical reagents (Kim et al., 2016). A significant drawback of this method is the production of significant amounts of minerals, which, if not sufficiently eliminated, may prevent microbial development and the fermentation of ethanol in later phases (Liu et al., 2019). Alternative methods such as ionic liquids (ILs, which consist of a high proportion of organic cations and a smaller quantity of inorganic anions), deep eutectic solvents (which are combinations of Bronsted and Lewis acids and bases), and organosolv processes (utilizing organic solvents like ethanol, methanol, butanol, and acetone) are also viable options (Galbe and Wallberg, 2019). Ionic liquid pretreatment effectively dissolves lignin and disrupts the structural integrity of lignocellulosic biomass, presenting a viable pretreatment strategy. Conversely, dilute ammonia pretreatment involves treating biomass with ammonia at low concentrations under mild conditions, facilitating the removal of lignin and hemicellulose while being more environmentally sustainable. Biological pretreatment represents an environmentally sustainable and economically viable alternative; this method employs microorganisms, including rot fungi, to decompose lignin and hemicelluloses present in lignocellulosic biomass, enhancing their susceptibility to enzymatic hydrolysis. A key benefit of this approach is the absence of chemical recycling requirements, resulting in no release of harmful substances into the environment. Nevertheless, it is associated with certain limitations, such as the necessity for prolonged reaction periods because of the sluggish rates of breakdown and the possible loss of a sizable amount of biomass during the procedure. The selection of a pretreatment method is influenced by various factors, including the type of biomass, the intended end product, economic viability, and environmental impact.

8.2.2 Hydrolysis

Hydrolysis is another crucial process that can be categorized into enzymatic hydrolysis and acidic hydrolysis. Enzymatic hydrolysis plays a vital role in the lignocellulosic biomass-based bioethanol production process. This process utilizes enzymes to decompose complex carbohydrates, including cellulose and hemicellulose, into simpler carbohydrates that can subsequently undergo fermentation to produce ethanol. Enzymatic hydrolysis is characterized by a reduced environmental impact and lower formation of inhibitors, whereas acid hydrolysis is noted for its rapidity. In the case of enzymatic hydrolysis, the lignocellulosic biomass necessitates a diverse array of enzymes, including cellulases (such as cellobiohydrolases, endo-glucanases, and β-glucosidases), hemicellulases (including endo-xylanases, β-xylosidases, and xyloglucanases), and enzymes that degrade lignin (Agrawal et al., 2021). The selection of the appropriate enzyme is also crucial, as it depends on the specific substrate being processed. This process employs solubilized enzymes to decompose insoluble substrates, resulting in a system of heterogeneous reactions that are affected by various parameters, which include the content of lignin and hemicellulose, the crystallinity of cellulose, the level of polymerization, the number of pores within the substrate, and the available surface area (Zhao et al., 2021). During the enzymatic hydrolysis of lignocellulosic biomass, enzymes facilitate the transformation of polymeric carbohydrates into small sugar molecules under mild and controlled operational conditions, ensuring optimal pH and temperature, which allows for the cleavage of bonds between sugar molecules and the release of simple sugars. This enzymatic process persists until a substantial portion of cellulose and hemicellulose is converted into sugars. This approach demonstrates high efficiency, resulting in significant sugar recovery while minimizing the formation of inhibitors and reducing the potential for corrosion. During hydrolysis, cellulose is converted into glucose, while hemicellulose is broken down into five-carbon sugars such as arabinose and xylose. Hemicellulose hydrolysis also produces acetic acid as a byproduct, which inhibits the growth of microorganisms and ethanol production, representing a significant drawback of the hydrolysis process. Additionally, the process is energy-intensive due to the presence of lignin, which occupies reactor space and necessitates increased mixing to ensure a uniform suspension. To mitigate enzyme costs, recycling strategies are implemented, allowing for the recovery, purification, and reuse of enzymes in subsequent processing batches. The resulting hydrolysate is then fermented, wherein microorganisms, typically yeast or bacteria, metabolize the sugars to produce ethanol (Robak and Balcerek, 2018). While enzymatic hydrolysis is essential for lignocellulosic bioethanol production, challenges remain, including the optimization of enzyme efficiency, reduction of enzyme costs, and management of potential inhibitors. Ongoing advancements in enzyme technology and process engineering are crucial for enhancing the yield and sustainability of bioethanol derived from lignocellulosic sources. Conversely, acid hydrolysis, which is a commonly used method, may necessitate different temperatures and concentrations to achieve higher yields. Strong acid-catalyzed hydrolysis at lower temperatures can lead to enhanced sugar recovery. However, this method is hindered by high expenses of production related to the recovery of acid, its disposal, and further recycling. Moreover, this process carries the possibility of sugar monomers deteriorating because of the acidic environment, which can adversely impact subsequent fermentation processes.

8.2.3 Fermentation and Extraction of Bioethanol

The fermentation process is a vital component in the conversion of sugars obtained from lignocellulosic biomass and other feedstocks into bioethanol. This process utilizes microorganisms, predominantly yeast or bacteria, to transform sugars into ethanol and carbon dioxide. Initially, hydrolysate is prepared, which consists of a blend of hexose and pentose sugars extracted from cellulose and hemicellulose. The transformation of glucose into ethanol is relatively straightforward; however, the conversion of other sugars presents greater challenges. The fermentation of various sugars necessitates distinct microbial populations, each requiring specific optimal growth conditions (Kucharska et al., 2018). The selection of the microorganism is contingent upon the specific type of sugar present in the hydrolysate. For instance, *Saccharomyces cerevisiae* is frequently employed for the fermentation of glucose, whereas engineered strains may be utilized for xylose and other sugar types. During sugar fermentation, it is common for microorganisms to preferentially metabolize one sugar type, typically glucose, over others. For instance, *Saccharomyces cerevisiae*, a widely utilized fermentation microorganism, lacks the ability to naturally ferment xylose (Jahanbakhshi and Salehi, 2019). Since *Saccharomyces cerevisiae* has multiple genetic components and is a tool for synthetic biology, it is the preferred host in metabolic engineering techniques for biofuel pathways. Several engineering techniques have been used to expand the range of substrates and products produced by *S. cerevisiae*.

Although some microorganisms can metabolize pentose sugars, the glucose derived from cellulose frequently inhibits the breakdown of these sugars (Zhao et al., 2021). These factors collectively hinder the development and management of fermentation processes utilizing lignocellulosic biomass as a feedstock, while incomplete conversions and sluggish enzyme reactions further complicate the process and diminish ethanol yields. Furthermore, microorganisms that can ferment these sugars often exhibit low tolerance to the conditions within bioreactors, primarily due to the accumulation of toxins. So, there is a pressing need for investigation into fermentative microorganisms to identify efficient and resilient microbial consortia or to develop genetically modified strains (Culaba et al., 2022). Once the appropriate microorganism is introduced to the hydrolysate, fermentation ensues, during which the microorganisms metabolize the sugars through a series of biochemical pathways. The fermentation process persists until the majority of the available sugars have been converted into ethanol, a duration that may extend over several days. Subsequent to fermentation, the process includes the recovery and purification of the fermented broth, which contains ethanol. Various fermentation methods documented in the literature include separate hydrolysis and fermentation (SHF) process, simultaneous saccharification and fermentation (SSF) process, simultaneous saccharification and co-fermentation (SSCF) process, and consolidated bioprocessing (CBP) process (Thompson et al., 2020). The SHF process, which is the most widely employed two-stage method, involves conducting enzymatic hydrolysis separately from fermentation. This separation allows enzymes to operate at higher temperatures while fermentation microorganisms function at moderate temperatures, optimizing their performance. Despite the advantages of SHF in lignocellulosic biomass processing, it faces challenges such as lower yields, elevated production costs, and contamination issues (Tavva et al., 2016; Lugani et al., 2020). In contrast, SSF facilitates the concurrent saccharification of cellulose and fermentation of monomeric sugars within the same reactor. The SSCF method integrates hydrolysis and saccharification in a single unit while simultaneously conducting fermentation. The consolidated bioprocessing (CBP) approach is a cost-effective strategy that combines enzyme production, hydrolysis, and fermentation in one vessel for bioethanol production. The latter three integrated processes (SSF, SSCF, and CBP) were developed to address the limitations of SHF, as they tend to exhibit reduced contamination, simpler process design, and improved yields (Vaid et al., 2018; Ravanal et al., 2019; Lugani et al., 2020; Edeh, 2021). Thus, fermentation is integral to the transformation of sugars from lignocellulosic biomass into a valuable and sustainable energy source. Ongoing advancements in biotechnology and strain optimization are consistently enhancing the fermentation process for bioethanol production. The choice of raw materials significantly influences the overall production cost, as their composition directly impact both pretreatment expenses and the yield of fermentable sugars. The effectiveness of pretreatment, hydrolysis, concentration of sugars in the hydrolysate, chosen microorganism, fermentation processes, along with the characteristics of lignocellulosic biomass (LCB), plays a crucial role in the competitive production of bioethanol (Qiao et al., 2022).

Following the process of monomeric sugar fermentation, the subsequent phase typically involves processing the fermenting broth to obtain ethanol. This extraction process generally necessitates a reduction in the broth's water content, facilitating the production of anhydrous ethanol. However, this method is complicated by the azeotropic characteristics of the ethanol-water mixture. To address this challenge, distillation at different boiling points is performed. The issue of the azeotropic solution can be mitigated by the addition of a separating agent that alters the relative volatility of the primary component. To extract the ethanol from the fermentation broth, a number of methods are used, such as chemical dehydration, membrane distillation, vacuum distillation, extractive distillation, and adsorption distillation. Among these conventional techniques, extractive distillation, liquid-liquid extraction, and azeotropic distillation are often used. Notably, extractive distillation is recognized as the predominant technique for large-scale operations. Nevertheless, innovative methods such as pervaporation and salt distillation are increasingly being explored for their potential advantages, particularly in terms of reduced energy consumption.

8.3 MAJOR OBSTACLES IN BIOETHANOL PRODUCTION

Lignocellulose-based bioethanol production process encounters numerous obstacles, such as ensuring a reliable and cost-effective supply of biomass, enhancing the efficiency of pretreatment processes, improving enzyme performance, and optimizing the fermentation of mixed sugars. The variation among waste feedstock content is one of the main obstacles. Variations in the properties of different waste sources may need modifications to processing techniques. Preprocessing and treatment are necessary for a lot of waste products in order to get rid of contaminants, lower recalcitrance, and improve the lignocellulose conversion rate. Ionic solutions, thermal explosion, organosolv, and acid/enzymatic treatments are examples of pretreatment techniques (Amiri and Karimi, 2015). The enzymatic hydrolysis process is characterized by its slow pace and high costs, making it imperative to address the inhibitory compounds generated during pretreatment. Additionally, challenges related to product inhibition and the attainment of elevated ethanol concentrations persist. The processes involved in bioethanol production are notably intensive in terms of water and energy, raising concerns regarding sustainability and economic feasibility. Effective waste management is critical, as the handling and disposal of toxic byproducts can pose significant environmental challenges. Trash, as feedstock, necessitates a thorough investigation of environmental and regulatory aspects, especially when it comes to municipal solid trash. It is imperative to tackle possible impurities and pollutants in order to guarantee the security and efficacy of the bioethanol output. The transition from laboratory-scale methodologies to commercial-scale production introduces further complexities, particularly in terms of process optimization, cost efficiency, and maintaining consistent product quality. Navigating the regulatory landscape concerning environmental sustainability, safety, and biofuel standards can be intricate and varies across different regions. The economic viability of bioethanol production is particularly challenged by competition from fossil fuel markets. Furthermore, the presence of alternative renewable energy sources, such as biodiesel, hydrogen, and electric vehicles, complicates efforts to capture a share of the transportation fuel market. Addressing these multifaceted challenges necessitates continuous research, development, and innovation in bioethanol production technologies, alongside collaborative efforts among researchers, industry stakeholders, and policymakers to foster a more sustainable and economically viable bioethanol industry.

8.3.1 Environmental Resilience

The production of lignocellulosic biomass is associated with a reduced carbon footprint in comparison to other biomass types and fossil fuels; however, its environmental implications can differ based on various factors, including the type of feedstock, cultivation techniques, harvesting strategies, and processing methods. It is frequently regarded as a more ecologically sustainable option relative to other biomass sources and fossil fuels. Nonetheless, it is crucial to evaluate the specific environmental consequences within diverse contexts. Greenhouse gas emissions are generally counterbalanced by the carbon dioxide absorbed during the growth of biomass, rendering the process carbon-neutral. Nevertheless, changes in land use and emissions from transportation can exacerbate overall emissions. The interplay between land use and biodiversity is vital, as the transformation of natural habitats into biomass production areas can disrupt local ecosystems and diminish biodiversity. Therefore, sustainable biomass production should prioritize the utilization of marginal or degraded lands to mitigate these adverse effects. Additionally, water consumption is a significant aspect of environmental impact assessments, as certain biomass crops may demand substantial water resources, potentially leading to water scarcity (Stenzel et al., 2021). Sustainable practices in biomass production strive to minimize or eliminate the reliance on pesticides and fertilizers through methods such as integrated pest management and nutrient recycling. Furthermore, soil health may be compromised by biomass production, especially when intensive cultivation practices are employed. Adopting strategies like crop rotation and no-till farming can contribute to the preservation of soil health. Enhancing transportation logistics and incorporating renewable energy sources within processing facilities can significantly mitigate environmental impacts. The implementation of advanced biomass conversion technologies plays a crucial role in reducing emissions. Effective management of waste and residues is vital to prevent environmental contamination. Additionally, biosecurity protocols are necessary to avert the introduction of invasive species or diseases linked to specific biomass crops.

8.3.2 Financial Stability

The economic feasibility and cost efficiency of producing lignocellulosic biomass and converting it into biofuels, especially bioethanol, are essential considerations in evaluating its potential as a sustainable energy alternative. Various elements, including feedstock expenses, conversion technologies, operational expenditures, economies of scale, market dynamics, governmental incentives and regulations, research and development efforts, financial uncertainties, environmental compliance, and the generation of co-products, can significantly influence the economic viability of this process (Bender et al., 2022). The costs associated with feedstock are determined by factors such as the type of feedstock utilized, its availability, and its proximity to processing facilities. While advanced technologies, including biochemical and thermochemical methods, may necessitate substantial initial capital outlays, they often yield greater efficiency and higher product outputs. Operational costs encompass labor, energy consumption, maintenance, and process optimization. The market price of bioethanol is subject to fluctuations based on regional demand, government interventions, and competition from alternative fuels. Government support in the form of incentives, subsidies, and tax breaks can enhance the economic viability of lignocellulosic bioethanol production, making it more competitive with fossil fuels. Furthermore, investments in research and development can lead to improved production efficiencies and cost reductions. Financial risks, stringent environmental regulations, and the effective use of co-products generated during bioethanol production can also contribute to enhanced cost-effectiveness.

8.3.3 Social Stability

The production of lignocellulosic bioethanol from waste materials has profound social implications, notably in terms of job creation, community engagement, and social accountability. This sector has the potential to generate employment across various segments of the value chain, including the collection of feedstocks, processing, logistics, and the operation of production facilities, thereby contributing positively to both local and regional economies. Additionally, it encourages the emergence of indirect job opportunities within the retail and service industries, as well as in the supply chain for goods and services essential to the bioethanol sector. Engaging local communities is vital, as it fosters participation in feedstock procurement, outreach initiatives, and collaborative efforts. Establishing partnerships with farmers, forest proprietors, and waste management entities can lead to the creation of economic prospects. Furthermore, bioethanol production facilities have the capacity to inform communities about the environmental and economic advantages of the industry, thereby cultivating constructive relationships and addressing any concerns that may arise. Social responsibility encompasses the commitment to uphold safety and health standards for both employees and residents, honoring the rights of indigenous populations, and actively participating in consultation and consent processes. Economic growth can be facilitated by diversifying local economies, diminishing reliance on a singular industry, and generating tax revenues that can be allocated to local infrastructure. Moreover, investing in workforce development and engaging stakeholders is essential for fostering trust and accountability within the industry, which in turn ensures sustainable employment opportunities and the maintenance of positive community relations (Praveenkumar et al., 2014).

8.4 THE UTILIZATION OF ARTIFICIAL INTELLIGENCE IN THE PROCESSES OF FEEDSTOCK PRETREATMENT AND HYDROLYSIS

The application of artificial intelligence in the pretreatment and hydrolysis processes has garnered attention, particularly in the context of enzymatic hydrolysis (EH), which poses challenges in bioethanol production due to the diversity of enzymes involved. Recent studies have demonstrated that machine learning (ML) technologies, specifically random forest (RF) algorithms and ANN, can effectively optimize EH in conjunction with fermentation processes. Over the past few years, machine learning has been applied in a number of biofuel research domains. A number of instances of such research in the scientific literature include the optimization of process variables for the anaerobic breakdown of corn stalk (Dong and Chen, 2019), predicting the chemical and physical characteristics of biodiesel depending on its content of fatty acids (Alviso et al., 2020), and viability of ML algorithms for determining biodiesel quality (Moayedi et al., 2020). With this optimization and accomplishment, the effects of feedstock, heat, and enzyme concentration on the ethanol generation from sugarcane biomass were successfully predicted. The author states that the results were in line with the experimental findings and draws conclusions about the modeling's efficacy for practical bioethanol production (Fischer et al., 2017). In order to forecast the manufacturing of bioethanol from lignocellulose, empirical information was also simulated (Smuga-Kogut et al., 2021). ANN and RF algorithms were also used in the modeling and forecasting of the output variable (bioethanol content) from these biomass sources using ionic liquids as the input variables. These algorithms were chosen because they could not only forecast the bioethanol content but also make it easier to choose the right ionic liquids for the procedure. They emphasized that using a coefficient of determination (R2) value of 0.961, their machine learning (ML) algorithms in hybrid modeling for multiple phase hydrolysis along with fermentation were efficient and in accordance with the experimental findings to forecast the bioethanol concentration. Phromphithak et al. (2021) used 520 entries of data from the scientific literature to simulate the cellulose enrichment factor (CEF) and solid recovery (SR) utilizing support vector machines (SVM), gradient boosting (GB), and random forests (RF) with 45 different types of feedstocks and 80 different kinds of solvents. It was demonstrated that RF outperformed the other ML algorithms in terms of CEF prediction, while SR (% w/w) and CEF were better served by RF. Response surface methodology (RSM) and artificial neural networks (ANN) were compared in a different study to forecast the various elements in the production hydrolysis stage for oligosaccharide combinations from sugar beet pulp. RSM is a statistical method that is mostly linked to randomness principles. A feedforward ANN called a multilayer perceptron (MLP) is characterized by a number of neurons cooperating to provide a set of outputs for intricate nonlinear processes. The input, hidden, and output layers make up the three layers of this perceptron. Reaction duration, cellulase activity to polygalacturonase activity ratio, and polygalacturonases to solid ratio all influence the enzymatic hydrolysis of sugar beet pulp (Astray et al., 2016), consequently made the decision to use the specified statistical procedures to evaluate these characteristics in connection to oligosaccharides. Several neurons were varied in several training and hidden layers in this strategy. The comparative assessment's conclusions demonstrated that the MLP is a useful instrument for simulating the pulp's enzymatic hydrolysis to produce oligosaccharides. In order to maximize the amount of reducing sugars throughout the enzymatic hydrolysis phases of producing bioethanol from water hyacinth biomass, Das et al. (2015) evaluated RSM and ANN approaches. This evaluation was carried out based on a comparison of these methods for an ideal enzymatic saccharification. With an average prediction error of 3.08 and an optimal prediction error of 0.95, respectively, MLP outperformed RSM once more. The efficacy of a hybrid model (PSO-ANN) and RSM for the generation of xylose and glucose was investigated using the regression/coefficient of determination (R2) as a benchmark. In order to increase the projected outputs of lignin-based biomass, these AI systems were utilized to model its pretreatment and enzymatic hydrolysis. Analysis showed that, when compared to RSM, the hybrid model accurately represented the steps of sugar generation with the same yields of xylose and glucose. The RSM model's R2 values were 0.8901 and 0.8439, while the hybrid model's values were 0.9939 and 0.9479 for glucose and xylose (Ethaib et al., 2018). The model findings (glucose concentrations) from an expertly trained artificial neural network (ANN) for the generation of bioethanol have been compared with the experimental findings from the pretreatment, hydrolysis, and hydrolysis of sugarcane bagasse using diluted acid and combination diluted acid ozonolysis. According to reports, MLP, a trained ANN, has high estimating skills and agrees well with experimental data (Gitifar et al., 2013). Low starting biomass and acid concentrations, high enzyme concentrations, and a 72-hour enzymatic hydrolysis length are the appropriate operational parameters for increased glucose production throughout the pretreatment and hydrolysis of sugar bagasse as predicted and determined by Tovar et al. (2018) with the assistance and efficiency of an MLP. Furthermore, Praveenkumar et al. (2014) used MLP to predict the enzymatic hydrolysis stage of the manufacture of bioethanol from sugarcane bagasse. β-glucosidase and cellulase loads were coupled by the neural network, which effectively predicted and optimized the glucose yield and concentration. In addition, Vani et al. (2015) discussed the precision and effectiveness of the MLP model utilizing the Levenberg–Marquardt backpropagation algorithm in forecasting the impact of substrate particle size, biomass loading, and reaction duration on the generation of glucose and xylose throughout the enzymatic hydrolysis of rice straw biomass extraction. Aruwajoye et al. (2022) used ANN, RF, and decision tree regression (DTR) to study the fermentable sugar level as well as a combined severity factor (CSF), which indicates the effectiveness of the pretreatment approach. They built algorithms using 49 experimental data and employed soaking temperature, soaking time, sterilization time span, HCl level, and solid load as descriptor variables. It was discovered that the best machine learning technique differed based on the output variable under investigation. Moodley et al. (2019) and Lee et al. (2020) investigated how the experimental circumstances affected the amount of fermentable sugars. They discovered that the best operating conditions were a 30-minute ultrasound treatment at 192.5 W and 48.2° C, which produced 356 mg/g of biomass. Onay (2022) provided algorithms for estimating the amount of carbohydrates, while Chen et al. (2022), Gitifar et al. (2013), and Plazas Tovar et al. (2018) predicted the level of glucose as the output variable. Similarly, Chang et al. (2011) used inputs from both the pretreatment and enzyme-mediated hydrolysis stages to predict enzymatic digestibility (%), while Ethaib et al. (2018) investigated both glucose and xylose as fermentable sugars. Using a variety of ML techniques, Charte et al. (2017) examined the recovery of solids and glucose concentration.

8.5 AI AND ITS APPLICATION IN BIOETHANOL PRODUCTION

Numerous studies have modeled fermentation, the last stage of the three-step process that produces bioethanol from LCB, either independently or in conjunction with the earlier stages. For instance, Suresh et al. (2020) and Smuga-Kogut et al. (2021) used the variables from three successive processes as the descriptors in their modeling of the manufacture of bioethanol. Using ANN to study the effect of hydrolysis, Suresh et al. (2020) used the concentration of *S. cerevisiae*, the ultrasonication period, and the HCl and α-amylase as input variables. The researchers discovered that ANN outperformed RSM in simulating the synthesis of bioethanol. Three distinct machine learning techniques were employed by Smuga-Kogut et al. (2021): ANN, RF, and a hybrid model combining the two. The hybrid model yielded more accurate findings. Fischer et al. (2017) and Vinitha et al. (2023) also looked at the impact of fermentation and hydrolysis parameters on bioethanol yield. Ahmadian-Moghadam et al. (2013) [14] investigated the efficacy of a multilayer perceptron (MLP) in predicting bioethanol concentration by utilizing fermentable sugar, as well as yeast, as input variables within a batch fermentation. The study involved training the neural network with 30 datasets and subsequently validating it against an established database. It was reported that the results of this comparison were consistent with the existing data. A strong correlation was observed between the input variables and the output variable (bioethanol concentration), which was evidenced by low error rates and high R^2 values. The findings led to the conclusion that MLP serves as an accurate, straightforward, and effective tool for modeling cost-efficient bioethanol production. Furthermore, the industrial bioethanol production process was modeled and simulated using a multilayer feedforward neural network in conjunction with particle swarm optimization (PSO) algorithms. PSO is acknowledged as a robust optimization technique, particularly when the modeling of a specific output variable is of paramount importance. By utilizing 3,400 data points from a fermentation unit of a Brazilian company, the MLP was optimized with PSO to improve production levels on an industrial scale. The coefficient of determination, which reached 0.91, underscored the capability of these integrated models and algorithms to accurately predict concentrations and effectively enhance bioethanol production levels. This was further validated by a 10% increase achieved through this innovative approach on an industrial scale (Pereira et al., 2020). In evaluating the fermentation phase for bioethanol production from watermelon waste (the output variable), three distinct quantities of yeast and varying agitator speeds in the fermenter (input variables) were tested. The modeling and prediction of outcomes were conducted using a well-established algorithm: the Levenberg–Marquardt algorithm for MLP and the adaptive neuro-fuzzy inference system (ANFIS) for the latter. The MLP employing the backpropagation training technique produced a mean square error (MSE) of 0.0089 and an R^2 value of 0.9895, in contrast to the adaptive neuro-fuzzy inference system (ANFIS), which recorded values of 0.3129 and 0.9993, respectively. These findings underscore the efficacy of both models in assessing production and forecasting bioethanol derived from waste materials (Jahanbakhshi and Salehi, 2019). Esfahanian et al. (2013) investigated the batch fermentation phase for bioethanol production from glucose utilizing the yeast species *Saccharomyces cerevisiae*. They modeled and optimized three input variables influencing production—temperature, pH, and glucose concentration—using RSM and MLP. Although the optimization results were relatively close, the MLP demonstrated superior precision compared to RSM in its predictions, as evidenced by R^2 values of 0.9975 and 0.9965 for MLP and RSM, respectively. The utilization of artificial intelligence in the critical phases of hydrolysis and fermentation has been examined by Sebayang et al. (2017), highlighting the optimization of efficiency, cost, and time through the application of backpropagation artificial neural networks for predicting sugars and their concentrations derived from enzymatic hydrolysis and fermentation processes. This analysis was grounded in the closed average values obtained, where the predicted reducing sugars were 175.94 g/L compared to an experimental value of 174.29 g/L, and the ethanol concentrations were 82.11 g/L versus 81.52 g/L, respectively. Talebnia et al. (2015) evolved and integrated two multilayer perceptron feedforward network models to forecast the course of time and amount of ethanol, applying these models to the enzyme-based hydrolysis and fermentation stages to represent the complete bioethanol production process from steam-treated rapeseed straw. Many groups solely used the bioethanol concentration as the output variable while modeling the fermentation process. Conversely, Selvakumar et al. (2018) looked into cellulase activity and bioethanol yield, Konishi (2020) examined cell growth and bioethanol production jointly, and Grahovac et al. (2016) examined the bioethanol content, yeast cell count, and sugar content during the alcoholic fermentation process.

8.6 DIFFICULTIES AND PROSPECTS FOR THE FUTURE

The sustainable production of lignocellulosic bioethanol using wastes has promising future prospects because present scientific and technological efforts are concentrated on improving the process's efficiency and sustainability. A circular bioeconomy strategy can be used to recycle lignocellulosic waste from industry and agriculture for the production of bioethanol, resulting in a closed-loop process that optimizes the value-added substances retrieved from waste. Studies and development are still being done to improve or develop suitable technologies that can efficiently and affordably produce bioethanol from lignocellulosic waste. This includes the development of successful and affordable pretreatment methods in addition to enzymatic hydrolysis, as well as fermentation processes. Moreover, a novel approach to produce zero waste through the conversion of lignocellulosic biomass to ethanol has been demonstrated. This process can promote durability and mitigate the adverse environmental impacts of manufacturing bioethanol. The production of environmentally friendly lignocellulosic bioethanol from waste is expected to benefit from additional studies and developments aimed at improving the method's effectiveness and longevity. A few potential futures that can enable sustainable production of bioethanol are innovations in technology, the circular bioeconomy, and waste-free biorefineries (Devi et al., 2022).

One of the main obstacles to ML uses for bioethanol production is the availability of a reasonably large amount of correct data, and this is expected to continue to occur in the not-so-distant future as well. A well-described dataset is necessary for machine learning. The needed information must first be present in the data (descriptors such as the material's physical and chemical qualities or operational conditions should explain certain important performance measurements). Second, for statistically sound conclusions, the dataset needs to be correct and big enough. One of the most difficult things about applications of machine learning in many domains is building an adequate and reliable dataset; this appears to be the situation for the manufacturing of lignocellulosic ethanol as well, and it is additionally one of the main drawbacks of existing uses. The process of making lignocellulosic ethanol may actually be more difficult than this, since more datasets are needed to create statistically significant models, as a greater number of descriptors are needed to accurately characterize such systems. Another factor contributing to the difficulty in producing bioethanol is the fact that different approaches (such as thermochemical or fermentation routes) or variations on the same route (such as sequential or simultaneous hydrolysis and fermentation) are explored for the production of ethanol, but none of them are thought to be the predominant method. The results from various process combinations change because the descriptors (and occasionally the performance measurements) for dissimilar pathways differ. As a result, the availability of several processes and process configurations distributes the work among the different paths and keeps any one of them from accumulating enough data. Moreover, novel substances or first-time process steps generate distinct variables not documented in previous studies. The implementation of machine learning (ML), which depends on inferring relationships from the dataset, is made extremely difficult by all of these factors. ML analysis can only be applied to a limited number of data points with unique variables. For more complicated setups, such as simultaneous fermentation and hydrolysis, the data appears to be a bigger issue because more descriptors will be required to characterize the combined process, necessitating additional data entries. The non-standard nature of cellulosic raw materials, which produces varying products and yields, particularly in the pretreatment and hydrolysis stages, appears to be another difficulty (Raj et al., 2022). Assuming that all descriptors are well-defined and that there is an adequate amount of data to smooth out variations, this should not normally pose a problem for machine learning. Nevertheless, in this domain, datasets are usually small, and there is a significant degree of uncertainty (or at least variation) attached to the descriptors. However, there are also attempts being made to get beyond these obstacles, and more are likely to come. The goal of one of these initiatives is to apply the machine learning models and analysis created for some domains to other related industries through a process known as transfer learning (Kaya and Hajimirza, 2019). Even though these are difficult to put into practice, they might be helpful for the production of lignocellulosic ethanol as well. For instance, models and experiences developed for the more established field of corn sugar fermentation should offer some insights for the machine learning analysis of the fermentation step in the production of lignocellulosic ethanol. Another way to generate a dataset for machine learning (ML) analysis is through the use of computational tools, particularly density functional theory (DFT), which is widely used in material science. Researchers can more easily share data thanks to the standard nature of the data produced in this way; in fact, numerous databases, including Material Project, OQMD, AFLOWLIB, and Computational Material Repository (Jain et al., 2016; Kirklin et al., 2015; Curtarolo et al., 2012; Landis et al., 2012), were created expressly for this purpose. However, most of such software programs and databases are designed for crystalline and small molecules; with present computer capabilities, it may not be feasible to compile the large number of information records required for an operation like fermentation. However, some sort of information sharing techniques can still be employed to maximize the advantage of machine learning, since bigger datasets with additional attributes always offer more precise and in-depth details for ML research. One approach to accomplish this is to create standardized testing and reporting techniques in conjunction with field researchers, allowing for the combination of data from multiple experimental efforts to provide a sizable volume of uniformly collected data. With the astounding rate of advancement in computational tools and algorithms, DFT and other computational tools may eventually be used in this discipline to generate data and comprehend the process. A different approach for small datasets is to reduce the number of descriptors (dimensionality reduction). This can be achieved by integrating them into a smaller new descriptor set (feature extraction) or removing insignificant descriptors (feature selection) in order to reduce the number of descriptors required (Alpaydin, 2020). In the meantime, recent years have seen the investigation of novel machine learning algorithms and techniques for tiny datasets (Zhang and Ling, 2018; Feng et al., 2019; Ma et al., 2020). Future developments are expected to see this trend continue and further the study of lignocellulosic biofuels. Lastly, one of the primary drawbacks (and criticisms) of the present ML applications is the black box character of ML models, which has been challenged in recent years by the idea of explainable ML (Suvarna et al., 2022). Like transfer learning, this idea is difficult to put into practice, but it has a lot of interest as it aims to clarify the rationale behind the outcomes produced by ML models. Because it makes it easier to understand the relationships between the descriptors and how they affect the result, this approach might be more advantageous for complex systems like lignocellulosic ethanol. It also makes it possible to reduce the number of descriptors—that is, the size of the dataset—by removing those that are not significant, which facilitates utilization of small datasets.

8.7 CONCLUSION

Improving the environmentally friendly manufacture of ethanol utilizing lignocellulosic waste materials is a commitment to solving the intricate problems of disposal, long-term viability, and mitigating climate change, rather than just seeking a substitute fuel source. The world is facing a pressing need to switch from using limited fossil fuels to more sustainable forms of energy, and lignocellulosic biomass is a rich resource that is both abundant and diversified, with the potential to play a significant part in the energy landscape of the future. We have examined the topic's complexity in this review, starting with the acknowledgment of bioethanol as an environmentally benign and sustainable fuel source. Its importance comes from its ability to not only decrease emissions of greenhouse gases but also convert waste products into useful energy sources, reducing the need for landfills and incineration. In this sense, sustainability includes both the production process's economic viability and environmental elements. Despite being the most plentiful biomass source, turning LCB to ethanol remains an extensive procedure that requires a lot of steps due to the characteristics of LCB. Particularly for enzymatic and microbiological hydrolysis, the structural makeup of lignocellulose biomass produces a hesitant tendency for hydrolysis. Extensive research has been conducted utilizing various techniques to convert biomass from lignocellulose into biofuel, specifically bioethanol. With the exception of biological pretreatment, the primary obstacles to all pretreatment procedures are the generation of inhibitory chemicals such as furfural and furfural derivative substances, as well as hemicellulose solubilization. The primary obstacle to scaling up (industrial scale production) is the issue of poor yield, inadequate productivity, and an extended period of residence for microbial degradation and hydrolysis. In order to effectively use and convert the plentiful resources, it is necessary to isolate and identify the right microorganisms, choose the right raw material (biomass), and optimize the process parameters (temperature, pH, aeration, particle size, time, substrate concentration, and inoculum concentration). The intricacy of the bioethanol production process includes microbe tolerance, pretreatment effectiveness, biomass supply, and enzyme cost, among many others. These problems are not insurmountable; rather, they call for creative thinking, advances in research, and constant technological innovation. In order to produce sustainable ethanol from biomass from lignocellulosic waste materials and promote renewable energy sources, the study recommends more research on enzymatic pretreatment methods, microbial strain improvement, creative bioreactor design, green solvent integration, and assessment of life cycle. Green technology, circular bioeconomies, and the creation of more eco-friendly and productive processes are the hallmarks of the future. The bioethanol sector is changing, as we have seen, with the goal of increasing yield, decreasing waste generation, and optimizing each stage of the production processes. These initiatives are not only admirable but also essential to the long-term viability and expansion of the sector. It is an effort to achieve a more sustainable and environmentally friendly future to improve the sustainable manufacture of ethanol from biomass from lignocellulosic waste materials, rather than just an academic or industrial activity. It's an acknowledgment that our energy decisions can be in line with our environmental obligations. Even if there are still difficulties, there is optimism regarding the future that bioethanol will play a significant role in our green energy landscape because of the persistent pursuit of answers and the cooperation of researchers, industry leaders, and politicians. We need to move on with this environmentally beneficial bioethanol manufacturing process from lignocellulosic waste because it will help us achieve our goal of safer, greener, and more sustainable energy in the future.

REFERENCES

Agrawal, R., Verma, A., Singhania, R.R., Varjani, S., Di Dong, C., & Patel, A.K. (2021). Current understanding of the inhibition factors and their mechanism of action for the lignocellulosic biomass hydrolysis. *Bioresource Technology,* 332, 125042.

Ahmadian-Moghadam, H., Elegado, F.B., & Nayve, R. (2013). Prediction of ethanol concentration in biofuel production using artificial neural networks. *American Journal of Modelling and Optimization,* 1, 31–35.

Alpaydin, E. (2020). *Introduction to Machine Learning*, 4 (ed). The MIT Press, Cambridge, Massachusetts, London, England.

Alvira, P., Tomas-Pejo, E., Ballesteros, M., & Negro, M.J. (2010). Pretreatment technologies for an efficient bioethanol production process based on enzymatic hydrolysis: A review. *Bioresource Technology,* 101(13), 4851–4861.

Alviso, D., Artana, G., & Duriez, T. (2020). Prediction of biodiesel physicochemical properties from its fatty acid composition using genetic programming. *Fuel,* 264, 116844.

Amiri, H., & Karimi, K. (2015). Improvement of acetone, butanol, and ethanol production from woody biomass using organosolv pretreatment. *Bioprocess and Biosystems Engineering,* 38(10), 1959–1972.

Aruwajoye, G.S., Faloye, F.D., Kassim, A., Saha, A.K., & Kana, E.G. (2022). Intelligent modelling of fermentable sugar concentration and combined severity factor (CSF) index from pretreated starch-based lignocellulosic biomass. *Biomass Conversion and Biorefinery,* https://doi.org/10.1007/s13399-022-03013-y.

Astray, G., Gullón, B., Labidi, J., & Gullón, P. (2016). Comparison between developed models using response surface methodology (RSM) and artificial neural networks (ANNs) with the purpose to optimize oligosaccharide mixtures production from sugar beet pulp. *Industrial Crops and Products,* 92, 290–299.

Bannor, B.E., & Acheampong, A.O. (2019). Deploying artificial neural networks for modeling energy demand: international evidence. *International Journal of Energy Sector Management,* 14(2), 285–315.

Bender, L.E., Lopes, S.T., Gomes, K.S., Devos, R.J.B., & Colla, L.M. (2022). Challenges in bioethanol production from food residues. *Bioresource Technology Reports,* 19, 101171.

Bonechi, C., Consumi, M., Donati, A., Leone, G., Magnani, A., Tamasi, G., & Rossi, C. (2017). Biomass: An overview. *Bioenergy Systems for the Future,* 3–42.

Brandt, A., Gräsvik, J., Hallett, J.P., & Welton, T. (2013). Deconstruction of lignocellulosic biomass with ionic liquids. *Green Chemistry,* 15(3), 550–583.

Chang, C.W., Yu, W.C., Chen, W.J., Chang, R.F., & Kao, W.S. (2011). A study on the enzymatic hydrolysis of steam exploded napiergrass with alkaline treatment using artificial neural networks and regression analysis. *Journal of the Taiwan Institute of Chemical Engineers,* 42(6), 889–894.

Charte, F., Romero, I., Pérez-Godoy, M.D., Rivera, A.J., & Castro, E. (2017). Comparative analysis of data mining and response surface methodology predictive models for enzymatic hydrolysis of pretreated olive tree biomass. *Computers & Chemical Engineering,* 101, 23–30.

Cheah W.Y., Sankaran R., Show P.L., Tg. Ibrahim T.N.B., Chew K.W., Culaba A., & Chang J-S. (2020). Pretreatment methods for lignocellulosic biofuels production: Current advances, challenges and future prospects. *Biofuel Research Journal,* 25, 1115–1127.

Chen, W.H., Lo, H.J., Aniza, R., Lin, B.J., Park, Y.K., Kwon, E.E., Sheen, H.K., & Grafilo, L.A.D.R. (2022). Forecast of glucose production from biomass wet torrefaction using statistical approach along with multivariate adaptive regression splines, neural network and decision tree. *Applied Energy,* 324, 119775.

Culaba, A.B., Mayol, A.P., San Juan, J.L.G., Vinoya, C.L., Concepcion, R.S., 2nd, Bandala, A.A., Vicerra, R.R.P., Ubando, A.T., Chen, W.H., Chang, J.S. (2022). Smart sustainable biorefineries for lignocellulosic biomass. *Bioresource Technology,* 344(Part B), 126215.

Curtarolo, S., Setyawan, W., Wang, S., Xue, J., Yang, K., Taylor, R.H., Nelson, L.J., Hart, G.L.W., Sanvito, S., Buongiorno-Nardelli, M., Mingo, N., & Levy, O. (2012). AFLOWLIB.ORG: A distributed materials properties repository from high-throughput ab initio calculations. *Computational Materials Science,* 58, 227–235.

Das, S., Bhattacharya, A., Haldar S., et al. (2015). Optimization of enzymatic saccharification of water hyacinth biomass for bioethanol: Comparison between artificial neural network and response surface methodology. *Sustainable Materials and Technologies*, 3, 17–28.

Devi, A., Bajar, S., Kour, H., Kothari, R., Pant, D., & Singh, A. (2022). Lignocellulosic biomass valorization for bioethanol production: A circular bioeconomy approach. *Bioenergy Research*, 5, 1820.

Dimos, K., Paschos, T., Louloudi, A., Kalogiannis, K.G., Lappas, A.A., Papayannakos, N., Kekos, D., & Mamma, D. (2019). Effect of various pretreatment methods on bioethanol production from cotton stalks, *Fermentation*, 5, 5.

Dong, C., & Chen, J. (2019). Optimization of process parameters for anaerobic fermentation of corn stalk based on least squares support vector machine. *Bioresource Technology*, 271, 174–181.

Edeh, I. (2021). Bioethanol production: An overview. *IntechOpen*, 1–24.

Esfahanian, M., Nikzad, M., Najafpour, G., & Ghoreyshi, A.A. (2013). Modeling and optimization of ethanol fermentation using saccharomyces cerevisiae: response surface methodology and artificial neural network. CICEQ, 19(2) 241–252.

Ethaib, S., Omar, R., Mazlina, M.K.S., Radiah, A.B.D., & Syafiie, S. (2018). Development of a hybrid PSO-ANN model for estimating glucose and xylose yields for microwave-assisted pretreatment and the enzymatic hydrolysis of lignocellulosic biomass. *Neural Computing and Applications*, 30(4), 1111–1121.

Feng, S., Zhou, H., & Dong, H. (2019). Using deep neural network with small dataset to predict material defects. *Materials & Design*, 162, 300–310.

Fischer, J., Lopes, V.S., Cardoso, S.L., Coutinho Filho, U., & Cardoso, V.L. (2017). Machine learning techniques applied to lignocellulosic ethanol in simultaneous hydrolysis and fermentation. *Brazilian Journal of Chemical Engineering* 34(1), 53–63.

Galbe, M., & Wallberg, O. (2019). Pretreatment for biorefineries: A review of common methods for efficient utilisation of lignocellulosic materials. *Biotechnology for Biofuels and Bioproducts*, 12, 294.

Gitifar, V., Eslamloueyan, R., & Sarshar, M. (2013). Experimental study and neural network modeling of sugarcane bagasse pretreatment with H_2SO4 and O_3 for cellulosic material conversion to sugar. *Bioresource Technology*, 148, 47–52.

Grahovac, J., Jokic, A., Dodic, J., Vucurovic, D., & Dodic, S. (2016). Modelling and prediction of bioethanol production from intermediates and byproduct of sugar beet processing using neural networks. *Renewable Energy*, 85, 953–958.

Griffin, D.W., & Schultz, M.A. (2012). Fuel and chemical products from biomass syngas: A comparison of gas fermentation to thermochemical conversion routes. *Environmental Progress & Sustainable Energy*, 31(2), 219–224.

Gupta, K., & Chundawat, T.S. (2020). Zinc oxide nanoparticles synthesized using *Fusarium Oxysporum* to enhance bioethanol production from rice-straw. *Biomass Bioenergy*, 143, 105840.

Jahanbakhshi, A., & Salehi, R. (2019). Processing watermelon waste using *Saccharomyces cerevisiae* yeast and the fermentation method for bioethanol production. *Journal of Food Process Engineering*, 42(7), e13283.

Jain, A., Hautier, G., Ong, S.P., & Persson, K. (2016). New opportunities for materials informatics: Resources and data mining techniques for uncovering hidden relationships. *Journal of Materials Research*, 31(8), 977–994.

Jorgensen, H., Kristensen, J.B., & Felby, C. (2007). Enzymatic conversion of lignocellulose into fermentable sugars: challenges and opportunities. *Biofuels, Bioproducts and Biorefining*, 1(2), 119–134.

Kaya, M., & Hajimirza, S. (2019). Using a novel transfer learning method for designing thin film solar cells with enhanced quantum efficiencies. *Scientific Reports*, 9(1), 5034.

Kim, J.S., Lee, Y.Y., & Kim, T.H. (2016). A review on alkaline pretreatment technology for bioconversion of lignocellulosic biomass. *Bioresource Technology,* 199, 42-48.

Kirklin, S., Saal, J.E., Meredig, B., Thompson, A., Doak, J.W., Aykol, M., Rühl, S., & Wolverton, C. (2015). The open quantum materials database (OQMD): assessing the accuracy of DFT formation energies. *NPJ Computational Materials,* 1(1), 15010.

Konishi, M. (2020). Bioethanol production estimated from volatile compositions in hydrolysates of lignocellulosic biomass by deep learning. *Journal of Bioscience and Bioengineering,* 129(6), 723–729.

Kucharska. K., Hołowacz, I., Konopacka-Łyskawa, D., Rybarczyk, P., & Kamiński, M. (2018). Key issues in modeling and optimization of lignocellulosic biomass fermentative conversion to gaseous biofuels. *Renewable Energy,* 129, 384–408.

Landis, D.D., Hummelshoj, J.S., Nestorov, S., Greeley, J., Dulak, M., Bligaard, T., Norskov, J.K., & Jacobsen, K.W. (2012). The Computational Materials Repository. *Computing in Science & Engineering,* 14(6), 51–57.

Larose, D.T., & Larose, C.D. (2014). *Discovering Knowledge in Data: An Introduction to Data Mining*, 2 (ed). Wiley, Hoboken, New Jersey.

Lee, K.M., Zanil, M.F., Chan, K.K., Chin, Z.P., Liu, Y.C., & Lim, S. (2020). Synergistic ultrasound-assisted organosolv pretreatment of oil palm empty fruit bunches for enhanced enzymatic saccharification: An optimization study using artificial neural networks. *Biomass Bioenergy.* 139, 105621.

Li, W., Ghosh, A., Bbosa, D., Brown, R., & Wright, M.M. (2018). Comparative techno-economic, uncertainty and life cycle analysis of lignocellulosic biomass solvent liquefaction and sugar fermentation to ethanol. *ACS Sustainable Chemistry & Engineering Journal,* 6(12), 16515–16524.

Liu, C.G., Xiao, Y., Xia, X.X., Zhao, X.Q., Peng, L., Srinophakun, P., & Bai, F.W. (2019). Cellulosic ethanol production: Progress, challenges and strategies for solutions. *Biotechnology Advances,* 37(3), 491–504.

Lugani, Y., Rai, R., Prabhu, A.A., et al. (2020). Recent advances in bioethanol production from lignocelluloses: A comprehensive review with a focus on enzyme engineering and designer biocatalysts. *Biofuel Research Journal*, 28, 1267–1295.

Ma, R., Colon, Y.J., & Luo, T. (2020). Transfer learning study of gas adsorption in metal-organic frameworks. *ACS Applied Materials & Interfaces Journal,* 12(30), 34041–34048.

Moayedi, H., Aghel, B., Foong, L.K., & Bui, D.T. (2020). Feature validity during machine learning paradigms for predicting biodiesel purity. *Fuel*, 262, 116498.

Moodley, P., Rorke, D.C., & Kana, E.B.G. (2019). Development of artificial neural network tools for predicting sugar yields from inorganic salt-based pretreatment of lignocellulosic biomass. *Bioresource Technology,* 273, 682–686.

Onay, M. (2022). Sequential modelling for carbohydrate and bioethanol production from Chlorella saccharophila CCALA 258: A complementary experimental and theoretical approach for microalgal bioethanol production. *Environmental Science and Pollution Research International,* 29(10), 14316-14332.

Pereira, R.D., Badino, A.C., & Cruz, A.J.G. (2020). Framework based on artificial intelligence to increase industrial bioethanol production. *Energy and Fuels,* 34(4), 4670–4677.

Perlack, R.D., Wright, L.L., Turhollow, A., & Graham, R.L., et al. (2005). Biomass as feedstock for a bioenergy and bioproducts industry: The technical feasibility of a billion-ton annual supply. USDA and DOE Joint Report 2005, DOE/GO-102005-2135 or ORNL/TM-2005/66.

Phromphithak, S., Onsree, T., & Tippayawong, N. (2021). Machine learning prediction of cellulose-rich materials from biomass pretreatment with ionic liquid solvents. *Bioresource Technology,* 323, 124642.

Praveenkumar, R., Suresh, K., Chozhavendhan, S., & Bharathiraja, B. (2014). Comparative analysis of saccharification of cassava sago waste using Aspergillus niger and Bacillus sp. for the production of bio-ethanol using *Saccharomyces cerevisiae. International Journal of Chemical Technology Research,* 6, 5090–5094.

Qiao, J., Cui, H., Wang, M., Fu, X., Wang, X., Li, X., & Huang, H. (2022). Integrated bio-refinery approaches for the industrialization of cellulosic ethanol fuel. *Bioresource Technology,* 360, 127516.

Raj, T., Chandrasekhar, K., Naresh Kumar, A., Rajesh Banu, J., Yoon, J.J., Kant Bhatia, S., Yang, Y.H., Varjani, S., & Kim, S.H. (2022). Recent advances in commercial biorefineries for lignocellulosic ethanol production: current status, challenges and future perspectives. *Bioresource Technology,* 344(Pt B), 126292.

Ravanal, M.C., Camus, C., Buschmann A.H., et al. (2019). Production of bioethanol from brown algae. In, - *Advances in Feedstock Conversion Technologies for Alternative Fuels and Bioproducts.* Woodhead Publishing, United Kingdom. 69–88.

Robak, K., & Balcerek, M. (2018). Review of second generation bioethanol production from residual biomass. *Food Technology and Biotechnolgy*, 56(2), 174–187.

Sebayang, A.H., Masjuki, H.H., Ong, H.C., et al. (2017). Optimization of bioethanol production from sorghum grains using artificial neural networks integrated with ant colony. *Industrial Crops and Products,* 97, 146–155.

Selvakumar, P., Kavitha, S., & Sivashanmugam, P. (2018). Optimization of process parameters for efficient bioconversion of thermo-chemo pretreated Manihot esculenta Crantz YTP1 stem to ethanol. *Waste Biomass Valorization,* 10(8), 2177–2191.

Smuga-Kogut, M., Kogut, T., Mark ew z, R., & Słowik, A. (2021). Use of machine learning methods for predicting amount of bioethanol obtained from lignocellulosic biomass with the use of ionic liquids for pretreatment. *Energies*, 14(1), 243.

Stenzel F., Greve P., Lucht W., Tramberend S., Wada Y., & Gerten D. (2021). Irrigation of biomass plantations may globally increase water stress more than climate change, *Nature Communications,* 12, 1512.

Suresh, T., Sivarajasekar, N., Balasubramani, K., Ahamad, T., Alam, M., & Naushad, M. (2020). Process intensification and comparison of bioethanol production from food industry waste (potatoes) by ultrasonic assisted acid hydrolysis and enzymatic hydrolysis: statistical modelling and optimization. *Biomass Bioenergy,* 142, 105752.

Suvarna, M., Jahirul, M.I., Aaron-Yeap, W.H., Augustine, C.V., Umesh, A., Rasul, M.G., Günay, M.E., Yildirim, R., & Janaun, J. (2022). Predicting biodiesel properties and its optimal fatty acid profile via explainable machine learning. *Renewable Energy,* 189, 245–258.

Talebnia, F., Mighani, M., Rahimnejad, M., & Angelidaki, I. (2015). Ethanol production from steam exploded rapeseed straw and the process simulation using artificial neural networks. *Biotechnology and Bioprocess Engineering,* 20(1), 139–147.

Tavva, S.S.M.D., Deshpande, A., Durbha S.R., et al. (2016). Bioethanol production through separate hydrolysis and fermentation of *Parthenium hysterophorus* biomass. *Renewable Energy*, 86, 1317–1323.

Thompson, T.M., Young, B.R., & Baroutian, S. (2020). Pelagic Sargassum for energy and fertiliser production in the Caribbean: A case study on Barbados. *Renewable Sustainable Energy Review,* 118, 109564.

Tovar, L.P., Rivera, E.C., Mariano, A.P., Maciel, M.R.W., & Filho R.M. (2018). Prediction of overall glucose yield in hydrolysis of pre-treated sugarcane bagasse using a single artificial neural network: Good insight for process development. *Journal of Chemical Technology and Biotechnology,* 93(4), 1031–1043.

Tursi, A. (2019). A review on biomass: Importance, chemistry, classification, and conversion. *Biofuel Research Journal,* 6, 962–979.

Vaid, S., Nargotra, P., & Bajaj, B.K. (2018). Consolidated bioprocessing for biofuel-ethanol production from pine needle biomass. *Environmental Progress & Sustainable Energy*, 37, 546–552.

Vani, S., Sukumaran, R.K., & Savithri, S. (2015). Prediction of sugar yields during hydrolysis of lignocellulosic biomass using artificial neural network modelling. *Bioresource Technology,* 188, 128–135.

Velazquez, J., Rodriguez-Jasso, R., Colla, L., Galindo, A., Cervantes, D., Aguilar, C., Fernandes, B., & Ruiz, H. (2018). Microalgal Biomass pretreatment for bioethanol production: A review. *Biofuel Research Journal*, 5, 780–791.

Vinitha, N., Vasudevan, J., & Gopinath, K.P. (2023). Bioethanol production optimization through machine learning algorithm approach: Biomass characteristics, saccharification, and fermentation conditions for enzymatic hydrolysis. *Biomass Conversion and Biorefinery,* 13, 7287–7299.

Wang, Z., Peng, X., Xia, A., Shah, A.A., Huang, Y., Zhu, X., Zhu, X., & Liao, Q. (2022). The role of machine learning to boost the bioenergy and biofuels conversion. *Bioresource Technology,* 343, 126099.

Xu, Z., & Huang, F. (2014). Pretreatment methods for bioethanol production. *Applied Biochemistry and Biotechnology,* 174(1), 43–62.

Zhang, J., Bao, J., Wang, N., Wang, W., Wang, X., & Zhu, Z. (2010). Biodetoxification of toxins generated from lignocellulose pretreatment using a newly isolated fungus, *Amorphotheca resinae* ZN1, and the consequent ethanol fermentation. *Biotechnology for Biofuels,* 3, 26.

Zhang, Y., & Ling, C. (2018). A strategy to apply machine learning to small datasets in materials science. *NPJ Computational Materials,* 4(1), 25.

Zhao, F., Yi, Y., & Lü, X. (2021). Chapter 4: Essential process and key barriers for converting plant biomass into biofuels. *Advances in 2nd Generation of Bioethanol Production*. pp. 53–70.

Zheng, Y., Pan, Z., & Zhang, R. (2009). Overview of biomass pretreatment for cellulosic ethanol production. *International Journal of Agricultural and Biological Engineering,* 2(3), 51–68.

Zoghlami, A., & Paes, G. (2019). Lignocellulosic biomass: understanding recalcitrance and predicting hydrolysis. *Frontiers in Chemistry*, 7, 874.

9 Current Status of Artificial Intelligence-based Biofuel Research

Arpan Das

9.1 INTRODUCTION

Over time, the availability and demand for energy in almost every sector have been out of equilibrium due to population growth and growth in the economy. The automobile sector's consumption of energy has increased significantly from 23 percent in 1971 to 29 percent in 2017, compared to the global final energy consumption by sector (Ahamed et al., 2021). In order to reduce reliance on fossil fuels, experts are increasingly looking for alternatives that are renewable as the release of greenhouse gases has primarily raised concerns about warming the planet. Among these options, bioenergy stands out as the largest renewable energy source, accounting for 50 percent of the worldwide renewable energy market and showing significant potential in tackling the challenges related to worldwide energy use and environmental issues (International Energy Agency, 2018). The contemporary industry has evolved numerous techniques to transform biomass into various energy forms, including solid (such as biochar), liquid (like bioethanol), and gaseous forms (such as biogas, biohydrogen, and syngas). Bioenergy's advantages have been extensively analyzed in the academic literature, including improving energy sustainability, boosting economic growth (Demirbas & Demirbas, 2007), enhancing local farming, and increasing the efficiency of natural resource use (Khishtandar et al., 2017). However, the widespread implementation of bioenergy systems faces obstacles pertaining to feedstocks for biomass, seasonal quality, challenges in biorefinery techniques, and process regulation, as well as challenges and complications in supply chains. To overcome these hurdles, improvement and research of bioenergy have received a lot of attention. Traditional modeling strategies, like optimizing processes and supply chains, have also been extensively studied in the academic field.

AI, or artificial intelligence, has seen a surge in interest lately. AI is about machines' capability to carry out tasks that resemble the intellect of humans. It can be applied through several computer science techniques, such as fuzzy logic, heuristic algorithms, and machine learning. In comparison to fields with thousands of AI-related publications, the use of AI in bioenergy systems is relatively scarce. However, studies have shown the possibility of AI in overcoming obstacles in the

 DOI: 10.1201/9781003571223-9

development of bioenergy. For instance, Castillo-Villar analyzed multiple studies employing metaheuristic algorithms to tackle challenges in the bioenergy supply chain (Castillo-Villar, 2014). Ardabili et al. (2020) explored the use of machine learning and deep learning in different fields of biofuel research. Previous evaluations on this topic are scattered, with most focusing on specific aspects of bioenergy systems (like the performance of biofuel-powered engines; Yusri et al., 2018) or a single AI technique (like Artificial Neural Networks [ANN]; Sewsynker-Sukai et al., 2017). A thorough analysis of existing and prospective AI applications across the whole bioenergy supply chain is crucial due to the wide variety of AI approaches, bioenergy products, conversion technologies, biomass types, and supply chain architectures.

9.2 SYSTEM OF BIOENERGY

Due to its significant impact on meeting energy needs and minimizing adverse environmental effects, bioenergy production plays a crucial role and is essential to sustainable development. However, the primary issue in bioenergy production is the low efficiency of current technologies, which leads to an economic imbalance. The development of bioenergy production on a commercial scale and the improvement of the efficiency of current bioenergy conversion procedures are primarily hindered by the challenges associated with bioenergy generation. According to Virkaj¨arvi et al. (2009), engineering research is necessary for the design, optimization, and scaling up of a conversion system that consists of the reactor as well as auxiliary and supporting components. Furthermore, the infrastructure for producing bioenergy must transition to high-tech processes in light of the growing need for energy. The bioenergy system is intricate and typically encompasses the cultivation, production, harvesting, pretreatment, bioenergy generation, distribution, and ultimate use of biomass. A broad range of biomass types can be transformed into bioenergy products, such as agricultural, animal, or industrial wastes, and blends. Each type of biomass exhibits variations in its chemical makeup and physical properties, including ash content, moisture level, chemical composition, energy content, density, and particle size, which influence the design and operation of biomass conversion processes. Consequently, previous studies have concentrated on the characterization and screening of biomass for energy applications. However, much of this data remains underutilized, and biomass research continues to heavily depend on conventional trial-and-error methods. Utilizing artificial intelligence could unlock the potential of this data to facilitate quicker and more efficient screening of biomass. There are two main categories of biomass conversion technologies: thermochemical and biochemical conversion. Thermochemical conversion involves the production of solid, liquid, and gas products through thermal processing of biomass. This includes technologies like gasification, pyrolysis, torrefaction, and hydrothermal carbonization (Zhang & Zhang, 2019). Biochemical conversion, on the other hand, transforms biomass into liquid (for instance, biodiesel and bioethanol) and gas (for example, biogas) products using small molecules, microorganisms, or enzymes. Various modeling strategies have been explored, including simulation and optimization models, which generally rely on an understanding of the interactions between biomass nature, process parameters, as well as the final yields and properties of the product. However, these relationships are not always fully understood. In contrast, many artificial intelligence techniques do not require prior knowledge of these interactions, potentially bridging the information gap and enhancing traditional modeling tactics. Many products may be created and applied to deliver energy in the forms of work, heat, and electricity based on the conversion technique (e.g., biofuel used in vehicle engines; Thrän et al., 2015). Biochar and other solid biofuels generated through pyrolysis and torrefaction can be burned to produce heat or co-fired with coal in conventional or combined heat and power (CHP) facilities (Nunes et al., 2014). With the potential to improve conversion economics, co-producing syngas and biochar in gasification has drawn more interest recently (Yao et al., 2018). Various processes can be employed to make liquid biofuels, such as fermentation-produced bioethanol, transesterification-produced biodiesel, pyrolysis-produced bio-oil, and Fischer-Tropsch synthesis-produced biofuel. For application in engines and electric generators, liquid biofuels provide an alternative to fossil fuel-based diesel, petrol, and jet fuel. Gaseous biofuels are often defined as gasification syngas or methane-rich gas (biogas) from the process of anaerobic digestion (Guo et al., 2015). According to IEA Bioenergy (2009), biogas and syngas may be converted into common fuel types or burned directly to produce electricity on-site. To guarantee the overall economic viability and supply chain resilience, bioenergy supply chains (BSC) must frequently have exquisite designs and precise operations due to the significant unpredictability surrounding biomass availability and quality. While several strategies have been investigated, few prior studies have adequately addressed those difficulties (Lan et al., 2020). Prior reviews have discovered the primary areas of study requiring further investigation. These include modeling issues, such as the high computational expenses and the significant number of variables required, as well as methodological challenges, such as the challenges associated with including stakeholder perspectives and possible disputes, socioeconomic issues, and environmental constraints.

9.3 ARTIFICIAL INTELLIGENCE

Among the difficulties in the context of various biomass types and transformation options for biomass-to-biofuels are selecting the best technology for each feedstock and optimizing process parameters to achieve financial viability, sustainable development, and high-quality products. It is quite challenging to attain the correctness of mechanical models in the instance of complex, nonlinear systems, including innovative fermentation procedures involving several microbial populations. As a result, the accuracy of biofuel output and quality forecast when process parameters are improved using classical modeling is in doubt. Conventional optimization methods used in the waste-to-biofuels sector are predicated on a quantitative understanding of the connections between the properties of biomass, operational variables, and the yield and qualities of biofuel. Artificial intelligence (AI) fills in certain gaps and enhances mathematically optimized models, but it is more effective than conventional approaches for predicting crucial variables for optimal biochemical as well as thermochemical process performances. AI algorithms lack an understanding of the connections between inputs and outcomes. The AI approaches that have been effective in the domain of recovering biomass for energy can be divided into four primary groups: heuristics, machine learning, hybrid approaches, and symbolic AI. These categories are based on the techniques and tools used, as well as specific to biofuel types that may be generated. ANN, Adaptive Neuro-Fuzzy Inference System (ANFIS), Random Forest (RF), and Support Vector Machine (SVM) are some of the artificial intelligence (AI) methods that can provide the most precise estimates of the outcome yield and composition for the thermochemical transformation of biomass. When these algorithms are coupled with kinetic and thermodynamic modeling of the processes, biomass decomposition can be even better controlled (Zhu et al., 2019; Safarian et al., 2019; Aghbashlo et al., 2019). Symbolic AI reflects the workings of human cognition by utilizing logic deduction and symbols to express cognition (Haugeland, 1985; Smolensky, 1987). Numerous uses for symbolic artificial intelligence (AI) may be found in the domain of process system design. Several of these applications are related to bioenergy systems, including boilers (Romeo & Gareta, 2009) and FL-based regulation for power plants that rely on biomass (Jurado et al., 2002). According to Zheng et al. (2013), Heuristics are random methods of searching that are used using experience and learning-based methodologies. Swarm intelligence and evolutionary algorithms are the two primary categories of heuristics. Genetic algorithms (GA), differential evolution, and evolutionary strategies are illustrations of evolutionary algorithms (Vikhar, 2017). Several techniques are used in swarm intelligence, including ant colony optimization (ACO; Chakraborty & Kar, 2017) and particle swarm optimization (PSO). Prior research has shown how effective algorithms are at locating close to ideal solutions for challenging issues such as BSC scheduling and planning. RBF, for instance, has a quicker training process and is less susceptible to noise in training data than a normal FNN. However, because it needs a high number of hidden nodes, it cannot be used in all situations (Mohd Ali et al., 2015). Another branch of AI is machine learning, which does not require explicit programming and may learn and advance via experience with certain tasks. The two most often used machine learning approaches are connectivism and statistical learning (Russell & Norvig, 2010). The advanced convolutional neural network, feedforward neural network (FNN), radial-basis functional network (RBF), recurrent neural network (RNN), and other techniques are included in connectivism (Jha et al., 2017). Another name for these AI methods is "Artificial Neural Network (ANN)." Among the most effective artificial intelligence methods for resolving very difficult issues in a variety of fields, including the generation of biofuels, are ANN algorithms. By establishing connections between any process's data input and output—without being aware of the connection between them—they are intended to mimic the way humans learn (Adeyemo and Enitan, 2011). ANNs behave similarly to human neurons in this regard. The underlying workings (hidden layers) of ANN algorithms are regarded as "black box" algorithms, meaning that they are highly challenging to comprehend and explain (Castrounis, 2019). The number of hidden layers, neurons in each layer, and/or connections between neurons may all be increased to make ANNs more complicated (Castrounis, 2019). According to several suggestions, ANNs are the best method for accurately predicting the behavior of various linear and nonlinear processes that rely on the input data for certain output functions (Almomani, 2020). Statistical approaches are the foundation of statistical learning. SVM, random forests (RF), and Bayesian networks (BN; Dounis, 2010) are examples of representative approaches. The basic ideas and architecture of these machine learning approaches vary, which affects how well they work. RBF, for instance, has a quicker training process and is less susceptible to noise in training data than a normal FNN. However, because it needs a high number of hidden nodes, it cannot be used in all situations (Mohd Ali et al., 2015).

9.4 AI TOOLS FOR PREDICTING THE QUALITIES OF BIOMASS FEEDSTOCKS

The quality of bio-based products and the feasibility of biomass conversion are directly impacted by the qualities of biomass. To forecast the features of biomass feedstock, applications of AI are reviewed in this work. These applications may be connected to traditional simulations of processes for biomass conversion or AI-based models. AI presents a viable substitute to minimize costs and duration for characterizing biomass. In this regard, García et al. (2013) employed a thermogravimetric assessment (TGA) with ASTM standard D7582-15 to measure the amount of ash/moisture (ASTM D7582-15, 2015) for proximate evaluation, an oxygen bomb calorimeter with ASTM standard D5865-13 for higher heating value (HHV). Conventional methods of analysis cost money and take a lot of time.

9.4.1 AI-Powered Biomass Conversion Applications

9.4.1.1 Thermochemical Conversion Systems

Renewable feedstock includes harvested plants, residues of crops, wood, algae, fatty acids, consumable vegetable oils, sewage waste, and leftover food. In order to address the robustness of lignocellulosic, microalgae, and macroalgae biomass and to make the following procedures accessible and economical, pretreatment approaches are essential. The term "pretreatment" describes processes applied to biomass feedstock prior to the primary conversion to break down the lignified cell wall and get past the inherent layer of resistance, improving the efficiency of further conversion into bioproducts. Biomass can be converted into liquid, gas, or solid biofuels through thermochemical, chemical, or biological processes, or by combining these methods. For example, biofuels can be produced more efficiently and with higher yields when the proper biomass feedstock is prepared for biological utilization through thermal preprocessing. By using thermochemical processes, a variety of biomass types may be broken down to create liquid and solid fuels as well as synthetic gas, which can then be converted into liquid fuels. When biomass is generated and utilized as a substitute for coal, it is referred to as "bio-coal" because it has better physicochemical qualities (density of matter, ashes, volatile components, hydrophobicity, and grindability) at low temperatures (less than 300°C) and thermal procedures such as torrefaction and hydrothermal carbonization (Cahyanti et al., 2020). Thermochemical conversion technique uses a variety of techniques, including pyrolysis, gasification, liquefaction, and combustion, to turn feedstock into fuel. The initial step of all thermochemical operations that result in solid, liquid, and gaseous end products with no aid of oxygen is called pyrolysis. In gasification, biomass undergoes semi-degradation (oxidation) to produce syngas; oxygen is provided in a controlled manner to facilitate this process (Sadhwani et al., 2013). In the liquefaction process, biomass is processed through a chemical procedure using solvents such as alcohol, water, and acetone, resulting in liquid products. In the course of combustion, organic material is thermally transformed with the aid of any oxidant, usually oxygen, producing CO_2 and H_2O. This is where biomass undergoes complete oxidation. Two of the most extensively developed thermochemical methods for conversion are the pyrolysis and gasification processes (Aslani et al., 2018). Both processes thermally break down the biomass, but they vary in periods of residence (for pyrolysis 10–20 min and for the gasification process 1–30 min) and temperatures (for pyrolysis 400–700°C and for gasification more than 700°C) (Sharma et al., 2015). In addition to producing heat and energy, biomass thermochemical processing (gasification or pyrolysis) at temperatures above 300°C also yields chemical precursors for a number of different industries. Cellulose, hemicelluloses, and lignin are transformed into biochar, bio-oil, and gaseous biofuels through pyrolysis at temperatures as high as 500°C. The conversion process is dependent on the properties of the biomass and the operational parameters that are used, such as inert gas type and flow rate, holding duration, temperature, and heating level. A combination of CO, H_2, CO_2, and CH_4 is produced when biomass is gasified at temperatures higher than 700°C, including supercritical water-based gasification and plasma-based gasification. The primary gasification product, synthetic natural gas, has several applications in the energy industry and serves as a starting point for the synthesis of liquid fuels, including methanol. From the dawn of human civilization, direct biomass burning has been a common practice and is maintained today, particularly in poor nations, despite significant pollution from CO, NO_X, SO_2, and particulate matter. AI demonstrates a unique capacity to assist with conventional pyrolysis modeling techniques. Process-based predictions at the workflow level depend on input data, such as the outcome and attributes, which are often acquired through experiments (Lan et al., 2019). Numerous studies have employed artificial intelligence (AI) algorithms to forecast product yields and attributes (such as HHV and carbon content) based on feedstock characterization data and/or operating circumstances (such as temperature, residence time, and heating rate). Without costly and time-consuming tests, such a prediction enables the rapid creation of process simulation frameworks (Cheng et al., 2020). The final product yield with the content of the outcomes of pyrolysis can also be predicted at the process level using kinetic simulations based on basic reaction procedures and reaction kinetics. However, experimental data are required to estimate a large number of kinetic parameters (Hameed et al., 2019). For predicting the logarithmic types of the three characteristics needed by kinetic models (pre-exponential factor, reaction order, and activation energy), one investigation trained FNN algorithms using compositional analysis information (cellulose, hemicellulose, and lignin concentrations; Sunphorka et al., 2017). By adding the temperature frequency as an extra input factor, Aghbashlo et al.

(2019) and Xing et al. (2019) altered this model. The precise composition of the pyrolysis byproducts may be predicted via the extended kinetic model, enabling additional process simulations with Aspen Plus software (Peters et al., 2017). One intriguing use of AI is to speed up the optimization of processes and shorten the solution time of reaction network models for pyrolysis (Hough et al., 2017). These investigations lead to the idea that AI methods may speed up the creation of kinetic models as well as simulations for pyrolysis. The capacity to examine various biomass kinds and incorporate full elements of biomass is a further benefit of AI that has been shown in earlier research. The primary input variables in thermochemical and biochemical transformation procedures are the installation's technical and operational specifications as well as the data on biomass characterization (specifically, sugar levels, enzyme levels, carbon and nitrogen proportions, quantity of the inoculum or yeast, etc.). Data on the kind and amount of catalyst, pretreatment procedures, and the microbial loading for biochemical reactions might occasionally also be utilized as input variables. Numerous AI experiments covering various biomass kinds have been examined, using biomass composition data as input variables. For instance, Karaci et al. (2016) predicted the proportion of hydrogen-rich gas in the final product by combining various substrate and catalyst kinds, catalyst amount, and the temperature of pyrolysis. Most of the research has made use of biomass characterization data from several analytical techniques, including lignocellulosic component analysis, proximate evaluation, and ultimate analysis. The incorporation of various biomass components enables prior AI research to quantitatively examine the effects of biomass compositions that are difficult to explore by conventional pyrolysis models and are frequently investigated by tests using contribution evaluation or sensitivity assessments (Liao et al., 2019; Zhu et al., 2019). Utilizing TGA data, certain investigations developed algorithms that might be utilized to cut down on both time and costs associated with TGA. The thermal response of biomass during pyrolysis was studied using TGA, an analytical method that evaluates the quantity of samples as they are treated in specified temperature ranges. The largest datasets (more than 1000 samples) among the research assessed in this area are those that use TGA data. Although in a study, Naqvi et al. (2019) employed mixed rice husk with sewage waste as their feedstock and set their mixing ratio to be one of the input parameters of FNN, the majority of studies concentrated on a single type of biomass, like walnut exterior, olive oil leftovers, forest leftovers, and sewage waste. The majority of AI research on gasification focuses on forecasting the chemical makeup of syngas, and the most commonly utilized output factor is H_2 content. The significance of syngas composition, particularly the H_2/CO ratio, for downstream operations like direct syngas combustion or Fisher-Tropsch synthesis to turn syngas into biofuel explains this (Sahoo et al., 2012). Several technology-specific factors are included in gasification investigations, along with reaction time and temperature. These include the equivalency proportion (which is the proportion of air required for the gasification process to air required for the full combustion of biomass feedstock; Asadullah, 2014), the steam to biomass proportion (for steam gasification), and the calcium oxide (CaO) to biomass proportion (for CaO catalyzed gasification; Rezk et al., 2019). An effective use of AI is for the dynamic evaluation, control, and improvement of gasification, which is extremely challenging to accomplish using conventional gasification modeling techniques like kinetic and thermodynamic equilibrium models. Since the thermodynamic equilibrium methodology depends on presumptions like perfect chemical mixing and constant temperature, it can only be used with a specific gasifier under specific operating parameters (Safarian et al., 2019). According to Safarian et al. (2019), kinetic models have the potential to be highly accurate, but their accuracy necessitates dependable comprehension of intricate processes, including microscopic evolution of particle distributions and gas–solid–particulate fluid flows. AI doesn't need to comprehend complicated phenomena to use data that is changing or even in real time. These studies often have big data sets (more than 2350 data samples). For instance, one study predicted the syngas temperature in real time for a co-current fixed bed gasifier under a variety of operating settings using a dynamic neural network (DNN) (Mikulandrić et al., 2016). The authors came to the conclusion that DNN outperforms several linear regression models in terms of accuracy. AI has been applied to gasification optimization in other studies. Based on the final analytical results of biomass feedstock, Kumar et al. employed FNN to identify the best operating parameters (Kumar et al., 2018). In order to enhance the quality of syngas, Rameshkumar & Mayilsamy (2014) trained an ANFIS model to forecast the tar content following gasification. The ability to incorporate reactor information is another possible benefit of AI over conventional gasification models. As previously mentioned, kinetic models take a long time to create (albeit they can anticipate product compositions at different reactor sites), while thermodynamic equilibrium models presume that the reactor is zero-dimensional (Safarian et al., 2019). Ruiz et al. noted that one possible avenue for future study to enhance gasification performance is to examine the effects of different types of gasifiers (Ruiz et al., 2013).

9.4.2 Biochemical Conversion Technologies

9.4.2.1 Transesterification

The main issues with using triglycerides to substitute diesel fuel include polyunsaturation, weak volatility, and excessive viscosity. Nevertheless, by creating vegetable-based oil derivatives using various techniques such as dilution, microemulsion, pyrolysis, and transesterification, these issues can be lessened. The most widely used process for producing biodiesel is transesterification, through which fatty biomass like vegetable and animal waste, algal lipids which are rich in triglycerides, are reacted with an organic alcohol, such as ethanol or methanol, in an acidic, alkaline, or enzyme-catalyzed medium to produce biofuel with at least 96 weight percent methylated or ethylated fatty acid esters (FAME or FAEE biodiesel) (Trejo-Zárraga et al., 2018). An alkali-catalyzed technique is employed if the oil's FFA concentration is less than 1 weight percent. However, a two-stage transesterification procedure that entails acid esterification and alkali transesterification is recommended for oil with a greater percentage of FFA. The two main byproducts of the transesterification are glycerin and the methyl ester of fatty acids, which is biodiesel with a high energy density (38–45 MJ/kg; Hoekman et al., 2012). The majority of research encompassed inputs related to assisted processes in addition to the transesterification process input variables. The pressure for supercritical transesterification method, the mixing intensity for mechanical stirring-assisted transesterification, and the sound waves for ultrasound-aided transesterification were among them (Baghban, 2019; Guo & Baghban, 2017; Sajjadi et al., 2017; Naderloo et al., 2017). According to these investigations, the methanol-to-oil proportion and other transesterification process characteristics have a greater impact on biodiesel yield than aided process parameters, but the former are still relevant (Karimi, 2017; Kusumo et al., 2017). Future optimization and scaling up of sophisticated transesterification procedures will benefit greatly from this information. Furthermore, two studies integrated process modeling with AI. To maximize the transesterification process, Nicola et al. (2010) integrated Aspen Plus simulation with GA. Another work predicted the output material flows of the transesterification simulation using the data produced by the Aspen Plus simulation to create soft sensors using boosting, an ensemble machine learning (ELM) technique (Ahmad et al., 2019). Genetic engineering using various methods (such as CRISPR) to get over the limitations of biochemical conversion and microbial resilience is another exciting AI application (Ng et al., 2017). Although they are still in their infancy, the use of genetic engineering to enhance the microalgae's metabolic mechanisms for biofuel production (Radakovits et al., 2010) and genetic machine learning (Libbrecht & Noble, 2015) seems promising. This field is always changing; further investigation and study are required to determine how and when AI may be applied to offer practical insights.

9.4.2.2 Hydrolysis and Fermentation

Bio-alcohols are produced through fermentation in aerobic conditions, followed by distillation and water retention using crops that contain sugar and starch, algal biomass, and cellulosic and lignocellulosic biomass, mostly generated from wood and wood waste. Fuel-grade bioethanol is a viable substitute for gasoline. It can be utilized as a primary fuel in combination with additional additives, or as a blend of 5 percent ethanol dehydrated to 99.6 weight percent and 95 percent gasoline (Karimi et al., 2021; Kul and Ciniviz, 2020). While the goal of all AI research on enzymatic hydrolysis is to forecast sugar yields, several metrics, such as the yields of glucose or xylose, or total reduced sugar, were employed. Depending on the kind of enzyme, each study employed different enzyme data instead of feedstock characterization data for input variables (Astray et al., 2016). Reaction time was included in every study for hydrolysis regarding operating parameters. But longer reaction periods are not well modeled by conventional modeling techniques, such as mechanistic kinetic models created for enzyme hydrolysis (Jeoh et al., 2017). Prior artificial intelligence (AI) research on fermentation has focused on determining the final amount of sugar or bioethanol yield using input variables like initial sugar quantity, yeast quantity, and operational conditions. One study (Saraceno et al., 2010) used FNN to forecast the batch fermentation process's kinetic parameters based on operational circumstances, while another investigation (Imtiaz et al., 2013) employed an inverted neural network to regulate the fermenter's temperature. The majority of earlier AI research has examined the two processes independently, despite the fact that fermentation and enzymatic hydrolysis are often carried out one after another to produce bioethanol. The integration of trained hydrolysis and fermentation AI models is technically possible, but more data will be required, as the majority of the examined AI experiments employed the same AI technique, FNN. For the whole process—which includes the pretreatment, hydrolysis, and fermentation processes—to be optimized holistically, more information and case studies are required.

9.4.3 Anaerobic Digestion Combined with Dark Fermentation

Anaerobic digestion is a microorganism-assisted biochemical conversion process to break down the portion of organic compounds that are biodegradable without the presence of oxygen. The result is biogas, a gaseous biofuel primarily composed of one-third carbon dioxide and two-thirds biomethane by volume, with minor amounts of other gases (Sewsynker-Sukai et al., 2017; Ghatak and Ghatak, 2018). Anaerobic fermentation in darkness may create biohydrogen from any organic material that has a high protein, lipid, and carbohydrate count. Sugar and starch-rich energy crops are among the best sources of feedstock for the synthesis of biohydrogen. Anaerobic digestion is a method used in both the generation of biogas and biohydrogen. However, the procedures differ in that the biomass content, pretreatment techniques, bacterial consortiums, and operating conditions direct the biochemical processes toward the synthesis of either biomethane or biohydrogen. Biogas, an exciting biofuel product that combines CH_4 and CO_2, is produced by anaerobic digestion, which involves using microorganisms to break down biomass. The studies included feedstock characterization for input variables. The most commonly utilized factors for characterizing biomass are the carbon-to-nitrogen ratio and substrate concentration. Hydraulic retention time (HRT) is the most commonly used, and operating conditions were included in the majority of research (Mao et al., 2015; Shi et al., 2017). It should be noted that, in comparison to other biochemical conversion processes, anaerobic digestion AI research has a large data size, and some of them have used data from large-scale biogas facilities (De Clercq et al., 2019). A few research contrasted traditional and AI-based approaches. For anaerobic digestion, the Anaerobic Digestion Model 1 (ADM1), created by Batstone et al. (2002), has been utilized extensively. The authors of one study (Tan et al., 2018) found that ANFIS is more accurate than ADM1 at predicting the characteristics of biogas produced from the anaerobic digestion of palm oil mill effluent. Another study that focused on ANN in combination with GA and ACO as a potential substitute for computationally costly ADM1 came to similar conclusions (Beltramo et al., 2019). In order to forecast the biogas yield, Sathish et al. evaluated the effectiveness of ANN and the Response Surface Methodology (RSM), which combines statistical and mathematical methods (Sathish & Vivekanandan, 2016). They came to the conclusion that while both models are highly accurate, ANN outperforms RSM by a small margin. Given the organized nature of RSM, they suggested ANN for further research. A few new fields for AI applications are developing. One of the most promising technologies for producing bio-based H_2 is dark fermentation, and considerable work is required to increase the efficiency of H_2 synthesis (Łukajtis et al., 2018). The dark fermentation was modeled using FNN in different researches (El-Shafie, 2014; Sewsynker & Gueguim Kana, 2016), and the findings may help with more process enhancement and expansion. Anaerobic digestion, which has been utilized to remediate wastewater in recent decades, uses industrial effluents as feedstock (Rajeshwari et al., 2000). In one AI investigation, water quality metrics, including chemical oxygen demand and total suspended particles, were employed as input variables, and the feedstock was palm oil mill effluent with seed sludge (Tan et al., 2018). Additionally, because it increases methane production and framework durability, the use of zero-valent iron (ZVI) to enhance the anaerobic breakdown process is typically promoted while attending to the absorption cycle. Instead of using time-consuming and tedious in-house experiments, accurate ZVI-based anaerobic digestion reactor modeling helps anticipate the methane output capacity, optimize operational strategy, and gather industrial design reference information. Anaerobic breakdown of food waste is recognized as the most effective cycle for handling because it enables waste adjustment and improves fertilizer and biogas. AI is also helpful for the development and execution of anaerobic digestion.

9.5 TRANSPORTATION-GRADE BIOFUEL

According to Liew et al. (2014), biofuel has a huge potential to reduce GHG and other air pollutants from the transportation industry. Forecasting of biofuel characteristics and car engine performance has been the main focus of earlier AI research. Cetane quantity, a crucial predictor of fuel's combustion characteristics, was predicted by numerous AI studies (Faizollahzadeh Ardabili et al., 2019; Mostafaei, 2018). Other studies used the FAME content as well as chemical structure details of biodiesel and fuel mixing ratio to figure out the physicochemical properties like viscosity, thickness, and iodine value (Aminian & Zare Nezhad, 2018; Razavi et al., 2019). The developed AI models for cetane number might be combined with those production algorithms to enhance/optimize the transformation of biomass to produce high-quality biofuel, since the FAME quantity and yields are the output parameters of AI models for biomass transformation to biodiesel. Additionally, by forecasting the binary relationship between variables of local composition models with specified temperature and biofuel/solvent composition, AI can help simulate biofuel phase equilibrium behaviors (Reynel-Ávila et al., 2019). Two studies contrasted conventional analytical techniques with AI methodologies. In forecasting cetane numbers, one study evaluated the effectiveness of ANN and multiple linear regression (MLR) (Piloto-Rodríguez et al., 2013). The study found that while ANN and MLR demonstrated comparable prediction capabilities, ANN might be better at modeling complicated characteristics. In another work, the viscosity of biodiesel was measured using ANN in comparison to the conventional Knothe-Steidley and Ramirez-Verduzco methods (Meng et al., 2014). Their findings demonstrated that ANN outperforms the other two techniques in terms of kinematic viscosity prediction, especially when it comes to accounting for the contributions of tiny components. Therefore, comparative investigations using larger datasets are still required to validate those claims. The majority of research on engine performance used AI to forecast energy or emissions-related metrics, which can save money and time on engine inspection and calibration. A number of studies examined engine vibration (Çelebi et al., 2017) and noise levels, two crucial aspects of vehicle comfort. The majority of research employed FNN, but some studies used ELM, a learning scheme of FNN (Huang et al., 2004). These studies found that ELM-based approaches outperform SVM and ANN in predicting the performance and emissions of biofuel engines (Aghbashlo et al., 2016; Wong, et al., 2015). As engine fuel, the majority of AI studies examined biodiesel mixed with petroleum-based diesel; however, a number of studies examined blending biodiesel-diesel together with additional fuels, such as ethanol, hydrogen, and natural gas (Ghaderi et al., 2019, Çelebi et al., 2017). Similar input-output variables were employed in these investigations as in the AI models for blends of biodiesel and diesel. These AI models might facilitate the use of various biofuels in the future.

9.6 ARTIFICIAL INTELLIGENCE IN BIOENERGY SUPPLY CHAINS (BSC)

Designing and optimizing the supply chain is crucial for the large-scale production and use of bioenergy. AI offers a predictive method that enables quick evaluation and more efficient reduction of potential hazards or disruptive occurrences across the supply chain. A bioenergy-producing facility's entire management of the supply chain is mostly dependent on power plant architecture, logistic planning, and resource assessment. According to the complexities, variation, and uncertainty of BSCs, the majority of conventional BSC modeling techniques—which include simulation and optimization—are computationally demanding (Ghaderi et al., 2016). Because it helps manage complex databases and offers real-time monitoring capabilities that may assist users with efficient raw material collection and use, bioenergy distribution, and cost-benefit analysis, Geographic Information Systems (GIS) hold promise in handling the bioenergy supply chain (Senocak and Guner, 2022). GIS technology is popularly utilized to determine the best site for a plant by estimating the terrain, kind of soil, feedstock accessibility, and transportation networks (Sahoo et al., 2018). In a paper published in 2021, Bharti et al. (2021) highlighted a number of GIS applications for estimating the bioenergy potential of residues. Another trend in recent years has been the rise of studies with mixed decision levels, such as mixed strategic-tactical and strategic-tactical-operational. Heuristic algorithms might assist in solving multi-decision-level BSC optimization issues in a suitable runtime because mixed decision levels typically signal higher complexity that may result in computational obstacles (Banos et al., 2011). According to earlier research, artificial intelligence (AI) holds promise for overcoming computational obstacles and offering nearly ideal solutions for BSC issues. Numerous studies coupled heuristic algorithms with conventional optimization techniques. Heuristics have been shown in several studies to reduce computing costs, particularly when resolving BSC models with a lot of variables and restrictions (Asadi et al., 2018; Kumar et al., 2015). According to Kumar et al. (2015), CPLEX took over three hours to solve the same BSC issue, whereas the heuristic approach—adaptive large neighborhood search (ALNS)—took 140 seconds. To draw a general conclusion about the benefit of heuristics, more research on other biomass systems is required. Although the GA outperforms PSO in optimizing the net present value of the specified BSC scenario (Rafiel & Zadeh, 2013), several studies that compared the computational time of various heuristics came to the conclusion that PSO performed better in reaching the near-optimal solution than other types of heuristics (Gómez-Gonzalez et al., 2013; Vera et al., 2010). The BSC uncertainties associated with biomass accessibility, biofuel demand, usage of land, and biomass expenses were addressed by a number of reviewed studies using fuzzy logic-based techniques (Arabi et al., 2019; Balaman & Selim, 2016; Khishtandar, 2019). Furthermore, two studies (Kim et al., 2018; Shu et al., 2017) employed simulation-based optimization after simulating individual behavior throughout BSC using agent-based modeling (ABM). To create standardized, generally recognized, and simple-to-implement AI selection processes and applications, more case studies are required. Further research into AI applications pertaining to the social and environmental facets of bioenergy systems is also required. In order to evaluate the sustainability of bioenergy on a regional level, a recent study (Rouleau & Zupko, 2019) provided a framework that used ABM modeling. This framework assessed sustainability from social, environmental, and economic perspectives. In order to assess the environmental effects of a product's life cycle, which encompasses the majority of BSC operations, this study recommended the use of standardized techniques such as Life Cycle Assessment (LCA). For biomass-based products, numerous life cycle assessments (LCAs) have been carried out. Additionally, the integration of LCA with BSC optimization has been investigated (Yue et al., 2014). Insufficient Life Cycle Inventory (LCI) data (such as usage of energy and air contaminants; Yao & Masanet, 2018) is a significant obstacle to doing Life Cycle Assessment (LCA) for bioenergy systems, particularly for new technologies or feedstock. Artificial Intelligence (AI) may be a viable solution to this problem. A number of studies generated missing LCI data using AI approaches like ANN (Liao et al., 2020; Song et al., 2017). A few studies produced spatial/temporal explicit LCI data for biomass production using machine learning techniques (Lee et al., 2020; Romeiko et al., 2020). To fully comprehend how well various AI approaches work to meet the data difficulties of LCA, more case studies are necessary. In terms of the social component, further study is required to determine whether particular social metrics are essential to BSC planning and design, as well as how AI might be used to assess, analyze, and produce these indicators.

9.7 CONCLUSION

In recent decades, there has been a lot of discussion about the use of biofuels as a greener substitute for traditional fuels. The use of inexpensive feedstock resources, minimal impact on the environment, sustainable production paths, and high return on investment are some of the significant advantages they offer. It was discovered that the usual approaches for producing various kinds of biofuels, including biochar, biodiesel, bioethanol, biogas, and biohydrogen, were chemical, thermochemical, and microbiological processes. Well-weighted optimization strategies are necessary for optimal ultimate outcomes and process sustainability because these are complicated processes where the choice of substrates, working conditions, or pretreatment protocols may significantly impact the efficiency of bioconversion. The article highlights how AI technologies may help the bioenergy industry become more efficient and sustainable. It also shows how effective AI can be in solving research, measurement, and modeling problems in the bioenergy sector. This study found that AI can be used to generate data that is costly, time-consuming, or difficult to measure directly. The four main types of AI algorithms utilized in biomass-to-biofuels conversion processes are hybrid approaches, machine learning, symbolic AI, and heuristics. Agricultural bioenergy producers can lower costs and enhance product quality by optimizing processes like biomass identification and production modeling with ANNs, SVMs, and other AI algorithms. To assist in advancing a more efficient and sustainable energy future, policymakers should think about funding AI research and development initiatives for the agro-bioenergy industry. A smart and sustainable digital future for agriculture and rural areas and the reduction of greenhouse gas emissions under the Carbon Neutralization Target are just two examples of the national and international environmental goals that could be supported by policies that promote the adoption of these technologies. Because it concentrated on agricultural bioenergy and ignored other possible domains where AI could be utilized, like energy storage, distribution, and grid management, this study has a narrow focus on applications of AI in agricultural bioenergy installations. Future research should focus on improving algorithms for artificial intelligence and mathematical modeling to make biomass-to-biofuels conversion processes more predictable. Additionally, more thorough experimental research is required to ascertain their efficacy based on certain process characteristics (e.g., substrate type, operating temperature, pretreatment technique, etc.). Big data analytics should be used to improve AI models, and more research should look at new AI application domains, including energy storage and grid control. To create more thorough and integrated AI applications that take into account the social, economic, and environmental aspects that affect the production and consumption of bioenergy—such as estimating the potential global bioenergy and optimizing the location of bioenergy plants—interdisciplinary research is also required. Furthermore, future studies should concentrate on creating models that can manage variation in data and uncertainty, adjust to shifting circumstances, and gain knowledge from experience while remaining interpretable for stakeholders in order to guarantee the resilience and comprehensibility of AI models.

REFERENCES

Adeyemo, J., & Enitan, A. (2011). Optimization of fermentation processes using evolutionary algorithms: A review. *Scientific Research and Essays*, 6(7), 1464–1472.

Aghbashlo, M., Shamshirband, S., Tabatabaei, M., Yee, P.L., & Larimi, Y.N. (2016). The use of ELM-WT (extreme learning machine with wavelet transform algorithm) to predict exergetic performance of a DI diesel engine running on diesel/biodiesel blends containing polymer waste. *Energy*, 94, 443–456.

Aghbashlo, M., Tabatabaei, M., Nadian, M.H., Davoodnia, V., & Soltanian, S. (2019). Prognostication of lignocellulosic biomass pyrolysis behavior using ANFIS model tuned by PSO algorithm. *Fuel*, 253, 189–198.

Ahamed, T.S., Anto, S., Mathimani, T., Brindhadevi, K., & Pugazhendhi. A. (2021). Upgrading of bio-oil from thermochemical conversion of various biomass–Mechanism, challenges and opportunities. *Fuel*, 287, 119329.

Ahmad, I., Ayub, A., Ibrahim, U., Khattak, M.K., & Kano, M. (2019). Data-based sensing and stochastic analysis of biodiesel production process. *Energies*, 12, 1–13.

Almomani, F. (2020). Prediction of biogas production from chemically treated co-digested agricultural waste using artificial neural network. *Fuel*, 280, 118573.

Aminian, A., & Zare Nezhad, B. (2018). Accurate predicting the viscosity of biodiesels and blends using soft computing models. *Renewable Energy*, 120, 488–500.

Arabi, M., Yaghoubi, S., & Tajik, J. (2019). A mathematical model for microalgae-based biobutanol supply chain network design under harvesting and drying uncertainties. *Energy*, 179, 1004–1016.

Ardabili, S., Mosavi, A., & Varkonyi-Koczy, A.R. (2020). Systematic review of deep learning and machine learning models in biofuels research. *Lecture Notes in Networks and Systems*, 101, 19–32.

Asadi, E., Habibi, F., Nickel, S., & Sahebi, H. (2018). A bi-objective stochastic location-inventory-routing model for microalgae-based biofuel supply chain. *Applied Energy*, 228, 2235–2261.

Asadullah, M. (2014). Barriers of commercial power generation using biomass gasification gas: A review. *Renewable and Sustainable Energy Reviews*, 29, 201–215.

Aslani, A., Mazzuca-Sobczuk, T., Eivazi, S., & Bekhrad, K. (2018). Analysis of bioenergy technologies development based on life cycle and adaptation trends. *Renewable Energy*, 127, 1076–1086.

Astray, G., Gullon, B., Labidi, J., & Gullon, P. (2016). Comparison between developed models using response surface methodology (RSM) and artificial neural networks (ANNs) with the purpose to optimize oligosaccharide mixtures production from sugar beet pulp. *Industrial Crops and Products*, 92, 290–299.

Baghban, A. (2019). Computational modeling of biodiesel production using supercritical methanol. *Energy Sources, Part A: Recovery, Utilization and Environmental Effects*, 41(1), 14–20.

Balaman, Ş.Y,. & Selim, H. (2016). Sustainable design of renewable energy supply chains integrated with district heating systems: A fuzzy optimization approach. *Journal of Cleaner Production*, 133, 863–885.

Banos, R., Manzano-Agugliaro, F., Montoya, F.G., Gil, C., Alcayde, A., & Gomez, J. (2011). Optimization methods applied to renewable and sustainable energy: A review. *Renewable and Sustainable Energy Reviews*, 15(4), 1753–1766.

Batstone, D.J., Keller, J., Angelidaki, I., Kalyuzhnyi, S.V., Pavlostathis, S.G., Rozzi, A., Sanders, W.T., Siegrist, H., & Vavilin, V.A. (2002). The IWA anaerobic digestion model no 1 (ADM1). *Water Science and Technology*, 45(10), 65–73.

Beltramo, T., Klocke, M., & Hitzmann, B. (2019). Prediction of the biogas production using GA and ACO input features selection method for ANN model. *Information Processing in Agriculture*, 6(3), 349–356.

Bharti, A., Paritosh, K., Mandla, V.R., Chawade, A., & Vivekanand, V. (2021). GIS application for the estimation of bioenergy potential from agriculture residues: an overview, *Energies*, 14(4), 898.

Cahyanti, M.N., Doddapaneni, T.R.K.C., & Kikas, T. (2020). Biomass torrefaction: an overview on process parameters, economic and environmental aspects and recent advancements. *Bioresource Technology*, 301, 122737.

Castillo-Villar, K.K. (2014). Metaheuristic algorithms applied to bioenergy supply chain problems: Theory, review, challenges, and future. *Energies*, 7(11), 7640–7672.

Castrounis, A. (2019). *AI for People and Business*. Ed. O'Reilly Media Inc., ISBN-10: 1492036579.

Celebi, K., Uludamar, E., Tosun, E., Yıldızhan, Ş., Aydın, K., & Ozcanlı, M. (2017). Experimental and artificial neural network approach of noise and vibration characteristic of an unmodified diesel engine fuelled with conventional diesel, and biodiesel blends with natural gas addition. *Fuel*, 197, 159–173.

Chakraborty, A., & Kar, A.K. (2017). Swarm intelligence: A review of algorithms. In S. Patnaik, X.-S. Yang, & K. Nakamatsu (Eds.), *Nature-Inspired Computing and Optimization: Theory and Applications* (pp. 475–494). Springer International Publishing.

Cheng, F., Luo, H., & Colosi, L.M. (2020). Slow pyrolysis as a platform for negative emissions technology: An integration of machine learning models, life cycle assessment, and economic analysis. *Energy Conversion and Management*, 223, 113258.

De Clercq, D., Jalota, D., Shang, R., Ni, K., Zhang, Z., Khan, A., et al. (2019). Machine learning powered software for accurate prediction of biogas production: A case study on industrial-scale Chinese production data. *Journal of Cleaner Production*, 218, 390–399.

Demirbas, A.H., & Demirbas, I. (2007). Importance of rural bioenergy for developing countries. *Energy Conversion and Management*, 48(8), 2386–2398.

Dounis, A.I. (2010). Artificial intelligence for energy conservation in buildings. *Advances in Building Energy Research*, 4(1), 267–299.

El-Shafie, A. (2014). Neural network nonlinear modeling for hydrogen production using anaerobic fermentation. *Neural Computing and Applications*, 24, 539–547.

Faizollahzadeh Ardabili, S., Najafi, B., & Shamshirband, S. (2019). Fuzzy logic method for the prediction of cetane number using carbon number, double bounds, iodic, and saponification values of biodiesel fuels. *Environmental Progress & Sustainable Energy,* 38, 584–599.

Garcia, R., Pizarro, C., Lavin, A.G., & Bueno, J.L. (2013). Biomass proximate analysis using thermogravimetry. *Bioresource Technology*, 139, 1–4.

Ghaderi, H., Pishvaee, M.S., & Moini, A. (2016). Biomass supply chain network design: An optimization-oriented review and analysis. *Industrial Crops and Products*, 94, 972–1000.

Ghaderi, M., Javadikia, H., Naderloo, L., Mostafaei, M., & Rabbani, H. (2019). Analysis of noise pollution emitted by moving MF285 Tractor using different mixtures of biodiesel, bioethanol and diesel through artificial intelligence. *Journal of Low Frequency Noise, Vibration and Active Control*, 38, 270–281.

Ghatak, M.D., & Ghatak, A. (2018). Artificial neural network model to predict behavior of biogas production curve from mixed lignocellulosic co-substrates. *Fuel*, 232, 178–189.

Gómez-González, M, López, A, & Jurado, F. (2013). Hybrid discrete PSO and OPF approach for optimization of biomass fueled micro-scale energy system. *Energy Conversion and Management,* 65, 539–545.

Guo, J., & Baghban, A. (2017). Application of ANFIS strategy for prediction of biodiesel production using supercritical methanol. *Energy Sources, Part A: Recovery, Utilization and Environmental Effects*, 39(17), 1862–1868.

Guo, M., Song, W., & Buhain, J. (2015). Bioenergy and biofuels: History, status, and perspective. *Renewable and Sustainable Energy Reviews*, 42, 712–725.

Hameed, S., Sharma, A., Pareek, V., Wu, H., & Yu, Y. (2019). A review on biomass pyrolysis models: Kinetic, network and mechanistic models. *Biomass and Bioenergy*, 123, 104–122.

Haugeland, J. (1985). *Artificial Intelligence: The Very Idea* (1st ed.). The MIT Press.

Hoekman, S.K., Broch, A., Robbins, C., Ceniceros, E., & Natarajan, M. (2012). Review of biodiesel composition, properties, and specifications. *Renewable and Sustainable Energy Reviews*, 16(1), 143–169.

Hough, B.R., Beck, D.A.C., Schwartz, D.T., & Pfaendtner, J. (2017). Application of machine learning to pyrolysis reaction networks: Reducing model solution time to enable process optimization. *Computers & Chemical Engineering*, 104, 56–63.

Huang, G.B., Zhu, Q.Y., & Siew, C.K. (2004). Extreme learning machine: A new learning scheme of feedforward neural networks. *IEEE International Conference on Neural Networks-Conference Proceedings*, 2, 985–990.

IEA Bioenergy. (2009). *Bioenergy: A Sustainable and Reliable Energy Source.* https://www.ieabi oenergy.com/blog/publications/main report-bioenergy-a-sustainable-and-reliable-energy-source-a-review-of-status-and-prospects/

Imtiaz, U., Assadzadeh, A., Jamuar, S.S., & Sahu, J.N. (2013). Bioreactor temperature profile controller using inverse neural network (INN) for production of ethanol. *Journal of Process Control,* 23, 731–742.

International Energy Agency. (2018). Market report series: Renewables 2018. OECD/IEA. https://www.iea.org/reports/renewables-2018.

Jeoh, T., Cardona, M.J., Karuna, N., Mudinoor, A.R., & Nill, J. (2017). Mechanistic kinetic models of enzymatic cellulose hydrolysis–A review. *Biotechnology and Bioengineering*, 114(7), 1369–1385.

Jha, S.K., Bilalovic, J., Jha, A., Patel, N., & Zhang, H. (2017). Renewable energy: Present research and future scope of Artificial Intelligence. *Renewable and Sustainable Energy Reviews*, *77*, 297–317.

Jurado, F., Ortega, M., Cano, A., & Carpio, J. (2002). Neuro-fuzzy controller for gas turbine in biomass-based electric power plant. *Electric Power Systems Research*, *60*(3), 123–135.

Karaci, A., Caglar, A., Aydinli, B., Pekol, S. (2016). The pyrolysis process verification of hydrogen rich gas (H-rG) production by artificial neural network (ANN). *International Journal of Hydrogen Energy,* 41, 4570–4578.

Karimi, M. (2017). Exergy-based optimization of direct conversion of microalgae biomass to biodiesel. *Journal of Cleaner Production*, 141, 50–55.

Karimi, S., Karri, R.R., Tavakkoli, Y.M., et al. (2021). Processes and separation technologies for the production of fuel-grade bioethanol: a review. *Environmental Chemistry Letters*, 19, 2873–2890.

Khishtandar, S. (2019). Simulation based evolutionary algorithms for fuzzy chance-constrained biogas supply chain design. *Applied Energy*, 236, 183–195.

Khishtandar, S., Zandieh, M., & Dorri, B. (2017). A multi criteria decision making framework for sustainability assessment of bioenergy production technologies with hesitant fuzzy linguistic term sets: The case of Iran. *Renewable and Sustainable Energy Reviews*, 77, 1130–1145.

Kim, S., Kim, S., & Kiniry, J.R. (2018). Two-phase simulation-based location-allocation optimization of biomass storage distribution. *Simulation Modelling Practice and Theory*, 86, 155–168.

Kul, B.S., & Ciniviz, M. (2020). Assessment of waste bread bioethanol-gasoline blends in respect to combustion analysis, engine performance and exhaust emissions of a SI engine. *Fuel*, 277, 118237.

Kumar, K., Javier, C., Lopez, C., Augusto, O., Sanchez, T., Gupta, A., et al. (2015). Integrated strategic and tactical optimization of animal-waste sourced biopower supply chains. *International Conference on Industrial Engineering and Systems Management,* 1367–1373.

Kumar, G.S., Gupta A.V.S.S.K.S., & Viswanadham M. (2018). Prediction of optimum parameters and product gas composition in gasification using ANNs. *AIP Conference Proceedings*, 2039.

Kusumo, F., Silitonga, A.S., Masjuki, H.H., Ong, H.C., Siswantoro, J., & Mahlia, T.M.I. (2017). Optimization of transesterification process for Ceiba pentandra oil: A comparative study between kernel-based extreme learning machine and artificial neural networks. *Energy*, 134, 24–34.

Lan, K., & Yao, Y. (2019). Integrating life cycle assessment and agent-based modeling: A dynamic modeling framework for sustainable agricultural systems. *Journal of Cleaner Production*, 238, 117853.

Lan, K., Park, S., & Yao, Y. (2020). Key issue, challenges, and status quo of models for biofuel supply chain design. In *Biofuels for a More Sustainable Future*. Elsevier Inc.

Lee, E.K., Zhang,W.J., Zhang, X., Adler, P.R., Lin, S., Feingold, B.J., Khwaja, H.A., & Romeiko, X.X. (2020). Projecting life-cycle environmental impacts of corn production in the US Midwest under future climate scenarios using a machine learning approach. *Science of The Total Environment*, 714, 136697.

Liao, M., Kelley, S., & Yao, Y. (2020). Generating energy and greenhouse gas inventory data of activated carbon production using machine learning and kinetic based process simulation. *ACS Sustainable Chemistry & Engineering*, 8, 1252–1261.

Liao, M., Kelley, S.S., & Yao, Y. (2019). Artificial neural network based modeling for the prediction of yield and surface area of activated carbon from biomass. *Biofuels Bioproduct and Biorefinery*. 13(4), 1015-1027.

Libbrecht, M.W., & Noble, W.S. (2015). Machine learning in genetics and genomics. *Nature Review Genetics*, 16(6), 321–332.

Liew, W.H., Hassim, M.H., & Ng, D.K.S. (2014). Review of evolution, technology and sustainability assessments of biofuel production. *Journal of Cleaner Production*, 71, 11–29.

Łukajtis, R., Hołowacz, I., Kucharska, K., Glinka, M., Rybarczyk, P., Przyjazny, A., & Kamiński, M. (2018). Hydrogen production from biomass using dark fermentation. *Renewable and Sustainable Energy Reviews*, 91, 665–694.

Mao, C., Feng, Y., Wang, X., & Ren, G. (2015). Review on research achievements of biogas from anaerobic digestion. *Renewable and Sustainable Energy Reviews*, 45, 540–555.

Meng, X., Jia, M., & Wang, T. (2014). Neural network prediction of biodiesel kinematic viscosity at 313 K. *Fuel*, 121, 133–140.

Mikulandrić, R., Bohning, D., Bohme, R., Helsen, L., Beckmann, M., & Lončar, D. (2016). Dynamic modelling of biomass gasification in a co-current fixed bed gasifier. *Energy Conversion and Management*, 125, 264–276.

Mohd Ali, J., Hussain, M.A., Tade, M.O., & Zhang, J. (2015). Artificial Intelligence techniques applied as estimator in chemical process systems: A literature survey. *Expert Systems with Applications*, *42*(14), 5915–5931.

Mostafaei, M. (2018). ANFIS models for prediction of biodiesel fuels cetane number using desirability function. *Fuel*, 216, 665–672.

Naderloo, L., Javadikia, H., & Mostafaei, M. (2017). Modeling the energy ratio and productivity of biodiesel with different reactor dimensions and ultrasonic power using ANFIS. *Renewable and Sustainable Energy Reviews*, 70, 56–64.

Naqvi, S.R., Hameed, Z., Tariq, R., Taqvi, S.A., Ali, I., Niazi, M.B.K., Noor, T., Hussain, A., Iqbal, N., & Shahbaz, M. (2019). Synergistic effect on co-pyrolysis of rice husk and sewage sludge by thermal behavior, kinetics, thermodynamic parameters and artificial neural network. *Waste Management*, 85, 131–140.

Ng, I.S., Tan, S.I., Kao, P., Chang, Y.K., & Chang, J.S. (2017). Recent developments on genetic engineering of microalgae for biofuels and bio-based chemicals. *Biotechnology Journal*, 12(10), 1–13.

Nicola, G.D., Moglie, M., Pacetti, M., & Santori, G. (2010). Bioenergy II: Modeling and multi-objective optimization of different biodiesel production processes. *International Journal of Chemical Reactor Engineering*, 8, 1–10.

Nunes, L.J.R., Matias, J.C.O., & Catalao, J.P.S. (2014). A review on torrefied biomass pellets as a sustainable alternative to coal in power generation. *Renewable and Sustainable Energy Reviews*, 40, 153–160.

Peters, J.F., Banks, S.W., Bridgwater, A.V., & Dufour, J. (2017). A kinetic reaction model for biomass pyrolysis processes in Aspen Plus. *Applied Energy*, 188, 595–603.

Piloto-Rodriguez, R., Sanchez-Borroto, Y., Lapuerta, M., Goyos-Perez, L., & Verhelst, S. (2013). Prediction of the cetane number of biodiesel using artificial neural networks and multiple linear regression. *Energy Conversion and Management*, 65, 255–261.

Radakovits, R., Jinkerson, R.E., Darzins, A., & Posewitz, M.C. (2010). Genetic engineering of algae for enhanced biofuel production. *Eukaryotic Cell*, 9(4), 486–501.

Rafiel, M., & Zadeh, M.S. (2013). Optimizing a typical biomass fueled power plant using genetic algorithm and binary particles swarm optimization. *International Journal on Technical and Physical Problems of Engineering*, 5(2), 15–21.

Rajeshwari, K.V., Balakrishnan, M., Kansal, A., Lata, K., & Kishore, V.V.N. (2000). State-of-the-art of anaerobic digestion technology for industrial wastewater treatment. *Renewable & Sustainable Energy Reviews*, 4(2), 135–156.

Rameshkumar, R., & Mayilsamy, K. (2014). Prediction of tar and particulate in biomass gasification using adaptive neuro fuzzy inference system. *Journal of Intelligent & Fuzzy Systems,* 27, 361–365.

Razavi, R., Bemani, A., Baghban, A., Mohammadi, A.H., & Habibzadeh, S. (2019). An insight into the estimation of fatty acid methyl ester based biodiesel properties using a LSSVM model. *Fuel,* 243, 133–141.

Reynel-Ávila, H.E., Bonilla-Petriciolet, A., & Tapia-Picazo, J.C. (2019). An artificial neural network-based NRTL model for simulating liquid-liquid equilibria of systems present in biofuels production. *Fluid Phase Equilibria*, 483, 153–164.

Rezk, H., Nassef, A.M., Inayat, A., Sayed, E.T., Shahbaz, M., & Olabi, A.G. (2019). Improving the environmental impact of palm kernel shell through maximizing its production of hydrogen and syngas using advanced artificial intelligence. *Science of the Total Environment*, 658, 1150–1160.

Romeiko, X.X., Lee, E.K., Sorunmu, Y., & Zhang, X. (2020). Spatially and temporally explicit life cycle environmental impacts of soybean production in the U.S. *Midwest. Environmental Science & Technology*, 54(8), 4758–4768.

Romeo, L.M., & Gareta, R. (2009). Fouling control in biomass boilers. *Biomass and Bioenergy*, *33*(5), 854–861.

Rouleau, M., & Zupko, R. (2019). Agent-Based Modeling for bioenergy sustainability assessment. *Landscape and Urban Planning*, 188, 54–63.

Ruiz, J.A., Juarez, M.C., Morales, M.P., Munoz, P., & Mendivil, M.A. (2013). Biomass gasification for electricity generation: Review of current technology barriers. *Renewable and Sustainable Energy Reviews*, 18, 174–183.

Russell, S., & Norvig, P. (2010). *Artificial Intelligence: A Modern Approach*. M. Hirsch (ed.) (3rd ed.). Pearson Education, Inc., New Jersey 07458.

Sadhwani, N., et al. (2013). Simulation, analysis, and assessment of CO_2 enhanced biomass gasification. *Computer Aided Chemical Engineering*, 32, 421–426.

Safarian, S. UnnÞórsson, R., & Richter, C. (2019). A review of biomass gasification modelling. *Renewable and Sustainable Energy Reviews*, 110, 378–391.

Sahoo, B.B., Sahoo, N., & Saha, U.K. (2012). Effect of H_2:CO ratio in syngas on the performance of a dual fuel diesel engine operation. *Applied Thermal Engineering*, 49, 139–146.

Sahoo, K., Mani, S., Das, L., & Bettinger, P. (2018) GIS-based assessment of sustainable crop residues for optimal siting of biogas plants. *Biomass and Bioenergy,* 110, 63–74.

Sajjadi, B., Asaithambi, P., Raman, A.A.A., & Ibrahim, S. (2017). Hybrid nero-fuzzy methods for estimation of ultrasound and mechanically stirring Influences on biodiesel synthesis through transesterification. *Measurement: Journal of the International Measurement Confederation*, 103, 62–76.

Saraceno, A., Curcio, S., Calabro, V., & Iorio, G. (2010). A hybrid neural approach to model batch fermentation of "ricotta cheese whey" to ethanol. *Computers and Chemical Engineering*, 34(10), 1590–1596.

Sathish, S., & Vivekanandan, S. (2016). Parametric optimization for floating drum anaerobic bio-digester using Response Surface Methodology and Artificial Neural Network. *Alexandria Engineering Journal*, 55(4), 3297–3307.

Senocak, A.A., & Guner Goren, H. (2022). Forecasting the biomass-based energy potential using artificial intelligence and geographic information systems: A case study, *Engineering Science and Technology, an International Journal, 26,* 100992.

Sewsynker, Y., & Gueguim Kana, E.B. (2016). Intelligent models to predict hydrogen yield in dark microbial fermentations using existing knowledge. *International Journal of Hydrogen Energy*, 41, 12929–12940.

Sewsynker-Sukai, Y., Faloye, F., & Bosco, G.K.E. (2017) Artificial neural networks: an efficient tool for modelling and optimization of biofuel production (a mini review). *Biotechnology & Biotechnological Equipment*, 31(2), 221–235.

Sharma, A., Pareek, V., & Zhang, D. (2015). Biomass pyrolysis: A review of modelling, process parameters and catalytic studies. *Renewable and Sustainable Energy Reviews*, 50, 1081–1096.

Shi, X.S., Dong, J.J., Yu, J.H., Yin, H., Hu, S.M., Huang, S.X., & Yuan, X.Z. (2017). Effect of hydraulic retention time on anaerobic digestion of wheat straw in the semicontinuous continuous stirred-tank reactors. *BioMed Research International*, Article ID 2457805, 1–6.

Shu, K., Schneider, U.A., & Scheffran, J. (2017). Optimizing the bioenergy industry infrastructure: Transportation networks and bioenergy plant locations. *Applied Energy*, 192, 247–261.

Smolensky, P. (1987). Connectionist AI, symbolic AI, and the brain. *Artificial Intelligence Review*, 1(2), 95–109.

Song, R., Keller, A.A., & Suh, S. (2017). Rapid life-cycle impact screening using artificial neural networks. *Environmental Science and Technology*, 51(18), 10777–10785.

Sunphorka, S., Chalermsinsuwan, B., & Piumsomboon, P. (2017). Artificial neural network model for the prediction of kinetic parameters of biomass pyrolysis from its constituents. *Fuel*, 193, 142–158.

Tan, H.M., Gouwanda, D., & Poh, P.E. (2018). Adaptive neural-fuzzy inference system vs. anaerobic digestion model No.1 for performance prediction of thermophilic anaerobic digestion of palm oil mill effluent. *Process Safety and Environmental Protection*, 117, 92–99.

Thrän, D., Dotzauer, M., Lenz, V., Liebetrau, J., & Ortwein, A. (2015). Flexible bioenergy supply for balancing fluctuating renewables in the heat and power sector–A review of technologies and concepts. *Energy Sustainability and Society*, *5*(1), 1–15.

Trejo-Zárraga, F., de Jesús Hernández-Loyo, F., Chavarría-Hernández, J.C., & Sotelo-Boyás, R. (2018). *Kinetics of Transesterification Processes for Biodiesel Production*. Biofuels–State of Development volume, Ed. Krzysztof Biernat.

Vera, D., Carabias, J., Jurado, F., & Ruiz-Reyes, N. (2010). A Honey Bee Foraging approach for optimal location of a biomass power plant. *Applied Energy*, 87, 2119–2127.

Vikhar, P.A. (2017). Evolutionary algorithms: A critical review and its future prospects. *Proceedings–International Conference on Global Trends in Signal Processing, Information Computing and Communication, ICGTSPICC*, 2016, 261–265.

Virkaj¨arvi, I., et al. (2009). Cellulosic ethanol via biochemical processing poses a challenge for developers and implementors. *Bio Resources*, 4(4), 1718–1735.

Wong, K.I., Vong, C.M., Wong, P.K., & Luo, J. (2015). Sparse Bayesian extreme learning machine and its application to biofuel engine performance prediction. *Neurocomputing*, 149, 397–404.

Xing, J., Wang, H., Luo, K., Wang, S., Bai, Y., & Fan, J. (2019). Predictive single-step kinetic model of biomass devolatilization for CFD applications: A comparison study of empirical correlations (EC), artificial neural networks (ANN) and random forest (RF). *Renewable Energy*, 136, 104–114.

Yao, Y., & Masanet, E. (2018). Life-cycle modeling framework for generating energy and greenhouse gas emissions inventory of emerging technologies in the chemical industry. *Journal of Cleaner Production*, 172, 768–777.

Yao, Z., You, S., Ge, T., & Wang, C.-H. (2018). Biomass gasification for syngas and biochar co-production: Energy application and economic evaluation. *Applied Energy*, 209, 43–55.

Yue, D., Slivinsky, M., Sumpter, J., & You, F. (2014). Sustainable design and operation of cellulosic bioelectricity supply chain networks with life cycle economic, environmental, and social optimization. *Industrial and Engineering Chemistry Research*, 53(10), 4008–4029.

Yusri, I.M., Abdul Majeed, A.P.P., Mamat, R., Ghazali, M.F., Awad, O.I., & Azmi, W.H. (2018). A review on the application of response surface method and artificial neural network in engine performance and exhaust emissions characteristics in alternative fuel. *Renewable and Sustainable Energy Reviews*, 90, 665–686.

Zhang, J., & Zhang, X. (2019). The thermochemical conversion of biomass into biofuels. *Biopolymer-Based Materials, and Bioenergy* (pp. 327–368). Elsevier Ltd.

Zheng, Y.J., Chen, S., Lin, Y., & Wang, W. L. (2013). Bio-inspired optimization of sustainable energy systems: A review. *Mathematical Problems in Engineering*, *2013*, 1–12.

Zhu, X., Li, Y., & Wang, X. (2019). Machine learning prediction of biochar yield and carbon contents in biochar based on biomass characteristics and pyrolysis conditions. *Bioresource Technology*, 288, 121527.

10 Life Cycle Assessment and Cost Analysis of Artificial Intelligence–based Biofuel Production

Malhar Santosh Kapadnekar, Shubham Sunil Behele, Mohd Haad Khan, Monika Kumari, Krishika Lakhmichand Budhani, Ramalingam Kayalvizhi, and Samuel Jacob

10.1 INTRODUCTION

10.1.1 Life Cycle of Biofuel Production

The life cycle of biofuel production begins with sourcing biomass from crops, agricultural residues, and waste. This biomass is collected and transported to facilities, where it undergoes conversion through processes like fermentation and gasification. Biofuels are sustainable alternatives to fossil fuels, reducing greenhouse gas emissions. Their production involves multiple stages: feedstock cultivation, biomass conversion, and fuel refining, each impacting the overall life cycle. Despite their environmental benefits, challenges in cost-effective production, processing techniques, and resource optimization persist. Advancements in these areas are crucial for widespread adoption. Biofuel production involves several stages, from feedstock cultivation to fuel processing, distribution, and end use. The life cycle of biofuel production begins with the selection of raw materials, which can include crops, waste, or other organic materials. These feedstocks undergo a conversion process—typically through biochemical or thermochemical methods—to produce biofuels like biodiesel or ethanol. The fuel is then refined, distributed, and utilized as a renewable energy source. Each stage of this process has environmental and economic impacts, including greenhouse gas (GHG) emissions, land and water use, and energy consumption. A life cycle assessment (LCA) helps quantify these impacts, evaluating the sustainability and efficiency of the biofuel production process. By understanding the entire life cycle, industries can improve resource utilization and minimize negative environmental consequences. This holistic approach also aids in making informed decisions about scaling biofuel technologies and integrating innovations like artificial intelligence (AI) for enhanced efficiency (Mu et al. 2017). The next step involves processing and refining the biomass into usable fuels such as ethanol,

DOI: 10.1201/9781003571223-10

biodiesel, or biogas. These processes consume energy and resources, impacting the overall environmental footprint, while transportation logistics play a key role in minimizing emissions. Once produced, biofuels are distributed for use in sectors like transportation and electricity generation. Although biofuels release carbon dioxide during combustion, this is offset by the carbon absorbed by biomass during its growth, making biofuels more carbon neutral compared to fossil fuels. Lastly, waste and by-products are managed, often reused or recycled, creating a more sustainable system. Addressing challenges like feedstock variability and land use is crucial for the long-term viability of biofuels. Highlighting the life cycle of biofuel production, this approach emphasizes the conversion of lignocellulosic biomass like agricultural residues into second-generation biofuels. The process begins with pretreatment to make sugars more accessible for fermentation, followed by biofuel production, which reduces greenhouse gas emissions. This approach minimizes competition for land and waste, offering a sustainable alternative to fossil fuels and first-generation biofuels (Prasad et al. 2016).

10.1.2 Role of Artificial Intelligence (AI) in Biofuel Production

AI, particularly machine learning (ML), is significantly offering advancements in biofuel technology through its ability to optimize and predict outcomes. AI applications range from predicting product yields and optimizing process conditions to evaluating the economic and environmental impacts of converting biomass into biofuels. AI tools, such as artificial neural networks (ANNs) and genetic algorithms (GAs), have emerged as promising methods for modeling and optimizing biofuel production. These AI techniques are particularly effective in managing the non-linear nature of bioprocesses, helping to predict complex interactions between inputs and outputs without prior knowledge of the process kinetics. ANNs, for example, simulate the learning process of the human brain, adapting and analyzing data to optimize biofuel yields, making them efficient for bioprocess research and development (Sewsynker-Sukai, Faloye, and Kana 2017). Recent studies have highlighted three key areas where ML is making a major impact: (i) discovering new materials for biofuel production, (ii) optimizing conversion processes like fermentation and pyrolysis, and (iii) aiding in decision-making for process improvements. Additionally, AI is being used for preliminary economic and environmental assessments, helping to identify the most efficient and sustainable pathways for biofuel production, ultimately enhancing both scalability and cost-effectiveness. AI and ML are revolutionizing biofuel production by offering advanced modeling techniques to optimize complex and non-linear processes like fermentation, gasification, and pyrolysis. These data-driven approaches enhance predictive capabilities, improving efficiency, scalability, and cost-effectiveness in biofuel technologies. AI/ML also helps optimize input parameters, ensuring better product yield and sustainability in biofuel production (Okolie 2024). This organized integration of AI in biofuel production is paving the way for enhanced efficiency and a more sustainable future.

10.1.3 Cost Analysis of Biofuel Production

Cost analysis of biofuel development examines the economic factors involved in production, including capital investment, operational costs, and feedstock expenses. Producing biofuels like ethanol or biodiesel requires substantial upfront investments in technology, infrastructure, and equipment, which vary based on the feedstock type and conversion technologies. Advanced biofuels from non-food biomass generally have higher capital costs due to more complex processes (Srivastava, Kharwar, and Mishra 2019). Operating costs cover labor, maintenance, and utilities, like energy and water, while feedstock expenses, often a significant part of the total cost, vary depending on availability and type. Crop-based biofuels face fluctuating feedstock prices influenced by agricultural markets. Cost analysis of biofuel development emphasizes the high expenses involved in producing biofuels compared to traditional fossil fuels. The manufacturing of ethanol, particularly using feedstocks like sugarcane bagasse, incurs significant costs, roughly Rs 95/L (Srivastava, Kharwar, and Mishra 2019). This includes expenses related to technology, infrastructure, and operational aspects. However, optimizing the production process through detailed cost analysis could lead to reductions in these expenses, thereby enhancing the efficiency and long-term sustainability of bioethanol production (Srivastava, Kharwar, and Mishra 2019).

Biofuel production faces several challenges, including the diversity of biomass feedstocks, the complexity of conversion technologies, and systems integration. Additionally, concerns about fuel quality, blending requirements, and environmental impacts like greenhouse gas emissions and life-cycle energy usage need to be addressed. Moving forward, research and development efforts are focused on improving biomass feedstocks, optimizing conversion processes, and integrating systems to enhance sustainability and efficiency in biofuel production (Hoekman 2009). The application of ML in bioproduction faces challenges such as low interpretability, limited compatibility with complex bioprocesses, and uncertainties in prediction. These limitations arise due to the intricate metabolic interactions involved in bioprocessing. Future efforts should focus on improving the interpretability and user-friendliness of ML models, as well as developing hybrid approaches that enhance prediction accuracy and adaptability to bioprocesses (Lim et al. 2023). It has been stated that scalability is also a major challenge in biofuel production, with large-scale implementation hindered by high energy demands and process inefficiencies. While AI tools and process simulations can optimize complex processes like fermentation and pyrolysis, these technologies still need refinement for real-world applications due to numerous variables and uncertainties. Moving forward, biofuel production should focus on efficient bioconversion processes, improving AI models for uncertainty prediction, and exploring consistent alternative feedstocks. Advances in recycling technologies and decarbonization strategies are also crucial for reducing biofuel production's carbon footprint, enhancing both sustainability and long-term viability. The main challenges for biofuel production include the high costs of production, inefficient harvesting techniques, and suboptimal processing methods. Future efforts should focus on optimizing mass culture, improving processing efficiency, and utilizing by-products to enhance sustainability (Mobin and Alam 2017).

This chapter highlights and offers an in-depth exploration of the rapidly evolving intersection of biofuel production and AI, focusing on both environmental and economic aspects. As biofuel production gains momentum globally, it has become crucial to evaluate its entire life cycle—from feedstock cultivation to final disposal—to ensure that the environmental benefits outweigh the costs. By emphasizing Life Cycle Assessment (LCA) and cost analysis, this chapter seeks to address these concerns while highlighting the transformative role of AI. Delving further, the chapter systematically explores the stages of biofuel production, ranging from feedstock cultivation to final disposal, with a focus on leveraging AI to enhance the precision and efficiency of these assessments. AI-driven tools offer deeper insights into uncertainties and variabilities within biofuel production processes, streamlining operations and optimizing resource usage. In addition, a detailed analysis of the cost implications of AI integration, showcasing how AI-enhanced simulations improve both environmental sustainability and economic feasibility, has been discussed in detail. The comparison between traditional biofuel production and AI-driven processes will demonstrate the return on investment (ROI) and long-term benefits, alongside addressing challenges like data variability and accuracy.

10.2 ESSENTIAL ASPECTS OF BIOFUEL PRODUCTION

10.2.1 Biofuel Technologies and Various Feedstocks Used

Currently, the total consumption of energy in the world has gone up to 14 billion tons of oil equivalent (ToE) (Jeswani, Chilvers, and Azapagic 2020). The leading cause of the increased GHG emissions is attributed to the continuous use of fossil fuels for the generation of energy. As of now, numerous environmental policies are prioritizing the reduction of GHG, and due to various sustainable advancements in crop engineering, bioprocessing, and fermentation technologies, biofuels prove to be viable alternatives to petroleum-based fuels. In general, the primary feedstocks used for producing bioethanol are sugar-based crops like sugarcane or sweet sorghum, whereas oil-containing crops like rapeseed or soybean serve as the feedstocks used in the production of biodiesel (Mat Aron et al. 2020). The major issue with using food crops as feedstocks in biofuel production is that they compete with food supplies. Moreover, due to differences in climatic conditions, agricultural practices, and available land mass, the supply of these food crops varies (NEULING 2017). Optimizing biofuel production using modern AI technologies would not only aid in the cost analysis during production but also create a strong and sustainable alternative to the currently used fossil fuels.

10.2.2 Use of AI in Biofuel Production

The use of AI in biofuel production greatly aids in reducing costs, making the overall process effective and sustainable. The different types of AI that are used to optimize biofuel production are shown in Figure 10.1 (Liao and Yao 2021).

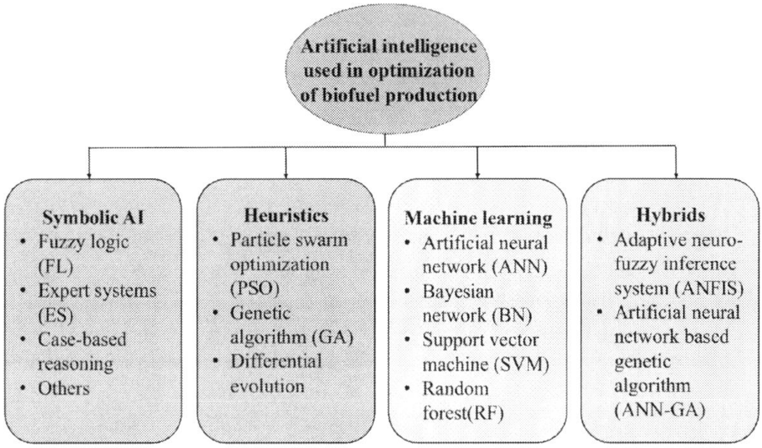

FIGURE 10.1 Artificial intelligence used for optimizing biofuel production.

ANN has various layers of web connections or nodes that imitate the neurons of the human brain (Mu et al. 2017). The transfer of data happens through a series of nodes wherein each node communicates with the adjacent one. The input variables include the primary resources like feedstock and catalysts, whereas the hidden layers include process variables like residence time or temperature. The feedstock composition and process control have an unknown impact, which can further impact the output variables. ANN identifies the optimal control function that improves the process results. Limitations of ANN include selecting a suitable network size and training the network (Owusu and Marfo 2023; Hansen and Mirkouei 2018). Symbolic AI utilizes symbols to represent cognition and logical deduction to imitate the process of human cognition (Hansen and Mirkouei 2018). Symbolism has been used to construct fuzzy logic (FL), expert systems (ES), and case-based reasoning (Liao and Yao 2021). Symbolic AI has various applications in process system engineering, some of which are related to bioenergy systems, such as fuzzy logic-based control for biomass-based power plants and boilers. Fuzzy logic modeling (FLM) is a type of AI that is ideal for biofuel production applications when precise numerical solutions are unavailable (Lerkkasemsan 2017). FLM replicates human reasoning by mathematically assigning a truth value. Usually, FLM is employed at lower-level control of equipment, except having a wide range of applications. FLM is ideal for system modeling, where mathematical models are ineffectual or unattainable. FLM has applications in the energy industry, like optimizing natural gas production, processing chemicals, and determining combustion engine performance with nanoparticles in diesel fuel. Similar methods could also be used to predict biofuel conversion process performance (Hansen and Mirkouei 2018).

An adaptive neuro-fuzzy inference system (ANFIS) integrates fuzzy logic and artificial neural networks to produce a hybrid intelligent system that draws advantages from both fuzzy logic and neural networks, and their efficiency is remarkably accurate (Mostafaei, Javadikia, and Naderloo 2016). The neuroadaptive learning mechanism operates similarly to neural networks, offering a learning strategy for fuzzy modeling. ANFIS can model the complex relationships between temperature, pressure, pH, and enzyme concentration to determine the desirable conditions for maximizing biofuel yield. ANFIS combines the interpretability of fuzzy logic with the learning capabilities of neural networks to provide an efficient and effective technique to optimize multiple parameters at the same time (Kumar 2020).

Nowadays, Convolutional Neural Networks (CNN) and Deep Belief Networks (DBN) are widely studied due to their established status (Hansen and Mirkouei 2018). Deep neural networks (DNNs) have numerous hidden layers between input and output, in contrast to ANNs that only have one layer (Filho and Viegas 2018). The advantages include improved datasets and training methods (Hansen and Mirkouei 2018). DNNs are ideal for huge datasets where ANNs are inadequate. DNN needs fewer nodes and is more effective than networks with a single hidden layer (Mu et al. 2017). Heuristics are stochastic search methods executed using learning-based methodologies and experience. There are two types of heuristics: evolutionary algorithms and swarm intelligence. Evolutionary algorithms comprise GA, differential evolution, and evolutionary techniques. Swarm intelligence algorithms involve particle swarm optimization (PSO) and ant colony optimization. Previous research has proved the great capability of applying heuristics to produce near-optimal solutions to complex issues like Biomass Supply Chains (BSC) scheduling and planning (Hansen and Mirkouei 2018; Owusu and Marfo 2023).

The incorporation of Internet of Things (IoT) and AI in biomass-to-biofuel conversion enables real-time monitoring of key parameters such as temperature, pressure, and chemical composition. Sensors placed at different points of the conversion process collect and transmit data to a centralized system, where AI algorithms analyze the data and make intelligent decisions. The advantages of using IoT and AI in biomass-to-biofuel conversion involve dynamic monitoring and control, resource optimization, reducing waste generation, and predictive modeling. Algorithms can learn from prior data and adapt to changing scenarios, increasing the conversion process efficiency and efficacy for better and sustainable outputs (Damian et al. 2024; Hussain et al. 2023; Ahmad et al. 2021b).

ML is another type of AI that can learn and develop based on experience for specific jobs without being specifically programmed (Mobin and Alam 2017). The two most used types of machine learning are connectivism and statistical learning (Hansen and Mirkouei 2018). Connectivism consists of techniques like feedforward neural networks (FNN), radial-basis functional networks (RBF), recurrent neural networks (RNN), and innovative convolutional neural networks. These AI approaches are also known as ANN. Statistical learning depends on statistical methodologies. Representative techniques include support vector machine (SVM), random forest (RF), and Bayesian networks (Hansen and Mirkouei 2018). These techniques differ in their fundamental principles and structure, leading to varied performance levels. ML has various applications in bioenergy and biofuel production, like catalyst utilization and agriculture. ML may boost sustainability by optimizing energy recovery from waste, reducing contamination, and minimizing negative environmental consequences associated with waste-based fuel production (Liao and Yao 2021).

TABLE 10.1
Applications of AI in the Optimization of Biofuel Production

Purpose	Application of AI	References
Biogas monitoring and estimating transmembrane pressure during production of bio-hydrogen in anaerobic membrane bioreactor	Internet of things (IoT) and adaptive neuro-fuzzy inference systems and artificial neural network (ANFIS)	Taheri et al. (2021)
Controlling fermentation time	ANN-GA combining strategy	Peng, Meng, and Ai (2013)
Parameters in fermentation	ANN-GA	Dave et al. (2021)

Using advanced learning rates to optimize operating capacity and production costs has aided in cost reductions in Brazilian sugarcane-based ethanol and maize ethanol production in the US. Moreover, they observed a substantial cost reduction in the range of 55–73% for bio-refineries (Daugaard et al. 2015). Using the same approach, Fischer-Tropsch diesel production costs were reduced from $4.42/gal to $2.00/gal. AI-enabled learning was effective across small-scale fuel production and large-scale gasification facilities. Additionally, the achieved learning rates surpassed industry experience and demonstrated economic superiority for AI optimization of biofuel (Daugaard et al. 2015). Applications of AI in the biochemical conversion of biomass are shown in Table 10.1.

10.3 LIFE CYCLE ASSESSMENT (LCA)

10.3.1 LCA Methodology

LCA is a useful and effective framework for evaluating impacts on the environment, natural resources, human beings, and so on. Attributional LCA is used for quantifying the present-day environmental implications of biofuels and bioproducts across the entire life cycle to identify the hotspots and strategies to minimize these implications. These results can be compared with their fossil fuel counterparts as well as to support decision-making for producers, consumers, and policymakers. On the other hand, consequential LCA assesses the changes in environmental implications occurring because of policies and decisions related to biofuels and bioproducts promotion. This often requires the inclusion of economic models to calculate the economic effects of policy decisions. Cradle-to-grave studies are inclusive of the whole life cycle, beginning from the cultivation of the biomass feedstock for biofuels to the transformation of the biomass to the product, its use, and final disposal in the case of bioproducts. Cradle-to-gate studies are used to compare biofuels or bioproducts from different feedstocks. Gate-to-gate studies may be used to focus on a particular novel transformation process (Osman et al. 2021).

10.3.2 Goal and Scope

The purpose of the LCA in biofuel production is to compare biofuels with conventional fuels or to determine the environmental impacts of different biofuel pathways. This sets the boundaries of the analysis, which incorporates the parts of the biofuel life cycle, such as feedstock cultivation, transportation, and fuel processing (Gheewala 2023). It also incorporates defining functional units and respective system boundaries. Functional units are an estimated description of the performance requirements that the product system fulfills and are directly linked with the functions of the product rather than with physical products (Osman et al. 2021).

10.3.3 Life Cycle Inventory

Life Cycle Inventory (LCI) for biofuel production is an organized collection of data on the inputs, like raw materials, energy, water, and the outputs, such as emissions and waste, at each step of the process. Feedstock production is the initial step in the process, requiring the use of energy, water, and fertilizers to cultivate crops for biofuel. The next step is transportation, in which the biomass is transported to processing facilities, therefore adding to emissions resulting from fuel consumption. In the conversion step, biomass is converted into biofuel via biochemical processes such as fermentation (for bioethanol) or thermochemical methods such as pyrolysis or gasification. Finally, the end-use stage considers the emissions and energy consumption during the combustion of biofuel, although biofuels typically generate

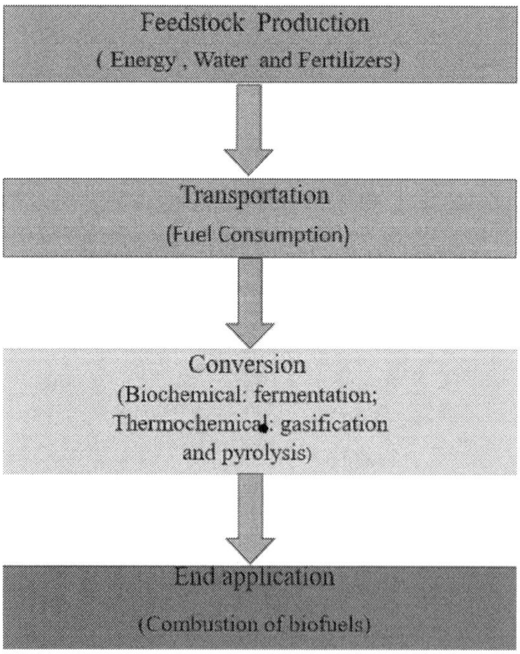

FIGURE 10.2 Biofuel production process.

fewer pollutants compared to fossil fuels. This thorough data gathering enables an accurate assessment of the environmental consequences of biofuel manufacturing, therefore facilitating the identification of strategies to mitigate its ecological imprint. The biofuel production process is shown in Figure 10.2.

10.3.4 Life Cycle Impact Assessment (LCIA)

LCIA is used to assess the direct and indirect inputs and releases at each step of a biofuel pathway, such as energy consumption, and environmental emissions like carbon dioxide, nitrogen oxides, and sulfur oxides (Gnansounou et al. 2009). LCIA is used to assess impact categories, such as climate change, eutrophication, ecotoxicity, human toxicity, photo-oxidant formation, acidification, stratospheric ozone depletion, depletion of biotic resources, and depletion of abiotic resources. The environmental impact can be measured at intermediate levels (midpoints) or damage levels (endpoints) after normalization and sometimes weighting approaches. Several frameworks in LCA help in the assessment of the impact of direct and indirect input, such as cradle-to-gate, cradle-to-grave, and well-to-wheel.

The cradle-to-gate approach concentrates only on the steps from the extraction of raw materials (cradle) to the point when the product exists at the factory gate (gate), omitting the phases of usage and disposal (Kokare, Oliveira, and Godina 2023). Concerning biofuel production, a cradle-to-gate LCA would determine the environmental impacts of feedstock cultivation, transportation, biofuel processing, and transportation but ignore the combustion of the biofuel in vehicles and its ultimate end use. This method is frequently employed when the goal is to draw a comparison between the production phase of biofuels and that of other fuels or products so that it can provide a clear view of the environmental impacts associated with production processes.

The cradle-to-grave approach is the most elaborate and extensive form of LCA, incorporating all steps of the product's life cycle, from raw material extraction (cradle) to the last disposal stage (grave) (Liu, Shafique, and Luo 2023). In biofuel production, this means determining everything from biomass cultivation, combustion, processing, and transportation of the biofuel to the end-of-life impacts. It also offers a comprehensive perspective of the biofuel's total environmental footprint, including combustion emissions when the fuel is burned in vehicles or other applications. The cradle-to-grave LCA is important for understanding the full environmental impact, especially when comparing biofuels to conventional fuels, as it captures both the pros and cons of the entire life cycle.

A well-to-wheel approach is commonly employed in the transportation industry to precisely assess the energy consumption and emissions resulting from the formation of fuel (well) and its subsequent use in vehicles (wheel). In the context of biofuels, this encompasses both the various phases of production (from the well to the tank) and the emissions resulting from the combustion of the ethanol in engines (from the tank to the wheel) (Gnansounou et al. 2009).

The present methodology offers a complete evaluation of the fuel's energy efficiency and emissions performance across its complete supply chain. The well-to-wheel approach is important for determining the overall benefits of biofuels, particularly regarding greenhouse gas emissions in comparison to fossil fuels.

10.3.5 Modification in LCA for AI Integration

Integration of AI in LCA for biofuel production helps in optimizing the process and predicting the environmental impacts at various stages of biofuel production, including soil preparation, production, feedstock management, and emissions. For optimizing the process, there are several ML tools such as ANN, ANFIS, Extreme Learning, SVM, and Extreme Learning Machine (ELM) (El Bouchefry and de Souza 2020). There are several stages in the biofuel production process where AI can easily be integrated without affecting the conventional method (Sewsynker-Sukai, Faloye, and Kana 2017). The first stage where AI can be included is the soil stage. In this stage, the SVM technique is the most widely used ML method, and the inputs we have to give are the soil characteristics, temperature, solar radiation, average precipitation, and wind speed. Due to this, we will get outcomes like future life cycle environmental impacts and biomass yield. The next stage is the feedstock stage; in this, the prerequisites are blending composition, mixing time, heating values, mixing speed, characteristics of fluorescence signals, temperature, and size dimensions of microalgae. The outputs are oxidizing stability, cetane number, yield, and characteristics of crude and hydrochar, viscosity, density, flash point, higher heating values, and fraction of methane obtained by employing various tools like ANN, RF, SVR, statistical regression model, multiple linear regression, and multiple non-linear regression for optimizing the process. The most important stage is the production stage, in which different AI tools are used based on the type of biofuel, such as biogas, biohydrogen, and miscellaneous. The application of ML in the life cycle of biofuel is shown in Figure 10.3.

For example, in biodiesel, the machine learning tool for quality estimation was the artificial neural network, followed by the regression model. The input variables were reaction duration, calcination temperature, flow rate, reaction temperature, and pressure. In yield estimation, the most common machine learning technology was ANN, followed by adaptive neuro-fuzzy inference systems. The inputs given are catalyst concentration, reaction duration, temperature, total volatile fatty acids (VFAs), and methanol-to-oil molar ratio in the effluent.

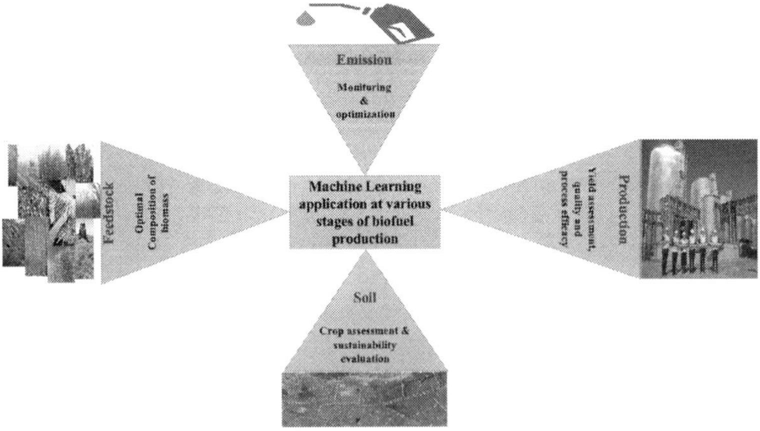

FIGURE 10.3 Application of machine learning in the life cycle of biofuels.

For quality and yield estimation, the GA-based ANN machine learning method was widely used, followed by GA-based SVM and ANFIS (Tufaner, Avşar, and Gönüllü 2017). The input variables are reaction temperature, stirring speed, methanol-to-oil molar ratio, reaction duration, and catalyst concentration. In the area of process efficiency estimates and optimization, artificial neural networks (ANN) emerged as the most effective machine learning tool, followed by the Adaptive Neuro-Fuzzy Inference System (ANFIS) together with various optimization tools of machine learning (Ahmad et al. 2021a). The input variables used here are concentration, water content, reaction duration, temperature, and methanol-to-oil molar ratio.

10.3.6 Real-time Monitoring

Real-time monitoring using AI in biofuel production aids in the dynamic optimization of important stages, which leads to process efficiency, environmental compliance, and resource management. Machine learning plays an essential role in estimating feedstock quality and adjusting parameters like moisture and biomass composition to forecast availability and prevent supply chain disruptions. During the production phase, AI tools continuously monitor variables such as temperature, pressure, and catalyst concentration, automatically adjusting them for optimal biofuel yields. Neural networks can optimize the methanol-to-oil ratio in biodiesel production. AI is also crucial in predictive maintenance, where algorithms analyze operational data to detect potential equipment failures before they occur; this leads to a reduction in downtime and maintenance costs. AI tool monitoring systems track emissions in real time, ensuring compliance with environmental regulations by adjusting processes to reduce greenhouse gases like CO_2 and NO_x. Finally, with the help of AI, it optimizes energy usage throughout the production process, determining energy needs and reducing waste by precise tuning operations at different stages. Despite the widespread adoption of production technology globally, there remains a need to enhance monitoring methods to improve efficiency. Online monitoring is essential for future-oriented bioenergy facilities; however, the parameters must be properly established. The investigations conducted show that pH is not an early indicator of instability, as it depends on organic loading due to its fundamental instability (Chiumenti et al. 2019). Alongside pH measurements, alkalinity assessments, crucial for the early identification of process instability, can be performed by real-time electrical conductivity measurements. This integration of AI across the biofuel life cycle leads to better control, efficiency, and environmental sustainability, ensuring optimized performance and reduced carbon footprints for biofuel producers.

10.3.7 Scenario Analysis

Scenario analysis in biofuel production processes is a crucial tool for assessing potential environmental, economic, and operational outcomes based on different assumptions. Concerning LCA, scenario analysis is widely used to investigate variations in feedstock types, processing technologies, transportation logistics, and end-use applications. This method allows for a comparison of different production pathways, such as environmental impacts (like GHG emissions) and their energy efficiency (Bichraoui-Draper et al. 2015). Scenario analysis can evaluate the environmental trade-offs between producing biofuels from algae versus other biomass, such as energy crops or agricultural residues. It also identifies the impact of integrating renewable energy sources into the production process or optimizing the supply chain to curb transportation emissions. Additionally, there are several dynamic LCA techniques that can further enhance scenario analysis by modifying variables such as energy compositions or regional resource accessibility over time. Scenario analysis helps decision-makers better understand risks and opportunities within biofuel production, helping to build strategies that can adapt to changing conditions while promoting sustainability and economic viability (Bichraoui-Draper et al. 2015; Chowdhury et al. 2023).

10.4 COST ANALYSIS IN BIOFUEL PRODUCTION

10.4.1 Traditional Biofuel Production

Ethanol production is increasing every year to meet the demand for it as a biofuel. Every year around 3,260 million liters of ethanol are produced. Recent fossil fuel depletion, rising environmental negative impacts, and petroleum reserves incurred for biofuel production (Hill et al. 2006; Murugesan et al. 2009). According to the US average of 2003–05, 1 ton of sugarcane yields 74 liters of ethanol, and 1 ton of sugar beet yields 94 liters of ethanol. The byproduct of sugarcane and sugar beet, i.e., sugar molasses, yields 263 liters of ethanol. Raw sugar, 1 ton, yields 513.2 liters to 534.4 liters of ethanol (Taghizadeh-Alisaraei, Motevali, and Ghobadian 2019).

Anaerobic digestion is used for biomethane production. It is done through the treatment of organic wastes by microbes. Various factors account for alkalinity, biogas composition, volatile fatty acid concentrations, and yield of methane that need to be monitored. Anaerobic digestion of food wastes is considered the most efficient technology for handling waste because it allows stabilization of waste that results in the yield of both fertilizer and biogas.

Artificial Neural Network—Genetic Algorithms (ANN-GA) hybrid approach and Response Surface Methodology (RSM) are used to improve the yield of methane (Li, Chen, and Wu 2019). One of the ingredients found in the anaerobic digestion of waste sludge is microplastic, which has an impact on the process by increasing the microplastic leachates, contents, and particle size. The presence of microplastics modifies the consequences of other anaerobic digestion contaminants as well as their course. Scanning electron microscopy (SEM), gas chromatography-mass spectrometry (GC–MS), Fourier transform infrared (FTIR), and Raman spectroscopy are a few of the techniques that AI offers to assess the content of these microplastics (Meena et al. 2021). The quickest method of measuring microplastics is generally visual identification under a microscope; however, the matching precision for microplastic particle size is lower. Spectral analysis and visual detection for precise microplastic properties are more beneficial (Lv et al. 2021).

Fermentation is used in biochemical conversion for energy and fuel generation. Two types of fermentation are used: solid-state and liquid-state fermentation

(Branco-Vieira et al. 2020). A solid-state is more challenging than a liquid one. Meena et al. (2021) developed an algorithm of ANN based on the level of pH. ANN is used for finding the different variables in the process that are necessary for optimum conditions and maximum yield. It was found that ANN gave better optimization results compared to RSM.

The increasing demand for biofuels periodically demands more production. The production process includes various parameters that need to be optimized to reduce the costs of production. Overall, AI helps in optimizing these parameters to produce biofuels, thus reducing costs. It also has positive environmental effects and is thus in line with the pollution policy. It helps in finding optimum resources and thus focuses on efficiency and reliability (Gumte et al. 2021).

10.4.2 Cost Components—Indirect and Direct

Biofuels have both direct and indirect land use. Direct effects can be tested by the amount of feedstock needed for one unit of bioenergy production. Indirect land transitions are harder to conclude. The various cost components in biofuel production are shown in Figure 10.4. Crop yields, location, and substitution on demand and supply have compelling aspects. Analysis by Elobeid et al. (2019) affirms the demand to acknowledge indirect land use change when checking the environmental impact of discrete biofuels.

10.4.3 Role of Artificial Intelligence

The global ML service, which was worth USD 2.4 billion in 2019, had become USD 15.47 billion in 2021 and is projected to ascend to USD 58.26 billion, USD 157.49 billion, and USD 305.62 billion in 2025, 2028, and 2030, respectively (Awogbemi and Kallon 2023). The architecture and enhancement of models and blueprints have confirmed reduced tasks and time, which allow for faster and more reliable decision-making. ML algorithms have empowered more accurate and adequate service delivery without the demand for human communication. ML algorithms can easily complete complicated tasks in the most ambiguous and difficult circumstances. Machine learning algorithms can handle enormous volumes of data, analyze trends,

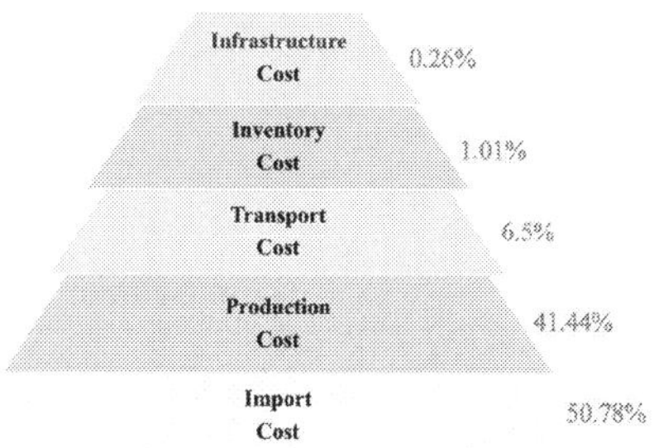

FIGURE 10.4 Cost components in biofuel production.

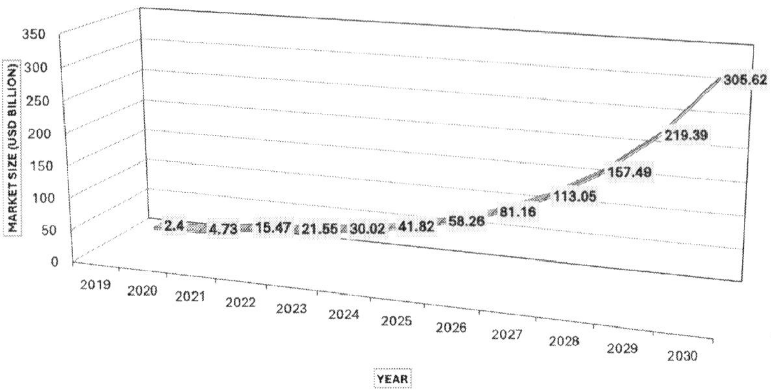

FIGURE 10.5 Machine learning service market share (billion USD). (Data taken from Awogbemi and Kallon 2023).

and use that information to make intelligent decisions with high certainty (Awogbemi and Kallon 2023). The increasing machine learning service market share is shown in Figure 10.5.

10.4.4 Comparing Traditional and Artificial Intelligence-based Biofuel Production Costs—Increased Efficiency and Productivity

In India, non-food resource-based biofuels are the most assured source of carbon-neutral renewable energy. These second-generation sources involve agro waste commodities such as wood, hay, and straw, among others, that do not compete with food resources. In biofuel production, the supply chain network is an important one for optimization and improving costs using AI. A mixed linear integer programming model was used for the 20% blending of biofuels with conventional fuels. It deals with the handling of different varieties of feed sources, products, and shipment alternatives. The basic process is almost the same, that is, treating lignocellulose biomass with carbohydrate hydrolysis agents followed by fermentation, with variation in temperature, pH, and enzyme usage based on specific biomass composition. Similarly, for biodiesel production, seeds are made to go through an expeller to obtain the oil and are subjected to a transesterification process to obtain biodiesel; here also, each process varies in terms of temperature, pH, and catalyst treatment based on biomass composition. Data-driven techniques are used to handle uncertainty in the biofuel supply chain; for instance, the Gaussian Mixture Model is an ML technique employed to determine the probability of individual data points of a given cluster. It can be used to alter the approach for determining the optimal cluster number into a Bayesian non-parametric method as well as for modeling data without any fixed number of fixed parameters. The DRIFT approach is another technique that is capable of handling fuzzy input and determining the capable cluster number through an artificial neural network-based transformation and evolutionary optimization of

FCM. The DRIFT technique can handle many different parameters, such as fuzziness, nonlinearity, high dimensionality, discontinuity, non-uniformity, or being scattered. Thus, the DRIFT approach and Gaussian mixture model are two novel methodologies used in data-driven techniques. This clarifies the implementation of DRIFT over other data-driven sampling techniques to handle uncertainty in biofuel supply chain modeling (Gumte et al. 2021).

10.4.5 ROI Analysis: Cost Reduction from AI Optimization

Infrastructure cost is a fixed investment cost. The others include supplier, retailer, manufacturer, transport, and distributor in a holistic manner, along with utility cost. Currently, the selling prices for bioethanol and biodiesel are ₹43.7/L and ₹52.8/L, respectively (Gumte et al. 2021). The distributor's storage cost is determined as 2% of the selling price for the specified term. The estimated cost for imported ethanol is 68% to 108% higher than indigenous, i.e., between ₹73.92/L and ₹91.52/L due to better quality shown by RON (research octane number), and a final value of ₹75/L is used (Ahmad et al. 2021b). For biodiesel, the imported cost is 5% higher than the indigenous value of ₹70/L. The price of fuel imported varies with a standard deviation of around 5%. The biomass supply chain network design framework handles product demand uncertainty. The novel supply chain model handles various resources of feed, products, transport, and technology options around varied periods to follow the Indian government's mandate of enabling the blending of 20% fuel-grade bioethanol with gasoline and biodiesel with diesel. While net present value has been calculated using Capex and Opex, the final revenue is generated by using GHG carbon credits (Gumte et al. 2021).

10.5 CHALLENGES OF AI-BASED LCA ASSESSMENT AND COST ANALYSIS

The production of biofuels has numerous technological obstacles that substantially impede advancement. The incorporation of artificial intelligence (AI) technologies greatly reduces these challenges. A considerable issue involves the utilization of non-consumable feedstocks, such as lignin-rich biomass and various biological wastes, requiring sophisticated AI-driven methodologies for effective processing. Furthermore, the complexities of manufacturing logistics need the application of AI algorithms to improve supply chain management. There is an urgent necessity to optimize pretreatment, enzyme hydrolysis, and fermentation technologies for energy efficiency with AI applications that can monitor and boost performance. The subject of co-product utilization is underexplored, presenting a potential for AI to improve resource recovery strategies. Furthermore, the establishment of biofuel standards and the improvement of distribution logistics are essential areas requiring AI frameworks to ensure compliance and efficiency. Moreover, the physical characteristics of biomass, such as humidity content and particle size, influence conversion efficiency, which AI may evaluate to forecast optimal processing conditions. The presence of inhibitors generated during pretreatment—such as detrimental chemicals that hinder microbial activity—highlights the necessity for AI-enhanced screening of microbial strains and the development of genetic engineering solutions to mitigate these effects. Integrating AI into these constraints could substantially enhance biofuel manufacturing techniques (Ambaye et al. 2021).

Indeed, the application of AI presents a remarkable chance to transform contemporary electrical infrastructure and the energy sector. Nonetheless, an obstacle persists that impedes the broader deployment of AI. Examples encompass constraints associated with the efficient tuning of network hyperparameters, data quality deficiencies, a shortage of skilled specialists and data science competencies, technical infrastructure problems, compliance obstacles, and legal security concerns, among others. Numerous technological issues exist, including the opacity of AI methodologies and the difficulties faced by AI systems in processing unstructured data. A substantial expenditure required for assessing power and technological acceleration in energy initiatives is also a concern for business owners. Sophisticated processing controller algorithms utilizing AI technologies can obscure the identification of defects and errors. The analysis and execution of AI rely on huge amounts of sensitive and occasionally personal data. It is rendered susceptible to significant problems, including identity theft and privacy breaches, by these uses. In recent years, rapid advancements have been made in AI, particularly through deep learning techniques utilized in ANN; yet uncertainty remains regarding the sustainability of this progress. Moreover, augmented computational capacity is frequently required for AI technologies to address more intricate issues (Ahmad et al. 2021a).

The challenges associated with integrating artificial intelligence (AI) into biofuel production must be addressed by considering several significant obstacles. A key limitation arises from the requirement for extensive datasets, which are often constrained to a limited range of experimental conditions. AI techniques, particularly machine learning (ML) models, rely heavily on large datasets to function effectively, making them suboptimal for small-scale experimental setups where training algorithms are insufficient. Additionally, the process of data acquisition and storage is both time-intensive and financially demanding. Furthermore, advanced deep learning models, which have shown promise in various industrial applications, are hindered by the unavailability of sufficiently large and comprehensive datasets within the current biofuel production framework. Consequently, these limitations pose substantial barriers to the optimization and scalability of AI-driven biofuel production systems (Hansen and Mirkouei 2018).

The application of artificial neural networks (ANNs) for modeling bioprocesses characterized by minimal data availability can pose significant challenges, often resulting in inadequate information for effective network training. Moreover, the deployment of ANNs for real-time monitoring and control of bioreactors used in biofuel production has been reported infrequently, highlighting a gap in the literature. Additional obstacles include the identification of optimal parameters that influence the model development phase, such as data partitioning and preprocessing, the appropriateness of the network architecture, and the rigor of model validation. These factors are critical in ensuring that the ANN models are robust and capable of accurately reflecting the complexities inherent in bioprocess dynamics (Sewsynker-Sukai, Faloye, and Kana 2017).

10.6 FUTURE PROSPECTS

Standardized and pragmatic protocols for algorithm selection and dataset size determination are formulated. Designing such processes necessitates a thorough comprehension of the implications and efficacy of diverse methodologies and training data collections to address specific bioenergy challenges. Optimizing gathered data, documentation, and dissemination in bioenergy sectors is essential. The accessibility and quality of data, primarily produced by independent experimental research and enhanced data sharing among diverse scientific and professional communities, are dependent on the successful implementation of AI approaches in the biofuel sector. Proposed ways to improve data accessibility and accuracy include the early involvement of AI professionals in testing, particularly in large-scale contexts. Furthermore, there is a necessity for readily available, extensively researched, and credible information to expand AI applications in facilitating decision-making about the development, operations, and enhancement of bioenergy systems. The capacity of AI to facilitate the sustainable advancement of biofuel networks is being investigated from a comprehensive viewpoint. Prior research has predominantly concentrated on a singular facet regarding environmental responsibility (e.g., economic) or a specific component of bioenergy technologies (e.g., biomass conversion). Analytic methods like LCA have served as a basis for comprehensive evaluation, and AI-integrated simulations could effectively tackle difficulties such as data scarcity. Appropriate and efficient methods for utilizing multiple artificial intelligence methods and current resources to facilitate comprehensive evaluation and enhancement of bioenergy systems for sustainability are required to be ascertained through further investigations (Liao and Yao 2021).

Future investigations into the integration of AI within biomass energy systems should focus on the development of sophisticated and comprehensive simulation methodologies. These advanced techniques should encompass not only the aforementioned parameters but also metrics such as payback time, LCA, and net profit, all consolidated onto a singular platform. Advancement could incorporate the Internet of Things (IoT) and the interconnection of bioenergy stations for enhanced communication and management of extensive bioresources, as well as technology for energy and chemical production (Meena et al. 2021).

The future prospects of AI in biofuel production are promising. AI can enhance the optimization of bioenergy conversion technologies by employing advanced algorithms such as GA and particle swarm optimization (PSO) to improve critical parameters like temperature, pH, and hydraulic retention time. The implementation of AI-driven modeling and simulation can facilitate a deeper understanding of system behavior, leading to quicker technological advancements and reduced experimental timelines. Moreover, integrating AI with renewable energy resources can promote self-sustainability in biofuel production systems, thereby reducing reliance on fossil fuels. The potential for AI to streamline the biomass supply chain, from collection to conversion, can further enhance economic feasibility and efficiency. As AI technologies evolve, they will likely play a critical role in addressing the existing bottlenecks in biofuel production, paving the way for a more sustainable and efficient bioenergy landscape (Banerjee 2023).

10.7 CONCLUSION

Incorporating AI in biofuel production represents a noteworthy progression toward realizing renewable energy objectives. AI technologies, including ML and neural networks, hold the promise of enhancing biofuel production processes, rendering them more efficient and environmentally conscious. By optimizing various stages—from feedstock cultivation to conversion processes—AI enhances resource efficiency, diminishes greenhouse gas emissions, and tackles issues such as feedstock variability and energy usage. Additionally, AI-driven cost analysis models enable the identification of opportunities for minimizing production expenses, potentially improving the economic competitiveness of biofuels relative to fossil fuels. Nonetheless, obstacles persist, particularly the necessity for extensive datasets and advanced AI models to manage the intricacies of biofuel production. As AI technologies advance, they are expected to be essential in addressing these challenges, facilitating the emergence of biofuels as a feasible and scalable substitute for traditional energy sources, thereby fostering a more sustainable and energy-efficient future.

REFERENCES

Ahmad, Iftikhar, Adil Sana, Manabu Kano, Izzat Iqbal Cheema, Brenno C. Menezes, Junaid Shahzad, Zahid Ullah, Muzammil Khan, and Asad Habib. 2021a. "Machine Learning Applications in Biofuels' Life Cycle: Soil, Feedstock, Production, Consumption, and Emissions." *Energies* 14 (16): 5072. doi:10.3390/en14165072.

Ahmad, Tanveer, Dongdong Zhang, Chao Huang, Hongcai Zhang, Ningyi Dai, Yonghua Song, and Huanxin Chen. 2021b. "Artificial Intelligence in Sustainable Energy Industry: Status Quo, Challenges and Opportunities." *Journal of Cleaner Production* 289 (March): 125834. doi:10.1016/j.jclepro.2021.125834.

Ambaye, Teklit Gebregiorgis, Mentore Vaccari, Adrián Bonilla-Petriciolet, Shiv Prasad, Eric D. van Hullebusch, and Sami Rtimi. 2021. "Emerging Technologies for Biofuel Production: A Critical Review on Recent Progress, Challenges and Perspectives." *Journal of Environmental Management* 290 (July): 112627. doi:10.1016/j.jenvman.2021.112627.

Awogbemi, Omojola, and Daramy Vandi Von Kallon. 2023. "Application of Machine Learning Technologies in Biodiesel Production Process: A Review." *Frontiers in Energy Research* 11 (February). doi:10.3389/fenrg.2023.1122638.

Banerjee, Nilanjana. 2023. "Biomass to Energy—an Analysis of Current Technologies, Prospects, and Challenges." *BioEnergy Research* 16 (2): 683–716. doi:10.1007/s12155-022-10500-7.

Bichraoui-Draper, Najet, Ming Xu, Shelie A. Miller, and Bertrand Guillaume. 2015. "Agent-Based Life Cycle Assessment for Switchgrass-Based Bioenergy Systems." *Resources, Conservation and Recycling* 103 (October): 171–78. doi:10.1016/j.resconrec.2015.08.003.

Bouchefry, Khadija El, and Rafael S. de Souza. 2020. "Learning in Big Data: Introduction to Machine Learning." In *Knowledge Discovery in Big Data from Astronomy and Earth Observation*, 225–49. Elsevier. doi:10.1016/B978-0-12-819154-5.00023-0.

Branco-Vieira, M., D. Costa, T.M. Mata, A.A. Martins, M.A.V. Freitas, and N.S. Caetano. 2020. "A Life Cycle Inventory of Microalgae-Based Biofuels Production in an Industrial Plant Concept." *Energy Reports* 6 (February): 397–402. doi:10.1016/j.egyr.2019.08.079.

Chiumenti, Alessandro, Francesco da Borso, Stefano Guercini, Andrea Pezzuolo, Michele Zanotto, Sandy Sgorlon, Gemini Delle Vedove, Fabiano Miceli, and Bruno Stefanon. 2019. "≪I≫The Impact of the Dairy Cow Diet on Anaerobic Digestion of Manure≪/I≫" In *2019 Boston, Massachusetts July 7- July 10, 2019*. St. Joseph, MI: American Society of Agricultural and Biological Engineers. doi:10.13031/aim.201900856.

Chowdhury, Raja, Nidia Caetano, Matthew J. Franchetti, and Kotnoor Hariprasad. 2023. "Life Cycle Based GHG Emissions from Algae Based Bioenergy with a Special Emphasis on Climate Change Indicators and Their Uses in Dynamic LCA: A Review." *Sustainability* 15 (3): 1767. doi:10.3390/su15031767.

Damian, Christopher Selvam, Yuvarajan Devarajan, Raja Thandavamoorthy, and Ravikumar Jayabal. 2024. "Harnessing Artificial Intelligence for Enhanced Bioethanol Productions: A Cutting-Edge Approach towards Sustainable Energy Solution." *International Journal of Chemical Reactor Engineering* 22 (7): 719–27. doi:10.1515/ijcre-2024-0074.

Daugaard, Tannon, Lucas A. Mutti, Mark M. Wright, Robert C. Brown, and Paul Componation. 2015. "Learning Rates and Their Impacts on the Optimal Capacities and Production Costs of Biorefineries." *Biofuels, Bioproducts and Biorefining* 9 (1): 82–94. doi:10.1002/bbb.1513.

Dave, Niyam, Thivaharan Varadavenkatesan, Raja Selvaraj, and Ramesh Vinayagam. 2021. "Modelling of Fermentative Bioethanol Production from Indigenous Ulva Prolifera Biomass by Saccharomyces Cerevisiae NFCCI1248 Using an Integrated ANN-GA Approach." *Science of The Total Environment* 791 (October): 148429. doi:10.1016/j.scitotenv.2021.148429.

Elobeid, Amani, Marcelo M.R. Moreira, Cicero Zanetti de Lima, Miguel Carriquiry, and Leila Harfuch. 2019. "Implications of Biofuel Production on Direct and Indirect Land Use Change: Evidence from Brazil." In *Biofuels, Bioenergy and Food Security*, 125–43. Elsevier. doi:10.1016/B978-0-12-803954-0.00007-3.

Filho, Alex Oliveira Barradas, and Isabelle Moraes Amorim Viegas. 2018. "Applications of Artificial Neural Networks in Biofuels." In *Advanced Applications for Artificial Neural Networks*. InTech. doi:10.5772/intechopen.70691.

Gheewala, Shabbir H. 2023. "Life Cycle Assessment for Sustainability Assessment of Biofuels and Bioproducts." *Biofuel Research Journal* 10 (1): 1810–15. doi:10.18331/BRJ2023.10.1.5.

Gnansounou, E., A. Dauriat, J. Villegas, and L. Panichelli. 2009. "Life Cycle Assessment of Biofuels: Energy and Greenhouse Gas Balances." *Bioresource Technology* 100 (21): 4919–30. doi:10.1016/j.biortech.2009.05.067.

Gumte, Kapil, Priyanka Devi Pantula, Srinivas Soumitri Miriyala, and Kishalay Mitra. 2021. "Achieving Wealth from Bio-Waste in a Nationwide Supply Chain Setup under Uncertain Environment through Data Driven Robust Optimization Approach." *Journal of Cleaner Production* 291 (April): 125702. doi:10.1016/j.jclepro.2020.125702.

Hansen, Samuel, and Amin Mirkouei. 2018. "Past Infrastructures and Future Machine Intelligence (MI) for Biofuel Production: A Review and MI-Based Framework." In *Volume 4: 23rd Design for Manufacturing and the Life Cycle Conference; 12th International Conference on Micro- and Nanosystems*. American Society of Mechanical Engineers. doi:10.1115/DETC2018-86150.

Hill, Jason, Erik Nelson, David Tilman, Stephen Polasky, and Douglas Tiffany. 2006. "Environmental, Economic, and Energetic Costs and Benefits of Biodiesel and Ethanol Biofuels." *Proceedings of the National Academy of Sciences* 103 (30): 11206–10. doi:10.1073/pnas.0604600103.

Hoekman, S. Kent. 2009. "Biofuels in the U.S.—Challenges and Opportunities." *Renewable Energy* 34 (1): 14–22. doi:10.1016/j.renene.2008.04.030.

Hussain, Zakir, M. Babe, S. Saravanan, G. Srimathy, H. Roopa, and Sampath Boopathi. 2023. "Optimizing Biomass-to-Biofuel Conversion." 191–214. doi:10.4018/978-1-6684-8238-4.ch009.

Jeswani, Harish K., Andrew Chilvers, and Adisa Azapagic. 2020. "Environmental Sustainability of Biofuels: A Review." *Proceedings of the Royal Society A: Mathematical, Physical and Engineering Sciences* 476 (2243). doi:10.1098/rspa.2020.0351.

Kokare, Samruddha, J.P. Oliveira, and Radu Godina. 2023. "Life Cycle Assessment of Additive Manufacturing Processes: A Review." *Journal of Manufacturing Systems* 68 (June): 536–59. doi:10.1016/j.jmsy.2023.05.007.

Kumar, Sunil. 2020. "Estimation Capabilities of Biodiesel Production from Algae Oil Blend Using Adaptive Neuro-Fuzzy Inference System (ANFIS)." *Energy Sources, Part A: Recovery, Utilization, and Environmental Effects* 42 (7): 909–17. doi:10.1080/15567036.2019.1602203.

Lerkkasemsan, Nuttapol. 2017. "Fuzzy Logic-Based Predictive Model for Biomass Pyrolysis." *Applied Energy* 185 (January): 1019–30. doi:10.1016/j.apenergy.2016.02.105.

Li Yue, Yinguang Chen, and Jiang Wu. 2019. "Enhancement of Methane Production in Anaerobic Digestion Process: A Review." *Applied Energy* 240 (April): 120–37. doi:10.1016/j.apenergy.2019.01.243.

Liao, Mochen, and Yuan Yao. 2021. "Applications of Artificial Intelligence-Based Modeling for Bioenergy Systems: A Review." *GCB Bioenergy* 13 (5): 774–802. doi:10.1111/gcbb.12816.

Lim, Seung Ji, Moon Son, Seo Jin Ki, Sang-Ik Suh, and Jaeshik Chung. 2023. "Opportunities and Challenges of Machine Learning in Bioprocesses: Categorization from Different Perspectives and Future Direction." *Bioresource Technology* 370 (February): 128518. doi:10.1016/j.biortech.2022.128518.

Liu, Fangjie, Muhammad Shafique, and Xiaowei Luo. 2023. "Literature Review on Life Cycle Assessment of Transportation Alternative Fuels." *Environmental Technology & Innovation* 32 (November): 103343. doi:10.1016/j.eti.2023.103343.

Lv, Lulu, Xiemin Yan, Limin Feng, Shiqi Jiang, Zifan Lu, Huifeng Xie, Shengli Sun, Jinjun Chen, and Chengyong Li. 2021. "Challenge for the Detection of Microplastics in the Environment." *Water Environment Research* 93 (1): 5–15. doi:10.1002/wer.1281.

Mat Aron, Nurul Syahirah, Kuan Shiong Khoo, Kit Wayne Chew, Pau Loke Show, Wei-Hsin Chen, and The Hong Phong Nguyen. 2020. "Sustainability of the Four Generations of Biofuels: A Review." *International Journal of Energy Research* 44 (12): 9266–82. doi:10.1002/er.5557.

Meena, Manish, Shubham Shubham, Kunwar Paritosh, Nidhi Pareek, and Vivekanand Vivekanand. 2021. "Production of Biofuels from Biomass: Predicting the Energy Employing Artificial Intelligence Modelling." *Bioresource Technology* 340 (November): 125642. doi:10.1016/j.biortech.2021.125642.

Mobin, Saleh, and Firoz Alam. 2017. "Some Promising Microalgal Species for Commercial Applications: A Review." *Energy Procedia* 110 (March): 510–17. doi:10.1016/j.egypro.2017.03.177.

Mostafaei, Mostafa, Hossein Javadikia, and Leila Naderloo. 2016. "Modeling the Effects of Ultrasound Power and Reactor Dimension on the Biodiesel Production Yield:

Comparison of Prediction Abilities between Response Surface Methodology (RSM) and Adaptive Neuro-Fuzzy Inference System (ANFIS)." *Energy* 115 (November): 626–36. doi:10.1016/j.energy.2016.09.028.

Mu, Dongyan, Roger Ruan, Min Addy, Sarah Mack, Paul Chen, and Yong Zhou. 2017. "Life Cycle Assessment and Nutrient Analysis of Various Processing Pathways in Algal Biofuel Production." *Bioresource Technology* 230 (April): 33–42. doi:10.1016/j.biortech.2016.12.108.

Murugesan, A., C. Umarani, R. Subramanian, and N. Nedunchezhian. 2009. "Bio-Diesel as an Alternative Fuel for Diesel Engines: A Review." *Renewable and Sustainable Energy Reviews* 13 (3): 653–62. doi:10.1016/j.rser.2007.10.007.

NEULING, ULF. 2017. "Review of Biofuel Production—Feedstock, Processes and Markets—Review Article." *Journal of Oil Palm Research* 29 (2): 137–67. doi:10.21894/jopr.2017.2902.01.

Okolie, Jude A. 2024. "Introduction of Machine Learning and Artificial Intelligence in Biofuel Technology." *Current Opinion in Green and Sustainable Chemistry* 47 (June): 100928. doi:10.1016/j.cogsc.2024.100928.

Osman, Ahmed I., Neha Mehta, Ahmed M. Elgarahy, Amer Al-Hinai, Ala'a H. Al-Muhtaseb, and David W. Rooney. 2021. "Conversion of Biomass to Biofuels and Life Cycle Assessment: A Review." *Environmental Chemistry Letters* 19 (6): 4075–4118. doi:10.1007/s10311-021-01273-0.

Owusu, Winnie A., and Solomon A. Marfo. 2023. "Artificial Intelligence Application in Bioethanol Production." *International Journal of Energy Research* 2023 (October): 1–8. doi:10.1155/2023/7844835.

Peng, Jiansheng, Fanmei Meng, and Yuncan Ai. 2013. "Time-Dependent Fermentation Control Strategies for Enhancing Synthesis of Marine Bacteriocin 1701 Using Artificial Neural Network and Genetic Algorithm." *Bioresource Technology* 138 (June): 345–52. doi:10.1016/j.biortech.2013.03.194.

Prasad, Aashish, Maria Sotenko, Thomas Blenkinsopp, and Stuart R. Coles. 2016. "Life Cycle Assessment of Lignocellulosic Biomass Pretreatment Methods in Biofuel Production." *The International Journal of Life Cycle Assessment* 21 (1): 44–50. doi:10.1007/s11367-015-0985-5.

Sewsynker-Sukai, Yeshona, Funmilayo Faloye, and Evariste Bosco Gueguim Kana. 2017. "Artificial Neural Networks: An Efficient Tool for Modelling and Optimization of Biofuel Production (a Mini Review)." *Biotechnology & Biotechnological Equipment* 31 (2): 221–35. doi:10.1080/13102818.2016.1269616.

Srivastava, Neha, Rahul Kumar Kharwar, and P.K. Mishra. 2019. "Cost Economy Analysis of Biomass-Based Biofuel Production." In *New and Future Developments in Microbial Biotechnology and Bioengineering*, 1–10. Elsevier. doi:10.1016/B978-0-444-64223-3.00001-1.

Taghizadeh-Alisaraei, Ahmad, Ali Motevali, and Barat Ghobadian. 2019. "Ethanol Production from Date Wastes: Adapted Technologies, Challenges, and Global Potential." *Renewable Energy* 143 (December): 1094–1110. doi:10.1016/j.renene.2019.05.048.

Taheri, Ensiyeh, Mohammad Mehdi Amin, Ali Fatehizadeh, Mashallah Rezakazemi, and Tejraj M. Aminabhavi. 2021. "Artificial Intelligence Modeling to Predict Transmembrane Pressure in Anaerobic Membrane Bioreactor-Sequencing Batch Reactor during Biohydrogen Production." *Journal of Environmental Management* 292 (August): 112759. doi:10.1016/j.jenvman.2021.112759.

Tufaner, Fatih, Yaşar Avşar, and Mustafa Talha Gönüllü. 2017. "Modeling of Biogas Production from Cattle Manure with Co-Digestion of Different Organic Wastes Using an Artificial Neural Network." *Clean Technologies and Environmental Policy* 19 (9): 2255–64. doi:10.1007/s10098-017-1413-2.

11 Artificial Intelligence–based Microalgal Biofuel Production

Future Prospects, Limitations, and Challenges

Bilal Ahmad Mir, Ritu Kumari, Aman Deep Raju, Rachana Singh, Sheo Mohan Prasad, Parul Parihar, and Shefali Gulliya

11.1 INTRODUCTION

In order to maintain a sustainable environment, government, scientists, policymakers, and researchers are trying to identify an alternative energy source due to the concerns related to depletion of fossil fuels, escalating oil prices, climate change, and environmental damage. One of the most eco-friendly alternatives to fossil fuels is biofuel manufactured from renewable resources that offers a practical solution for long-term ecological and economic sustainability. Algae are found in aquatic and terrestrial habitats and are capable of photosynthesis. They have the ability to thrive in harsh environment like high temperature, salinity, and high pH independently as well as symbiotically with other species (Barsanti et al., 2008). Algae are categorized into different classes like Rhodophyta, Chlorophyta, and Phaeophyta—and majorly categorized as macroalgae and microalgae based on size. While macroalgae are large and multicellular, microalgae are microscopic and single-celled. The microalgae, either prokaryotes or eukaryotes, are considered to be the major source of biofuels, cosmetics, medications, and dietary supplements (Das et al., 2011). Talking about biofuels like biodiesel, microalgae are considered the most important because of their high lipid content and also their quick multiplication and survivability in any natural as well as greenhouse conditions. Moreover, they do not compete for land resources and even grow on non-cultivable lands. Biofuels can be solid, liquid, or gaseous fuels made from organic sources. Despite their potential, first-generation biofuels brought up serious political, environmental, and financial issues.

Large-scale production of these biofuels competes with land for food production, and their production methods have worsened environmental damage, leading to

DOI: 10.1201/9781003571223-11

decreased public support. In response, attention shifted to second-generation biofuels, but their reliance on expensive, advanced technologies has made them commercially unfeasible (Brennan and Owende, 2010). Consequently, research has focused on third-generation biofuels, with microalgae being the primary source. Microalgae offer several advantages, including the ability to produce methane, biodiesel, and biohydrogen, as well as being 15 to 300 times more efficient in biodiesel production per unit area than traditional crops. With their short growth cycles and rapid reproduction, microalgae can be cultivated without the need for high-quality agricultural land.

11.1.1 Production of Biofuel from Microalgae

Freshwater and marine environments are home to single-cell tiny creatures known as microalgae. Microalgae are far more diverse than plants, with more than 300,000 species. Microalgae are better at converting solar energy into chemical energy in comparison to higher plants, and they can even grow in lab conditions in culture media having all the required nutrients and the required conditions. Table 11.1 shows different types of microalgae and their oil content.

11.2 DIFFERENT TYPES OF CULTIVATION TECHNIQUES

Microalgae serve as excellent alternatives for raw material because of their faster multiplicity capability and high efficiency in storing the bioproducts within the cells (Randrianarison and Ashraf, 2017). Moreover, they do not compete with other crops for resources because they do not require cultivable land and to increase their numbers no pesticides or fertilizers are required, which makes them more suitable (Khan et al., 2018). Microalgae can also be cultivated using wastewater from households and companies, such as palm oil milling, to aid in wastewater bioremediation (Tan et

TABLE 11.1
Various Microalage's Oil Contents in Relation to Dry Weight

S. No	Name of Microalgae	Percentage Dry Weight
1	*Chlorella* spp.	28–32
2	*Crypthecodinium cohnii*	20
3	*Dunaliella primolecta*	23
4	*Isochrysis* spp.	25–33
5	*Tetraselmis sueica*	15–23
6	*Cylindrotheca* spp.	16–37
7	*Botryococcus braunii*	25–75
8	*Phaeodactylum tricornutum*	20–30
9	*Schizochytrium* spp.	50–77
10	*Neochloris oleoabundans*	35–54

al., 2020). Despite these advantages, there are still many challenges facing microalgae growth. Through photosynthesis, the production of microalgae can reduce atmospheric carbon dioxide levels, reducing the greenhouse effect and global warming. However, because of their small cell size and poor biomass output when grown in liquid media, harvesting becomes exceedingly costly. To make microalgae more viable for industrial applications, efforts have been made to increase their growth rates to offset these challenges. While various tools and technologies have been developed to enhance microalgal production, ensuring their improved performance in large-scale manufacturing remains a significant challenge, even though they can be easily cultured in controlled laboratory environments. The following qualities should be present in a desirable microalgae culturing system: (1) efficient material transfer across liquid–gas barrier; (2) sufficient light source; (3) minimal contamination rate; (4) straightforward operation procedure; (5) high land efficiency; and (6) affordable overall construction and production cost. In general, photobioreactor and open pond are the two primary types that make up microalgae cultivation systems.

11.2.1 Open Pond Method for Microalgal Culture

One of the oldest and most basic techniques for producing microalgae on a large scale is open pond farming. Open ponds are a common feature in the business because of their comparatively reduced costs of development, upkeep, and operation. Open pond systems also have the advantages of being simple to set up, requiring little care, and being easily expanded. There are several varieties of open ponds, including man-made ones like raceways and circular ponds, as well as natural ones like lakes and ponds. According to Sun et al. (2016), microalgae can also occasionally be cultivated in tank-like containers. The main benefit of open pond systems is their affordability. The cell concentration of microalgae in natural water is relatively low, despite the enormous cultivation area, necessitating highly effective collection techniques. In addition, a number of obstacles affect the growth of microalgae in open ponds, including increased water turbidity, bank erosion that results in leaks, and precipitation runoff that modifies pH and salinity. Another big concern in open pond systems is contamination by bacteria and protozoa, which can make the products toxic and unusable. Growing microalgae that can flourish in extremely saline or alkaline conditions, where fewer pollutants may survive, is one current option.

Because of the system's openness, controlling growth parameters like temperature and light intensity becomes more challenging and may impact how quickly microalgal growth occurs. One of the earliest techniques for producing microalgae that humans have ever employed is harvesting microalgae from naturally existing sources. The first people to harvest microalgae are said to have been the Aztecs from Lake Texcoco in Mexico. The primary source of nutrients and ideal growing conditions for microalgae in this culture method is naturally existing bodies of water. Currently operating in Australia's Hutt Lagoon is one of the largest commercial microalgae farms. The facility produces 6 tons of β-carotene from Dunaliella annually using its 700-hectare ponds. The first artificial pond that was widely utilized for microalgae production was a circular pond. This method is made up of a revolving agitator in the middle of a circular culture tank. Normally, the tank has a diameter of 45 meters and a depth of 30–70 cm. The revolving agitator prevents algal biomass from sedimenting and helps to ensure efficient mixing, according to Show et al. (2017). However, the size of the pond limits the design because bigger ponds increase water resistance and place more strain on the agitator's mechanical components. Hamed (2016) draws attention to the fact that this system has a number of drawbacks as well, including high construction costs and significant energy consumption during the agitation process. Using this agricultural method, Taiwan and Japan have started to cultivate Chlorella for human consumption recently. One of the most popular designs for open-pond microalgae production is the raceway pond. The microalgae biomass is moved by paddlewheels and several 30-cm-deep closed-loop channels, which prevent sedimentation and equally distribute nutrients. Due to the high energy efficiency, open pond farming is considered best, where one paddlewheel produces 5-hectare algal biomass and this concept has been utilized by many farms like Sapphire Energy's Columbus Algal Biomass Farm in Columbus (Rogers et al., 2014; White and Ryan, 2015).

11.2.2 Photobioreactor

Growing phototrophic organisms, like microalgae, in a closed environment without exchanging resources with the outside world is done in a photobioreactor, a type of bioreactor. Open pond culture systems have a number of drawbacks that photobioreactors help to overcome. They offer a confined, regulated growing environment that permits the uncontaminated growth of a single strain of microalgae. Moreover, the controlled conditions improve the metabolic and nutritional efficiency that enhances the biomass accumulation per unit of substrate, but low scalability due to design defects creates an obstacle for the application of photobioreactors, thus making them less feasible for large-scale manufacturing (Gupta et al., 2015). Moreover, the stringent parameters of photobioreactors are linked to elevated initial and ongoing expenses. In photobioreactors, an airlift device or a mechanical pump is used to move the microalgae culture across the apparatus. Because of the potential for long tubes to cause changes in substrate and product concentrations throughout the system, one of the main problems with tubular photobioreactors is their low mass transfer efficiency. Despite this, *Chlorella* is produced commercially in Israel and *Haematococcus* in Germany using large-scale tubular photobioreactors (Acién et al., 2017). A sparger is used in vertical column bioreactors to pump air bubbles into the vertical cylindrical tube. This homogenizes the culture and makes it easier for the air and microalgae to exchange carbon dioxide and oxygen. Compared to other designs, this system has the highest gas–liquid mass transfer efficiency because the sparger creates smaller air bubbles, which increase the overall surface area and improve mass transfer efficiency (Mohan et al., 2019).

The straightforward design allows it to function more effectively and with less energy use. However, the microalgae do not receive enough light in the cylindrical container for optimal photosynthesis. Due to their expensive construction costs and difficult cleaning requirements, vertical column photobioreactors are hard to use commercially (Huang et al., 2017). Moreover, there are numerous large-scale experimental reactors for Guangdong, China's production of *Chlorella zofingiensis* (Huo et al., 2018a). The design of this photobioreactor has been used commercially. On the other hand, Vo et al. (2018) suggested that flat plate photobioreactor design can be utilized due to its transparent composition and depth that ranges from 1 to 5 cm. In the flat plate photobioreactor, culture is agitated with the help of an airlift system. This type of photobioreactor maximizes surface area for light exposure, and the recirculation airlift system reduces oxygen buildup, which may induce stress damage to microalgae cells. Innovative designs, including plastic sheets and twin-layer flat plates, are being implemented to improve the flat plate photobioreactors' efficiency (Vo et al., 2018).

11.3 HARVESTING TECHNIQUES OF MICROALGAE

The harvesting of microalgae is a critical part of their processing. Studies have shown that this specific phase accounts for 20–30% of total manufacturing costs due to its significant energy consumption and capital expenditures. All harvesting approaches seek to maximize the extraction of culture media from microalgae biomass to enhance subsequent downstream processes, such as the extraction of bioactive compounds. Several harvesting techniques employed for biomass collection include centrifugation, filtration, flotation, and flocculation (Singh and Patidar, 2018). There are some circumstances where combining two or more strategies is necessary to increase harvesting effectiveness even more.

11.3.1 Centrifugation

On the basis of the particle size as well as the density of components, centrifugation is used to separate microalgae colonies from the growing media (Soomro et al., 2016). Although this technique is quite effective in harvesting cells, it requires a lot of time and energy. Furthermore, centrifugation is not appropriate for situations where maintaining sensitive nutrients is crucial because the strong gravitational force utilized in the process can harm cells. The industry uses a variety of centrifugal devices, such as hydro-cyclones, imperforated and perforated basket centrifuges, disk stack centrifuges, and decanters (Singh and Patidar, 2018).

11.3.2 Filtration

The filtration method uses a semipermeable membrane, and this method harvests the trace level of microalgae and lets the liquid media pass through (Al Hattab et al., 2015), and this system can be adjusted as per the requirements of different microalgal species, including more sensitive ones that are vulnerable to shear damage. High cell concentrations from the media can be obtained with this technique. However, due to its high propensity for fouling and clogging, filters or membranes must be replaced frequently, which can dramatically raise processing expenses. In light of this reaction, a filter membrane composed of easily accessible and reasonably priced materials has been developed. Successfully developed stretch cotton-based filter membranes have harvesting efficiency ranging from 60 to 93 percent.

11.3.3 Floatation

In order to facilitate microalgae cells' ascent to the surface of the culture fluid for simple collection, flotation uses small bubbles that attach to the cells. Numerous advantages of the flotation system include its low cost, high processing capacity, and ease of operation. Furthermore, its efficiency in harvesting is relatively high. There are three main types of flotation systems, and to create the required air bubbles, each system requires a unique mechanism. By first saturating the culture with compressed air and subsequently releasing it into the surrounding environment, the dissolved air flotation technique creates air bubbles. Although this technique has been extensively utilized for treating wastewater, the chemicals and energy involved in it might make it costly. Dispersed air flotation, on the other hand, uses less energy than dissolved air flotation since it uses a sparger to create air bubbles. The third method, called electroflotation, creates tiny bubbles by electrolyzing an electrode in order to collect free-floating microalgae. This technique uses alternating current to enable simultaneous cell rupture and harvesting. However, this method can result in higher production costs due to its high energy consumption and frequent electrode replacements due to fouling (Singh and Patidar, 2018).

11.3.4 Flocculation

Unicellular, free-floating microalgae cells lose their surface charge during the flocculation process, which leads to the accumulation of bigger particles known as flocs (Muylaert et al., 2017). Chemical and biological flocculants are the two primary categories of flocculating agents. Chemical flocculants like salts of iron and aluminum are frequently utilized in industry because they are readily available and reasonably priced (Bracharz et al., 2018). Though the system is good, the flocculants are extremely toxic and not good for the environment, and thus there is a need for some process to eliminate them, which will further raise the production costing (Rinanti and Purwadi, 2018). On the other hand, bioflocculants are much safer and environmentally friendly. Additionally, they are less expensive to utilize and frequently don't need pretreatment before reusing culture material and processing microalgae further. Biopolymers, which can be created chemically or organically, including chitosan and acrylic acid, are the most common types of bio-flocculants used today. In comparison to the chemical flocculating agents, chitosan is said to achieve 90% cell recovery at lower dosages (Nguyen et al., 2019).

11.4 KEY PATHWAY FOR MICROALGAE-BASED BIOFUEL PRODUCTION

The current microalgae biorefinery method can produce a variety of high-value products in both fuel and non-fuel categories, including common biofuels such as bioethanol (or biobutanol), biodiesel, biogas, and biohydrogen.

11.4.1 Thermochemical Conversion

Wet microalgae biomass can be converted into bio-oil by the thermochemical liquefaction process at low temperatures and high pressures. This method does not require a catalyst; hence, it does away with the labor-intensive and time-consuming drying procedure. The process of turning lipids, carbohydrates, and proteins found in microalgae biomass into crude bio-oil is known as liquefaction. The natural structure and content of the microalgal biomass, pressure, temperature, reaction time, catalyst presence, and other variables can all have an impact on the process's efficiency (Hong et al., 2021). Pyrolysis is another often-used method that produces syngas, bio-oil, and charcoal by heating biomass at medium to high temperatures (350 to 700°C) without oxygen (Yang et al., 2019). The temperature range needed for fast pyrolysis is about 500°C, and the reaction only lasts for one second even in the presence of heated vapor. Up to 95% of the biodiesel produced by the technique is biodiesel (Venderbosch, 2019). When compared to slow pyrolysis, fast pyrolysis produces more bio-oil while using less total energy due to its rapid reaction. Fast pyrolysis, on the other hand, results in a thinner bio-oil than slow pyrolysis, which creates a thicker bio-oil. Therefore, for producing liquid biofuels on a larger industrial scale, the fast pyrolysis process is frequently more appropriate (Hoang et al., 2021).

Microalgae species called chlorella have a great deal of promise for pyrolysis to produce high energy content (Arvindnarayan et al., 2017). Additionally, using bio-oil made from biomass microalgae in conjunction with regular transportation gasoline doesn't require any major engine adjustments. Spirulina platensis produced 39% biochar, 23.8% bio-oil, and 19.2% gases when it was pyrolyzed at 350°C. Furthermore, nine microalgal species were hydrothermally liquefied between 28°C and 320°C operating temperatures, with a pressure of 0.7 MPa (Muñoz and Gonzalez-Fernandez, 2017). The transformation of lipid-rich nannochloropsis at 320°C produced the highest yield of biocrude. The studies on the thermal conversion of microalgae to biofuels are compiled in Table 11.2.

11.4.2 Biochemical Conversion of Microalgal Biomass into Biofuels

Anaerobic digestion is a productive biochemical method for converting biomass into biofuels. Bacteria utilize the process to transform the biological biomass of microalgae into biogas, composed chiefly of carbon dioxide and methane with small amounts of hydrogen sulfide. The anaerobic digestion process is often carried out at a constant temperature, which usually creates the ideal circumstances for bacterial proliferation. The temperature range corresponds to mesophilic temperatures of 42°C and thermophilic conditions of 55°C. Because anaerobic digesters are so popular and dependable in processing various biomass kinds, they are frequently designed using the characteristics of previous operating conditions (Muñoz & Gonzalez-Fernandez, 2017). Anaerobic digestion typically involves three steps: fermentation, hydrolysis, and methanogenesis. The complex polysaccharides in the biomass are first hydrolyzed to produce simple sugars. Subsequently, the sugars undergo fermentation, yielding volatile fatty acids, acetic acid, alcohol, and minute amounts of H_2 and CO_2.

TABLE 11.2
Thermal Decomposition of Microalgae

Microalgae Types	Thermochemical Method	Biofuel Types	Yield	References
Chlorella vulgaris	Gasification	Hydrogen	81.6%	Liu et al. (2017)
Scenedesmus sp.	Gasification	Hydrogen	81.6%	Gholkar et al. (2021)
Galdieria sulphuraria	Hydrothermal gasification	Hydrogen	45.7%mol	Ibrahim et al. (2020)
Spirulina	Hydrothermal gasification	Bio-oil	46.6 wt%	Liu et al. (2021)
Cyanobacteria	Pyrolysis	Bio-oil	41.1 wt%	Sotoudehniakarani et al. (2019)
Sargassum tenerrimum	Liquefaction	Bio-oil	33.0 wt%	Biswas et al. (2020)
Scenedesmus almeriensis	Microwave-induced pyrolysis	Syngas	93.8 vol%	Beneroso et al. (2013)
Spirulina, chlorella	Microwave-induced pyrolysis	Syngas	73.3 vol%	Hong et al. (2017)
U. prolifera	Pyrolysis	Bio-oil	41.3 wt%	Ma et al. (2020)

The gas mixture is converted into methane and CO_2 in percentages of 60–70 and 30–40, respectively, in the presence of methanogens. The energy content is 20–40% higher for the produced biogas in comparison to the biomass fuel. Wet biomass with an 80–90% moisture content is suitable for anaerobic digestion. Many studies have looked into using microalgae as the primary fuel for the anaerobic digestion process to produce biomethane. However, this method usually yields very subpar results since bacteria cannot decompose the outer cell structure of microalgae. The low carbon-to-nitrogen ratio further inhibits methane synthesis since free ammonia is available. Microalgae biomass, however, exhibits higher biomass yields when compared to biomass derived from terrestrial plants like Jatropha curcas (Trivedi et al., 2015). But, because the latter requires less feedstock, it is more economically viable.

Major biomass components, including sugars, starch, and cellulose, must first be broken down by bacteria or yeast for microalgae to be converted into bio-alcohol. Additional distillation is frequently necessary for higher quantities of diluted alcohol (Alves et al., 2020). The primary sources of carbohydrates are the cellulose found in organisms' cell walls and starch reserves. When bacteria are present, it can be challenging to ferment the sugars in biomass. Thus, to prepare these carbohydrates for further processing, they must first be hydrolyzed into monomeric sugars utilizing biological as well as chemical agents (Lin et al., 2019). Enzymatic hydrolysis is less advantageous, as it necessitates the costly and time-consuming preparation of microalgal biomass.

In the process of alcoholic fermentation, the simple sugars created during the first processing of biomass are then transformed into the intended products. Hydrolysis and fermentation can be carried out independently or concurrently in the same reaction chamber using two distinct methods for fermentation. For example, Chlorella vulgaris, which has a dry cell weight of 37% starch, may be converted to produce about 65% ethanol. In a different study, every gram of Chlamydomonas reinhardtii biomass yielded 0.235g of bioethanol by a distinct hydrolysis and fermentation process. When compared to other fermentation methods, anaerobic fermentation is thought to be a rather simple process for making bioethanol from microalgal biomass. *Chlorella vulgaris* is one of the favored strains of starch-rich microalgae that can increase ethanol yield by up to 65% conversion efficiency. Microalgae can also be dark fermented at 30°C to create a sizable bioethanol yield (Javed et al., 2019). Table 11.3 summarizes the biofuel produced by biochemical methods in different microalgal species.

11.5 TRANSESTERIFICATION METHOD FOR PRODUCING BIODIESEL FROM ALGAE

The process commonly used to produce biodiesel is called transesterification. This process uses thermochemical methods including hydrothermal liquefaction and pyrolysis to transform microalgal biomass into biodiesel. Renewable diesel is a biofuel made from biological resources by hydrotreating them, whereas biodiesel made by transesterification is made up of fatty acid methyl esters (FAME). It is discovered that the fuel properties of renewable biodiesel are more akin to those of petroleum

TABLE 11.3
Types of Microalgal Species That Are Utilized in Biofuel Production

Microalgae Types	Biochemical Method	Biofuel Types	Yield	References
Scenedesmus sp.	Thermophilic inoculum with acetate-butyrate	Biohydrogen	0.72 ml/g	Batista et al. (2014)
Chlorella sp.	Anaerobic sludge with butyrate-acetate	Biohydrogen	45 ml/g	Huo et al. (2018b)
Arthrospira platensis	Fermentation with butyrate acetate	Biohydrogen	85 ml/g	Bhatia et al. (2023)
Chlorella vulgaris	Anaerobic digestion after being pretreated by enzyme	Bio-methane	86.6–200.8 ml/day	Mahdy et al. (2015)
Scenedesmus sp.	Anaerobic digestion	Bio-methane	237 ml/g	Olsson et al. (2018)
Scenedesmus quadricauda	Anaerobic digestion	Bio-methane	222 ml/g	Nguyen (2019)
Spirulina platensis	Fermentation	Bio-ethanol	4.2 g/L	Rempel et al. (2019)
C. vulgaris ESP	Hydrolysate	Bio-ethanol	0.0761 g/g	Yu et al. (2021)

diesel, with sustainable biodiesel showing the greatest cetane number and a cloud point that is nearly the same as petroleum diesel. Transesterification is one of the most prevalent methods used in producing biodiesel that utilizes methanol along with a catalyst to convert lipids and triglycerides into fatty acid esters. According to Ma et al. (2020), the microalgal biomass that is extracted using oil or FAME is rich in carbs and perfect for the synthesis of bioethanol. Biodiesel, also known as long-chain fatty acid monoalkyl esters, is generally made by transesterifying lipids derived from sustainable biomass feedstocks. Through a sequence of reversible processes that liberate moieties of fatty acid methyl esters, the transesterification process transforms lipids and triglycerides into diglycerides, monoglycerides, and ultimately glycerol (Kumar et al., 2020). Reactive transesterification shows the direct conversion of lipid-containing microalgal biomass into FAME, as opposed to extractive transesterification, which creates FAME from the extracted lipids.

A variety of pre-treatment methods, including high-pressure homogenization, microwave irradiation, ultrasonic sonication, water bath treatment, osmotic shock, shaking mill, and Triton-X-100 treatment, can harm microalgae's cell wall during the lipid extraction process (Ramola et al., 2019). For example, solvothermal techniques were used to extract lipids from spirulina in a microwave for 30 minutes at 750 W and 60°C (Singh et al., 2019). There are two methods for transesterification of microalgal oil. The amount of free fatty acids (FFA) is decreased by a two-step process known as acid esterification. Using acid transesterification, the first step of transesterification reduced the FFA level of algal oil from 6.3% to 0.34% under ideal conditions of 1.5% H_2SO_4, a 12:1 methanol-to-oil ratio, 400 rpm, and 60°C for 90 minutes. 1% NaOH, a 9:1 methanol-to-oil ratio, 600 rpm, and 60°C for 70 minutes were the ideal conditions for base transesterification in the second stage, which produced a maximum biodiesel production of 90.6%. It is possible to perform in-situ transesterification reactions in two steps, but most of them are completed in one phase (Ghosh et al., 2017). After a 70-minute reaction at 90°C, the FAME recovery from the one-step in-situ technique was up to 60.89%, whereas the overall FAME recovery from the two-step in-situ procedure in Chlorella sorokiniana (UTEX 1602) reached up to 94.87 ± 0.86%. In-situ transesterification techniques are becoming more popular for biodiesel production from microalgae because they eliminate the expensive biomass drying and lipid extraction procedures (Ma et al., 2020).

11.6 ARTIFICIAL INTELLIGENCE IN MICROALGAL BIOFUEL PRODUCTION

Enhancing productivity, efficiency, and economic feasibility in this developing field can be achieved through a revolutionary approach: integrating artificial intelligence (AI) into the microalgal biofuel production process. Because of their quick growth rates and capacity to be cultivated on non-arable land and wastewater, microalgae are acknowledged for their potential as a sustainable source of biofuels (Jeswani et al., 2020). However, obstacles including high production costs and the difficulty of processing biomass have made algal biofuels less viable commercially (Culaba et al., 2020). Artificial intelligence (AI) technologies, namely machine learning (ML), are being utilized more frequently to maximize the production of microalgal biofuel at many stages, including strain selection, growing, and harvesting (Çoşgun et al., 2023; Alzahmi, 2024). Large-scale datasets can be analyzed by machine learning algorithms to determine the best growing conditions for certain microalgal strains, which improves lipid accumulation—a critical step in the generation of biofuel (Zhu et al., 2022). For example, recent research has shown that AI can greatly reduce the time and resources required for conventional breeding approaches by facilitating the selection of high-performing strains through predictive modeling (Alzahmi, 2024).

Moreover, AI-driven optimization of cultivation parameters, such as light intensity, nutrient supply, and CO_2 concentration, can lead to improved biomass yields and lipid content, thereby increasing the overall efficiency of biofuel production (Meng et al., 2021; Nagi et al., 2021). The concept of microalgal biorefineries further underscores the potential of AI in this sector. By integrating biofuel production with the extraction of high-value co-products, such as biochemicals and nutraceuticals, the economic feasibility of algal biofuel systems can be significantly enhanced (Kim et al., 2022; Tiwari & Kiran, 2018). AI can play a pivotal role in managing these complex biorefinery operations by optimizing resource allocation and process integration, ultimately leading to a reduction in waste and an increase in profitability (Cruce and Quinn, 2019; Figueroa-Torres et al., 2020). Additionally, the use of AI in monitoring and controlling bioreactor conditions can help mitigate issues related to algal phages and other biological threats that can compromise production (Kraft et al., 2019; Vaniyankandy et al., 2022). In conclusion, the application of artificial intelligence in microalgal biofuel production not only addresses the existing challenges of economic viability and efficiency but also paves the way for innovative solutions that can transform the biofuel landscape. By leveraging AI technologies, the microalgal biofuel industry can enhance its sustainability and contribute significantly to global energy needs while minimizing environmental impacts.

Artificial intelligence (AI) is revolutionizing microalgal biofuel production by enhancing efficiency, optimizing processes, and improving yield. The integration of AI technologies, such as machine learning and predictive modeling, is crucial for advancing microalgal biotechnology and bioenergy applications.

11.7 AI-DRIVEN OPTIMIZATION

Machine learning models, like Artificial Neural Networks (ANN), help in optimizing the crucial parameters like light intensity, CO_2 concentration, airflow rate, and carbon-to-nitrogen ratio in order to maximize the production of biomass and lipid content in microalgae (Kumar et al., 2024).

AI tools facilitate global sensitivity analysis, allowing researchers to identify the most impactful variables in microalgal cultivation, leading to significant increases in productivity (Kumar et al., 2024).

AI-driven optimization in microalgal biofuel production represents a significant advancement in the quest for sustainable energy sources. Microalgae are recognized for their potential to produce biofuels due to their rapid growth rates and ability to utilize CO_2, making them a viable alternative to fossil fuels. However, challenges such as low biomass productivity, high production costs, and inefficient harvesting methods have hindered their commercial viability (Zhu et al., 2023; Khan et al., 2023). Recent studies have highlighted the role of artificial intelligence (AI) in optimizing microalgal cultivation processes. For instance, AI can enhance strain selection and cultivation strategies, thereby improving biomass yield and lipid accumulation. Figueroa-Torres et al. (2020) demonstrated that integrating predictive modeling with experimental methodologies can significantly optimize microalgal growth dynamics, leading to increased starch and lipid formation. Furthermore, the application of AI-informed cultivation strategies allows for real-time adjustments based on environmental conditions, which can enhance productivity and reduce costs (Alzahmi, 2024). The optimization of growth conditions is critical for maximizing photosynthetic efficiency, the primary driver of microalgal biomass production. Hu et al. emphasized that improving light-to-biomass conversion efficiency is essential for increasing microalgal productivity (Hu et al., 2023). Techniques such as adjusting light intensity and duration, as well as employing artificial lighting systems, have been identified as key factors in enhancing growth rates (Krátký et al., 2023; Hawrot-Paw, 2024). Moreover, integrating genetic engineering approaches, such as metabolic engineering, can further enhance lipid accumulation by overcoming the trade-off between growth and lipid production (Zhu et al., 2022). In addition to optimizing cultivation conditions, the development of efficient harvesting and processing techniques is crucial for the economic feasibility of microalgal biofuels. Traditional harvesting methods are often energy-intensive and costly, which can negate the benefits of microalgal biomass (Zhu et al., 2018). Innovative approaches, such as using natural flocculants or optimizing flow rates in photobioreactors, can reduce harvesting costs and improve overall efficiency (Zhu et al., 2018; Banerjee et al., 2023). Furthermore, the concept of biorefineries, which allows for the co-production of biofuels and value-added chemicals, can enhance the economic viability of microalgal biomass (Figueroa-Torres et al., 2020; Meng et al., 2021).

The optimization of growth conditions is critical for maximizing photosynthetic efficiency, which is the primary driver of microalgal biomass production. Hu et al. emphasized that improving light-to-biomass conversion efficiency is essential for increasing microalgal productivity (Hu et al., 2023). Techniques such as adjusting light intensity and duration, as well as employing artificial lighting systems, have been identified as key factors in enhancing growth rates (Krátký et al., 2023; Hawrot-Paw, 2024). Moreover, the integration of genetic engineering approaches, such as metabolic engineering, can further enhance lipid accumulation by overcoming the trade-off between growth and lipid production (Zhu et al., 2022). In addition to optimizing cultivation conditions, the development of efficient harvesting and processing techniques is crucial for the economic feasibility of microalgal biofuels. Traditional harvesting methods are often energy-intensive and costly, which can negate the benefits of microalgal biomass (Zhu et al., 2018). Innovative approaches, such as using natural flocculants or optimizing flow rates in photobioreactors, can reduce harvesting costs and improve overall efficiency (Zhu et al., 2018; Banerjee et al., 2023). Furthermore, the concept of biorefineries, which allows for the co-production of biofuels and value-added chemicals, can enhance the economic viability of microalgal biomass (Figueroa-Torres et al., 2020; Meng et al., 2021). In conclusion, AI-driven optimization in microalgal biofuel production encompasses a multifaceted approach that integrates advanced modeling, genetic engineering, and innovative harvesting techniques. By addressing the challenges of low productivity and high costs, these strategies can significantly enhance the sustainability and economic feasibility of microalgal biofuels, paving the way for their broader commercial application.

11.7.1 Enhanced Cultivation Strategies

AI informs strain selection and cultivation methods, improving biomass yields and enabling the use of wastewater for cultivation (Alzahmi et al., 2024).

Machine learning assists in evaluating and managing uncertainties in microalgal biorefineries, thereby promoting sustainable biofuel production (Ravi et al., 2024).

The integration of artificial intelligence (AI) into microalgae cultivation strategies significantly enhances productivity and efficiency. AI technologies facilitate strain optimization, real-time monitoring, and predictive modeling, addressing key challenges in the industry.

11.7.2 Strain Optimization and Selection

AI-driven methods enable the identification of optimal microalgae strains for specific bioproducts, enhancing biomass yields (Alzahmi et al., 2024).

Machine learning techniques classify microalgae species with high accuracy, aiding in the selection of the most productive strains.

11.7.3 Real-time Monitoring and Control

Automated systems that utilize AI and IoT can monitor growth factors such as pH, temperature, and light intensity, ensuring optimal cultivation conditions (Khor et al., 2022).

Predictive models simulate algal growth under varying conditions, allowing for tailored cultivation strategies (Syed et al., 2024).

11.7.4 Innovative Cultivation Techniques

Semi-continuous algal cultivation designs informed by machine learning improve biomass productivity while reducing harvesting costs (Long et al., 2022). While AI presents transformative potential for microalgae cultivation, challenges such as high initial investment and the need for specialized knowledge may hinder widespread adoption. Nonetheless, the benefits of enhanced productivity and sustainability are compelling.

11.7.5 Future Perspectives

The combination of AI with hydrothermal liquefaction processes in biorefineries shows promise for producing high-yield biofuels, significantly improving energy recovery rates (Wu et al., 2013).

Despite advancements, the application of AI in microalgal biofuel production is still evolving, highlighting the need for standardized protocols to facilitate broader implementation (Ravi Rajan et al., 2024).

While AI presents transformative potential in microalgal biofuel production, challenges remain in standardizing methodologies and ensuring scalability for commercial applications.

11.8 IMPEDIMENTS, LIMITATIONS, AND FUTURE PROSPECTS OF ARTIFICIAL INTELLIGENCE IN MICROALGA-BASED BIOFUEL PRODUCTION

11.8.1 Impediments

Integrating artificial intelligence (AI) in microalga-based biofuel production faces several impediments that hinder its commercial viability. These challenges encompass economic, technical, and environmental factors that must be addressed for successful implementation.

11.8.2 Economic Constraints

Algal fuels are currently more expensive than petroleum-derived fuels, limiting their market competitiveness (Chisti, 2013).

The need for low-cost, concentrated carbon dioxide and nutrient recycling systems adds to production costs (Chisti, 2013).

11.8.3 Technical Challenges

The simultaneous optimization of biomass and lipid productivities remains a significant challenge, as these objectives often conflict (Kumar et al., 2024).

AI applications, while promising, require extensive data and robust models to effectively enhance algal productivity (Alzahmi et al., 2024).

11.8.4 Environmental Considerations

Freshwater scarcity poses a limitation, even for marine algae, necessitating sustainable water management practices (Chisti, 2013).

The environmental impact of large-scale algal cultivation must be carefully assessed to ensure sustainability (Yu et al., 2023).

Despite these challenges, advancements in AI and biotechnology present potential pathways to overcome these obstacles, indicating a need for continued research and innovation in this field.

11.8.5 Limitations

The integration of artificial intelligence (AI) in microalgae-based biofuel production presents several limitations that hinder its full potential. While AI can optimize processes, challenges remain in scalability, energy consumption, and cost-effectiveness.

11.8.6 Scalability Issues

AI models often require extensive data for training, which can be difficult to obtain at scale in microalgal production environments. The transition from laboratory-scale experiments to industrial applications can lead to discrepancies in model performance, limiting the effectiveness of AI solutions (Kumar et al., 2024).

11.8.7 Energy Consumption

The energy-intensive nature of microalgal cultivation and processing poses a significant challenge. AI can optimize certain processes, but the overall energy footprint remains high. For instance, while AI can enhance drying processes, the energy required for cultivation still impacts the sustainability of biofuel production.

11.8.8 Economic Viability

High production costs of microalgae and low energy recovery efficiency are significant barriers to commercialization, despite AI's potential to improve these metrics (Parakh et al., 2023). The need for integrated biorefinery approaches and cost-reduction strategies remains critical for the economic feasibility of AI-driven solutions (Parakh et al., 2023). In contrast, some researchers argue that continuous advancements in AI could eventually overcome these limitations, resulting in more efficient and sustainable methods for microalgal biofuel production.

11.8.9 Future prospects

The future prospects of artificial intelligence (AI) in microalgae-based biofuel production are promising, as AI technologies have the potential to significantly enhance the efficiency and sustainability of biorefinery processes. By integrating machine learning and optimization techniques, researchers aim to address the challenges associated with microalgal cultivation and biofuel production.

11.8.10 AI-Driven Optimization of Cultivation

Machine learning tools can evaluate and predict physiological factors affecting microalgal growth, enabling more efficient cultivation strategies (Ravi Rajan et al., 2024). AI-informed approaches have shown potential in optimizing strain selection and cultivation conditions, leading to increased biomass and lipid productivity (Alzahmi et al., 2024).

11.8.11 Multi-Objective Optimization

Advanced modeling techniques, such as Artificial Neural Networks (ANN), have been employed to simultaneously maximize both biomass and lipid yields, resulting in significant increases in productivity (Kumar et al., 2024). These models facilitate the identification of optimal process parameters, which are crucial for scaling up production in commercial settings (Parakh et al., 2023).

11.8.12 Economic and Environmental Sustainability

AI can help reduce production costs and improve energy recovery efficiency, thereby addressing key barriers to the commercialization of microalgal biofuels (Parakh et al., 2023). The integration of AI in biorefineries not only enhances biofuel production but also supports the recovery of valuable bioproducts, contributing to a circular economy (Dash et al., 2022). Although the application of AI in microalgal biofuel production is still in its early stages, its potential to revolutionize the industry is evident. Continued research and development are essential to fully realize these benefits and to address the challenges that remain.

In the past ten years, a rise in interest in the manufacture of third-generation biofuels from microalgal biomass has occurred. Several studies on microalgal cultivation and biofuel production have been conducted, and these studies, in collaboration with commercial enterprises, university institutions, government agencies, and public research groups, have helped in developing third-generation biofuels. However, substantial commercial adoption of microalgae-based biofuels is still some distance off due to the cost of third-generation biofuels being significantly greater than that of conventional fuels (Chowdhury and Loganathan, 2019). The costs of producing biofuels are now estimated primarily from historical data or small-scale experimental investigations. Although growth rate, dietary needs, lipid production, and energy consumption are important considerations, real data validation is sometimes lacking in the modeling of these cost estimations (Chia et al., 2018). Studies on microalgae production in open ponds are more numerous than those on photobioreactor systems, which are 2 to 2.5 times more expensive. Microalgae-based biofuel is not economically competitive with traditional gasoline sources, as indicated by price forecasts ranging from $0.44 to $8.76 per liter. Thus, it is imperative to lower the cost of microalgae biofuels by streamlining processes and enhancing production methods. Additionally, with methodical improvement of the production and conversion processes, these costs may drop soon (Chowdhury and Loganathan, 2019).

Much research has been conducted in the last few decades on the manufacture of biofuels from microalgae; however, only a limited number of pilot programs have been evaluated for potential uses in real-world settings, and commercialization has not yet yielded significant results. There isn't a documented case study of continuous industrial-scale microalgae-based biofuel production in the real world. Numerous challenges remain in the way of the commercialization and expansion of microalgae-based biofuels, including high capital and operating costs. Much work has been done up to this point on the growth of microalgae in photobioreactors and open systems. Among these concerns, the composition of the microalgal population highlights issues with the conversion efficiency of biofuels produced (Zabed et al., 2020). The increase in biomass is one of the primary barriers to microalgae growth in raceway ponds and photobioreactors. To further transform microalgae into bioethanol, more pretreatment procedures and a fruitful yeast fermentation process are required. To increase the current production of bioethanol, more research is needed to produce modified microalgae species with the capacity for rapid growth. Automation technology significantly boosts microalgae productivity and biofuel production, which reduces the overall cost of production, as demonstrated by a number of case studies (Branco-Vieira et al., 2020). The amount and quality of lipids that can be collected can vary greatly due to different climates and microalgae regeneration after a series of subcultures. Moreover, the two challenges, i.e., low biomass yield and lower concentration of volatile solid chemicals, are major issues in microalgae biofuel production.

According to some research, combining the generation of bioenergy with wastewater treatment can optimize microalgae biorefinery processes in novel ways (De

Bhowmick et al., 2019). However, there are a lot of challenges when cultivating microalgae utilizing wastewater. The growth of microalgal biomass is subject to uncontrollable variables due to the varying features of different effluent streams, including their origin, pretreatment methods, and nutrient levels. Microalgae can digest typical nutrients found in wastewater streams; however, their growth can be inhibited by particular nutrients and growth inhibitors. Reduced biomass productivity can also result from incompatible carbon-to-nitrogen ratios, which can impede ideal growth rates (Li et al., 2020). Adding more downstream processing steps would raise the overall cost of manufacturing, especially considering the increase in contaminants discovered in microalgae biomass. Other aspects like biomass concentration, harvesting, moisture extraction, fertilizer prices, and harvesting practices are all related to these difficulties. Although employing bioenergy produced from microalgae has several benefits, there are a number of obstacles preventing its broad use. These include choosing the right conversion methods, regulating seasonal variations in feedstock availability, gathering biomass, and identifying viable microalgae strains. Low dispersion of sunlight is one of these issues that negatively affects biomass production and microalgae development. The tiny cell size and low biomass concentration of microalgae greatly increase the cost of harvesting. The biomass's high moisture content, which calls for extra energy-intensive drying processes, is another significant problem. In order to assure ideal growing conditions, producing microalgae involves a significantly larger capital expenditure for facility development and maintenance than traditional farming. The development of biorefinery processing has been suggested as a way to overcome the difficulties involved in biomass farming and the extraction of proteins and lipids to create biofuels, possibly doing away with the stages of cultivation and harvesting that are now included in the production cycle. Nowadays, the process of turning lipids from microalgae into biodiesel involves several phases. Further chemicals and refining steps are required to guarantee that the output quality of biodiesel is suitable for engine operation.

For example, the more complex process of dark fermentation for biohydrogen production requires the removal of CO_2 before treatment can begin, resulting in reduced hydrogen emissions. The gas output must undergo further processing, such as drying and compression to a few thousand PSI, to achieve higher energy density. However, this compression process consumes substantial energy, which cannot be recovered. Additionally, various other challenges in biohydrogen generation need to be addressed, including the variation in production yields caused by metabolic differences among hydrogen-producing species. The challenge of reducing 5-hydroxymethylfurfural and furfural during the preprocessing stage remains unresolved; specific adsorbent materials are required to remove these undesirable pollutants. In light of the benefits provided by microalgae as a sustainable fuel source, they are regarded as a vital source of sustainable and carbon-neutral biomass feedstock. Third-generation biofuels can be produced from microalgae through various processes, including fermentation, hydro-treatment, anaerobic digestion, transesterification, and pyrolysis (Pourkarimi et al., 2019). While the cost-effectiveness of producing biofuel from microalgae biomass has improved, progress has been impeded by the current slow pace of development. Future commercial success in promoting the production of microalgae-based biofuels will depend on how effectively these challenges are addressed.

11.9 CONCLUSION

This review offers the most recent information on the state of microalgae-based biofuel research by carefully examining the advantages and present difficulties related to the several biofuels that can be created from microalgae. The ability of microalgae to fix atmospheric CO_2 is one of its main advantages. Another is that they significantly reduce pollution in the environment because wastewater is used as a nutrient source in microalgae production. Lastly, microalgae contribute to environmental sustainability as a whole. The high energy needs for digesting microalgae feedstock, which raises processing costs, were the biggest obstacle found, though. In order to produce molecules of an acceptable quality for the generation of biofuel, future research directions should concentrate on using recombinant gene technology to boost the biochemical capabilities of microalgae. Additionally, within the frameworks of sustainable development and the circular economy, more research should be done on the integration of microalgae biorefinery processing and the feasibility of producing biofuel from microalgae. Artificial intelligence (AI) has a lot to offer the production of microalgal biofuels in comparison to conventional techniques, both economically and ecologically. AI is a good substitute since it improves farming procedures, lowers costs, and encourages sustainability. Artificial Intelligence (AI) has the potential to produce microalgae biofuel while preserving the environment by cultivating the algae using wastewater, which reduces pollution and water consumption. Reducing greenhouse gas emissions can also be accomplished by using AI-powered devices to efficiently extract CO_2 from industrial emissions. Additionally, by promoting the development of a variety of algae strains and reducing land consumption, AI promotes environmental balance. AI offers various financial advantages in the production of microalgal biofuel. For instance, methods such as machine learning can increase productivity and reduce the cost of harvesting biomass, making biofuels more competitive with fossil fuels. Additionally, combining microalgal production with other industries makes it easier to produce high-value products and biofuels concurrently, increasing the industry's financial viability. AI also helps find the best nutritional conditions, which leads to much higher yields of starch and lipids and increases profitability. But even with AI's promising future in this area, expanding these technologies and generating steady profits still present difficulties. To fully reap these benefits, further research and funding are required.

REFERENCES

Acién, F. G., Molina, E., Reis, A., Torzillo, G., Zittelli, G. C., Sepúlveda, C., & Masojídek, J. (2017). Photobioreactors for the production of microalgae. In *Microalgae-Based Biofuels and Bioproducts* (pp. 1–44). Woodhead Publishing.

Al Hattab, M., Ghaly, A., & Hammouda, A. (2015). Microalgae harvesting methods for industrial production of biodiesel: Critical review and comparative analysis. *Journal of Fundamentals of Renewable Energy*, 5(2), 1000154.

Alves, J. L. F., da Silva Filho, V. F., Machado, R. A. F., & Marangoni, C. (2020). Ethanol enrichment from an aqueous stream using an innovative multi-tube falling film distillation column equipped with a biphasic thermosiphon. *Process Safety and Environmental Protection, 139*, 69–75.

Alzahmi, A. (2024). Enhancing algal production strategies: Strain selection, AI-informed cultivation, and mutagenesis. *Frontiers in Sustainable Food Systems*, 8.

Alzahmi, Amnah, Sarah, Daakour., David, R., Nelson., Dina, Al-Khairy., Jean-Claude, Twizere., Kourosh, Salehi-Ashtiani. (2024). 3. Enhancing algal production strategies: Strain selection, AI-informed cultivation, and mutagenesis. *Frontiers in Sustainable Food Systems*. doi: 10.3389/fsufs.2024.1331251

Arvindnarayan, S., Prabhu, K. K. S., Shobana, S., Kumar, G., & Dharmaraja, J. (2017). Upgrading of micro algal derived bio-fuels in thermochemical liquefaction path and its perspectives: A review. *International Biodeterioration & Biodegradation, 119*, 260–272.

Banerjee, S., Dasgupta, S., Atta, A., Das, D., Dayal, D., Malik, S., ... & Almutary, A. (2023). Flow rate optimization in a flat-panel photobioreactor for the cultivation of microalgae for mitigating waste gas. *Water, 15*(15), 2824.

Barsanti, L., Coltelli, P., Evangelista, V., Frassanito, A. M., Passarelli, V., Vesentini, N., & Gualtieri, P. (2008). Oddities and curiosities in the algal world. In *Algal Toxins: Nature, Occurrence, Effect and Detection* (pp. 353–391). Springer, Dordrecht.

Batista, A. P., Moura, P., Marques, P. A., Ortigueira, J., Alves, L., & Gouveia, L. (2014). Scenedesmus obliquus as feedstock for biohydrogen production by *Enterobacter aerogenes* and *Clostridium butyricum*. *Fuel, 117*, 537–543.

Beneroso, D., Bermúdez, J. M., Arenillas, A., & Menéndez, J. A. (2013). Microwave pyrolysis of microalgae for high syngas production. *Bioresource Technology, 144*, 240–246.

Bhatia, S. K., Banu, J. R., Singh, V., Kumar, G., & Yang, Y. H. (2023). Algal biomass to biohydrogen: Pretreatment, influencing factors, and conversion strategies. *Bioresource Technology, 368*, 128332.

Yang, C., Li, R., Zhang, B., Qiu, Q., Wang, B., Yang, H., ... & Wang, C. (2019). Pyrolysis of microalgae: A critical review. *Fuel Processing Technology, 186*, 53–72.

Biswas, B., Kumar, A., Fernandes, A. C., Saini, K., Negi, S., Muraleedharan, U. D., & Bhaskar, T. (2020). Solid base catalytic hydrothermal liquefaction of macroalgae: Effects of process parameter on product yield and characterization. *Bioresource Technology, 307*, 123232.

Bracharz, F., Helmdach, D., Aschenbrenner, I., Funck, N., Wibberg, D., Winkler, A., ... & Brück, T. B. (2018). Harvest of the oleaginous microalgae Scenedesmus obtusiusculus by flocculation from culture based on natural water sources. *Frontiers in Bioengineering and Biotechnology, 6*, 200.

Branco-Vieira, M., Mata, T. M., Martins, A. A., Freitas, M. A. V., & Caetano, N. S. (2020). Economic analysis of microalgae biodiesel production in a small-scale facility. *Energy Reports, 6*, 325–332.

Brennan, L., & Owende, P. (2010). Biofuels from microalgae: A review of technologies for production, processing, and extractions of biofuels and co-products. *Renewable and Sustainable Energy Reviews, 14*(2), 557–577.

Chia, S. R., Ong, H. C., Chew, K. W., Show, P. L., Phang, S. M., Ling, T. C., ... & Chang, J. S. (2018). Sustainable approaches for algae utilization in bioenergy production. *Renewable E, 129*, 838–852.

Chisti, Y. (2013 Sep. 10). Constraints to commercialization of algal fuels. *Journal of Biotechnology, 167*(3), 201–214. doi: 10.1016/j.jbiotec.2013.07.020.

Chowdhury, H., & Loganathan, B. (2019). Third-generation biofuels from microalgae: A review. *Current Opinion in Green and Sustainable Chemistry, 20*, 39–44.

Çoşgun, A., Günay, M., & Yıldırım, R. (2023). Machine learning for algal biofuels: A critical review and perspective for the future. *Green Chemistry, 25*(9), 3354–3373.

Cruce, J., & Quinn, J. (2019). Economic viability of multiple algal biorefining pathways and the impact of public policies. *Applied Energy*, 233–234, 735–746.

Culaba, A., Ubando, A., Ching, P., Chen, W., & Chang, J. (2020). Biofuel from microalgae: Sustainable pathways. *Sustainability, 12*(19), 8009.

Das, P., Aziz, S. S., & Obbard, J. P. (2011). Two phase microalgae growth in the open system for enhanced lipid productivity. *Renewable Energy, 36*(9), 2524–2528.

Dash, S., Parida, S., Sahoo, B., & Rath, B. (2022). The Promising Future of Microalgae as Biofuels and Valuable Bioproducts. In: Verma, P. (eds) *Micro-Algae: Next-Generation Feedstock for Biorefineries.* Clean Energy Production Technologies. Springer, Singapore.

De Bhowmick, G., Sarmah, A. K., & Sen, R. (2019). Zero-waste algal biorefinery for bioenergy and biochar: A green leap towards achieving energy and environmental sustainability. *Science of the Total Environment, 650*, 2467–2482.

Figueroa-Torres, G., Pittman, J., & Theodoropoulos, C. (2020). Optimising microalgal cultivation for biofuels production: The biorefinery paradigm.

Gholkar, P., Shastri, Y., & Tanksale, A. (2021). Renewable hydrogen and methane production from microalgae: A techno-economic and life cycle assessment study. *Journal of Cleaner Production, 279*, 123726.

Ghosh, S., Banerjee, S., & Das, D. (2017). Process intensification of biodiesel production from Chlorella sp. MJ 11/11 by single step transesterification. *Algal Research, 27*, 12–20.

Gupta, P. L., Lee, S. M., & Choi, H. J. (2015). A mini review: Photobioreactors for large scale algal cultivation. *World Journal of Microbiology and Biotechnology, 31*(9), 1409–1417.

Hamed, I. (2016). The evolution and versatility of microalgal biotechnology: A review. *Comprehensive Reviews in Food Science and Food Safety, 15*(6), 1104–1123.

Hawrot-Paw, M. (2024). Efficient production of microalgal biomass: Step by step to industrial scale. *Energies, 17*(4), 944.

Hoang, A. T., Ong, H. C., Fattah, I. R., Chong, C. T., Cheng, C. K., Sakthivel, R., & Ok, Y. S. (2021). Progress on the lignocellulosic biomass pyrolysis for biofuel production toward environmental sustainability. *Fuel Processing Technology, 223*, 106997.

Hong, W., Chen, J., Ding, Q., Gao, Y., Ye, L., Yin, Y., & Tu, S. (2021). Efficient thermochemical liquefaction of microalgae *Haematococcus pluvialis* for production of high quality biocrude with high selectivity over Fe/montmorillonite catalyst. *Journal of the Energy Institute, 97*, 73–79.

Hong, Y., Chen, W., Luo, X., Pang, C., Lester, E., & Wu, T. (2017). Microwave-enhanced pyrolysis of macroalgae and microalgae for syngas production. *Bioresource Technology, 237*, 47–56

Hu, J., Wang, D., Chen, H., & Wang, Q. (2023). Advances in genetic engineering in improving photosynthesis and microalgal productivity. *International Journal of Molecular Sciences, 24*(3), 1898.

Huang, Q., Jiang, F., Wang, L., et al. (2017). Design of photobioreactors for mass cultivation of photosynthetic organisms. *Engineering,3*(3), 318–329.

Huo, S., Kong, M., Zhu, F., Zou, B., Wang, F., Xu, L., Zhang, C., & Huang, D. (2018a). Mixotrophic Chlorella sp. UJ-3 cultivation in the typical anaerobic fermentation effluents. *Bioresource Technology, 249*, 219–225.

Huo, S., Wang, Z., Zhu, S., Shu, Q., Zhu, L., Qin, L., ... & Dong, R. (2018b). Biomass accumulation of Chlorella zofingiensis G1 cultures grown outdoors in photobioreactors. *Frontiers in Energy Research, 6*, 49.

Ibrahim, A. F., Dandamudi, K. P. R., Deng, S., & Lin, J. Y. (2020). Pyrolysis of hydrothermal liquefaction algal biochar for hydrogen production in a membrane reactor. *Fuel, 265*, 116935.

Javed, F., Aslam, M., Rashid, N., Shamair, Z., Khan, A. L., Yasin, M., ... & Bazmi, A. A. (2019). Microalgae-based biofuels, resource recovery and wastewater treatment: A pathway towards sustainable biorefinery. *Fuel, 255*, 115826.

Jeswani, H., Chilvers, A., & Azapagic, A. (2020). Environmental sustainability of biofuels: A review. *Proceedings of the Royal Society a Mathematical Physical and Engineering Sciences, 476*(2243).

Khan, M. I., Shin, J. H., & Kim, J. D. (2018). The promising future of microalgae: Current status, challenges, and optimization of a sustainable and renewable industry for biofuels, feed, and other products. *Microbial cell factories, 17*(1), 1–21.

Khan, S., Das, P., Quadir, M., Thaher, M., Mahata, C., Sayadi, S., ... & Jabri, H. (2023). Microalgal feedstock for biofuel production: Recent advances, challenges, and future perspective. *Fermentation, 9*(3), 281.

Khor, J. G., Lim, H. R., Chia, W. Y., & Chew, K. W. (2022). Automated cultivation system for microalgae: Growth factors and control. *Current Nutrition & Food Science, 18*(9), 776–779.

Kim, K., Yoo, J., & Lee, W. (2022). Recent advances in photobioreactor systems for sustainable and enhanced microalgal biofuel production. *Sustainable Energy & Fuels, 6*(24), 5459–5473.

Kraft, K., Alum, A., & Abbaszadegan, M. (2019). Environmental algal phage isolates and their impact on production potential for food and biofuel applications. *Journal of Applied Microbiology, 128*(1), 182–190.

Krátký, L., Jirout, T., & Belohlav, V. (2023). Economic feasibility study for artificial lighting of microalgal flat-panel photobioreactors. *International Journal of Environmental Science and Technology, 20*(11), 12089–12100.

Kumar, M., Sun, Y., Rathour, R., Pandey, A., Thakur, I. S., & Tsang, D. C. (2020). Algae as potential feedstock for the production of biofuels and value-added products: Opportunities and challenges. *Science of the Total Environment, 716*, 137116.

Kumar, R. R., Sarkar, D., & Sen, R. (2024). Simultaneously maximizing microalgal biomass and lipid productivities by machine learning driven modeling, global sensitivity analysis and multi-objective optimization for sustainable biodiesel production. *Applied Energy, 358*, 122597.

Li, S., Ji, L., Chen, C., Zhao, S., Sun, M., Gao, Z., ... & Fan, J. (2020). Efficient accumulation of high-value bioactive substances by carbon to nitrogen ratio regulation in marine microalgae Porphyridium purpureum. *Bioresource Technology, 309*, 123362.

Lin, R., Deng, C., Ding, L., Bose, A., & Murphy, J. D. (2019). Improving gaseous biofuel production from seaweed Saccharina latissima: The effect of hydrothermal pretreatment on energy efficiency. *Energy Conversion and Management, 196*, 1385–1394.

Liu, B., Wang, Z., & Feng, L. (2021). Effects of reaction parameter on catalytic hydrothermal liquefaction of microalgae into hydrocarbon rich bio-oil. *Journal of the Energy Institute, 94*, 22–28.

Liu, G., Liao, Y., Wu, Y., Ma, X., & Chen, L. (2017). Characteristics of microalgae gasification through chemical looping in the presence of steam. *International Journal of Hydrogen Energy, 42*(36), 22730–22742.

Long, B., Fischer, B., Zeng, Y., Amerigian, Z., Li, Q., Bryant, H., ... & Yuan, J. S. (2022). Machine learning-informed and synthetic biology-enabled semi-continuous algal cultivation to unleash renewable fuel productivity. *Nature Communications, 13*(1), 541.

Ma, C., Geng, J., Zhang, D., & Ning, X. (2020). Non-catalytic and catalytic pyrolysis of Ulva prolifera macroalgae for production of quality bio-oil. *Journal of the Energy Institute, 93*(1), 303–311.

Mahdy, A., Mendez, L., Ballesteros, M., & González-Fernández, C. (2015). Protease pretreated Chlorella vulgaris biomass bioconversion to methane via semi-continuous anaerobic digestion. *Fuel, 158*, 35 41.

Meng, T., Harun, R., & Kamaludin, M. (2021). Microalgal biorefinery: Challenge and strategy in bioprocessing of microalgae carbohydrate for fine chemicals and biofuel. *Journal of Applied Biology & Biotechnology* 9, 1–17.

Mohan, S. V., Rohit, M. V., Subhash, G. V., et al. (2019). Biofuels from algae. In: Pandey, A., Chang, J. -S., Soccol, C. R., Lee, D. -J., Chisti, Y., (eds.) *Algal Oils as Biodiesel.* Elsevier.

Muñoz, R., & Gonzalez-Fernandez, C. (eds.) (2017). *Microalgae-Based Biofuels and Bioproducts: From Feedstock Cultivation to End-Products.* Woodhead Publishing.

Muylaert, K., Bastiaens, L., Vandamme, D., & Gouveia, L. (2017). Harvesting of microalgae: Overview of process options and their strengths and drawbacks. *Microalgae-Based Biofuels and Bioproducts*, 113–132.

Nagi, G., Minhas, A., Gaur, S., Jain, P., & Mandal, S. (2021). Integration of algal biofuels with bioremediation coupled industrial commodities towards cost-effectiveness. *Frontiers in Energy Research, 9.*

Nguyen, T. D. P., Le, T. V. A., Show, P. L., Nguyen, T. T., Tran, M. H., Tran, T. N. T., & Lee, S. Y. (2019). Bioflocculation formation of microalgae-bacteria in enhancing microalgae harvesting and nutrient removal from wastewater effluent. *Bioresource Technology, 272*, 34–39.

Olsson, J., Forkman, T., Gentili, F. G., Zambrano, J., Schwede, S., Thorin, E., & Nehrenheim, E. (2018). Anaerobic co-digestion of sludge and microalgae grown in municipal wastewater: A feasibility study. *Water Science and Technology, 77*(3), 682–694.

Parakh, S. K., Tian, Z., Wong, J. Z. E., & Tong, Y. W. (2023). From microalgae to bioenergy: Recent advances in biochemical conversion processes. *Fermentation, 9*(6), 529.

Pourkarimi, S., Hallajisani, A., Alizadehdakhel, A., & Nouralishahi, A. (2019). Biofuel production through micro-and macroalgae pyrolysis–A review of pyrolysis methods and process parameters. *Journal of Analytical and Applied Pyrolysis, 142*, 104599.

Ramola, B., Kumar, V., Nanda, M., Mishra, Y., Tyagi, T., Gupta, A., & Sharma, N. (2019). Evaluation, comparison of different solvent extraction, cell disruption methods and hydrothermal liquefaction of Oedogonium macroalgae for biofuel production. *Biotechnology Reports, 22*, e00340.

Randrianarison, G., & Ashraf, M. A. (2017). Microalgae: A potential plant for energy production. *Geology, Ecology, and Landscapes, 1*(2), 104–120.

Kavitha, S., Yukesh Kannah, R., Gopalakrishnan, K., Yogalakshmi, K., Nandabalan., R., & Banu, J. (2024). 4. Microalgal biorefineries: Advancement in machine learning tools for sustainable biofuel production and value-added products recovery. *Journal of Environmental Management.* doi: 10.1016/j.jenvman.2024.120135

Rempel, A., de Souza Sossella, F., Margarites, A. C., Astolfi, A. L., Steinmetz, R. L. R., Kunz, A., Treichel, H., & Colla, L. M. (2019). Bioethanol from Spirulina platensis biomass and the use of residuals to produce biomethane: An energy efficient approach. *Bioresource Technology, 288*, 121588.

Rinanti, A., & Purwadi, R. (2018). Harvesting of freshwater microalgae biomass by Scenedesmus sp. as bioflocculant. In *IOP Conference Series: Earth and Environmental Science* (Vol. 106, No. 1, p. 012087). IOP Publishing.

Rogers, J. N., Rosenberg, J. N., Guzman, B. J., Oh, V. H., Mimbela, L. E., Ghassemi, A., ... & Donohue, M. D. (2014). A critical analysis of paddlewheel-driven raceway ponds for algal biofuel production at commercial scales. *Algal Research, 4*, 76–88.

Show, P., Tang, M., Nagarajan, D., et al. (2017). A holistic approach to managing microalgae for biofuel applications. *International Journal of Molecular Sciences, 18*(1), 215.

Singh, G., & Patidar, S. (2018). Microalgae harvesting techniques: A review. *Journal of Environmental Management, 217*, 499–508.

Singh, R., Kumar, A., & Chandra Sharma, Y. (2019). Biodiesel production from microalgal oil using barium–calcium–zinc mixed oxide base catalyst: Optimization and kinetic studies. *Energy & Fuels*, *33*(2), 1175–1184.

Soomro, R. R., Ndikubwimana, T., Zeng, X., et al. (2016). Development of a two-stage microalgae dewatering process: A life cycle assessment approach. *Frontiers in Plant Science, 7*, 113.

Sotoudehniakarani, F., Alayat, A., & McDonald, A. G. (2019). Characterization and comparison of pyrolysis products from fast pyrolysis of commercial *Chlorella vulgaris* and cultivated microalgae. *Journal of Analytical and Applied Pyrolysis*, *139*, 258–273.

Sun, Z., Liu, J., & Zhou, Z. G. (2016). Algae for biofuels: An emerging feedstock. In *Handbook of Biofuels Production* (pp. 673–698). Woodhead Publishing.

Syed, T., Krujatz, F., Ihadjadene, Y., Mühlstädt, G., Hamedi, H., Mädler, J., & Urbas, L. (2024). A review on machine learning approaches for microalgae cultivation systems. *Computers in Biology and Medicine*, 108248.

Tan, Jia Sen, Lee, S. Y., Chew, K. W., Lam, M. K., Lim, J. W., Ho, S. -H., & Show, P. L. (2020). A review on microalgae cultivation and harvesting, and their biomass extraction processing using ionic liquids. *Bioengineered*, *11*(1), 116–129.

Tiwari, A., & Kiran, T. (2018). Biofuels from microalgae. https://doi.org/10.5772/intechopen.73012

Trivedi, J., Aila, M., Bangwal, D. P., Kaul, S., & Garg, M. O. (2015). Algae based biorefinery—how to make sense?. *Renewable and Sustainable Energy Reviews*, *47*, 295–307.

Vaniyankandy, A., Ray, B., Karthikeyan, S., & Rakesh, S. (2022). Thermochemical conversion of algal based biorefinery for biofuel.

Venderbosch, R. H. (2019). Fast pyrolysis. *Thermochemical Processing of Biomass: Conversion into Fuels, Chemicals and Power*, 175–206.

Vo, H. N., Ngo, H. H., Guo, W., et al. (2018). A critical review on designs and applications of microalgae-based photobioreactors for pollutants treatment. *Science of the Total Environment,* 651, 1549–1568.

White, R. L., & Ryan, R. A. (2015). Long-term cultivation of algae in open-raceway ponds: Lessons from the field. *Industrial Biotechnology*, *11*(4), 213–220.

Wu, L. F., Chen, P. C., & Lee, C. M. (2013). The effects of nitrogen sources and temperature on cell growth and lipid accumulation of microalgae. *International Biodeterioration & Biodegradation, 85*, 506–510.

Yu, Z., Hou, Q., Liu, M. et al. (2023). From lab to application: Cultivating limnetic microalgae in seawater coupled with wastewater for biodiesel production on a pilot scale. *Water Resources, 229,* 119471

Yu, K. L., Lee, X. J., Ong, H. C., Chen, W. H., Chang, J. S., Lin, C. S., ... & Ling, T. C. (2021). Adsorptive removal of cationic methylene blue and anionic Congo red dyes using wet torrefied microalgal biochar: Equilibrium, kinetic and mechanism modeling. *Environmental Pollution*, *272*, 115986.

Zabed, H. M., Akter, S., Yun, J., Zhang, G., Zhang, Y., & Qi, X. (2020). Biogas from microalgae: Technologies, challenges and opportunities. *Renewable and Sustainable Energy Reviews*, *117*, 109503.

Zhu, L., Li, Z., & Hiltunen, E. (2018). Microalgae chlorella vulgaris biomass harvesting by natural flocculant: Effects on biomass sedimentation, spent medium recycling and lipid extraction. *Biotechnology for Biofuels*, *11*(1).

Zhu, S., Higa, L., Barela, A., Lee, C., & Du, Z. (2023). Microalgal consortia for waste treatment and valuable bioproducts. *Energies*, *16*(2), 884.

Zhu, Z., Sun, J., Fei, Y., Liu, X., & Lindblad, P. (2022). Enhancing microalgal lipid accumulation for biofuel production. *Frontiers in Microbiology*, *13*.

12 Artificial Intelligence and Machine Learning in Biofuels as Tools for Advancing Efficiency and Sustainability

Thiago José Machado, Anthony Andrey Ramalho Diniz, and Moisés Dantas dos Santos

12.1 INTRODUCTION

The biomass-based biofuels industry has seen significant advancements driven by digital transformation, particularly through the application of AI and ML. These technologies play crucial roles in optimizing processes, increasing operational efficiency, and promoting sustainability in the production of ethanol, biodiesel, jet fuel, and hydrogen. AI and ML are widely used in the exploration of biomass feedstocks and conversion pathways. For example, predictive models analyze biomass characteristics and optimize feedstock selection, ensuring efficient biofuel production. Advanced neural networks contribute to improving fermentation processes, increasing ethanol yield and minimizing waste such as vinasse. Notably, initiatives such as "Usina 4.0" in Brazil have employed real-time data streams and computational algorithms to optimize fermentation at industrial scale, generating productivity gains and reducing environmental impact.

In ethanol production, AI has revolutionized fermentation monitoring by automating adjustments to environmental conditions, ensuring higher alcohol yields and lower energy consumption (Ibarra-Junquera et al. 2007; Itto-Nakama et al. 2022; Kwon et al. 2024). It is also used to monitor critical parameters such as temperature, pH, oxygen levels, and concentrations of substrates and products in real time, enabling predictive analytics to optimize efficiency and minimize the production of by-products like glycerol and acetic acid. ML models predict fermentation failures, reducing waste and maximizing sugar conversion. Beyond industrial processes, AI supports agricultural planning with crop forecasts based on climate and soil data, ensuring a stable supply of raw materials. Integrations with IoT and smart sensors improve operational control (Lopez et al. 2019), while advanced research uses AI to develop genetically modified yeasts and catalysts. Companies such as Petrobras are

DOI: 10.1201/9781003571223-12

also expanding ethanol production through joint ventures, reflecting the growing role of biofuels in the global energy transition.

Biogas, another pillar of renewable energy, benefits from AI's ability to evaluate the biochemical methane potential (BMP) of diverse feedstocks, identifying efficient and cost-effective materials for production (Cinar et al. 2021). AI optimizes feedstock combinations to enhance process stability and gas yields, while predictive maintenance ensures efficient operation of biogas plants by minimizing downtime (Khatri & Khatri 2022). Additionally, AI improves biogas integration into energy grids by optimizing upgrading processes, such as carbon dioxide and impurity removal, to produce biomethane. This purified form of biogas can be injected into natural gas grids or used as a renewable fuel. AI-driven research is also advancing microbial consortia development (Long et al. 2021), tailoring them to specific feedstocks and conditions, further enhancing the efficiency of anaerobic digestion.

In the biodiesel industry, Brazil leads in production and leverages AI to integrate low-carbon innovations. Optimization of transesterification and feedstock selection reduces costs and enhances biodiesel quality, aligning with sustainability goals (Osman et al. 2024; Kyriklidis et al. 2025). AI-driven systems monitor and control chemical reactions, ensuring consistent quality and minimizing waste. Predictive models identify optimal reaction parameters—such as temperature, catalyst concentration, and reaction time—while evaluating feedstocks like waste cooking oil and soybean oil to determine cost-effective and sustainable options. Integration with IoT enables real-time monitoring of equipment and processes, improving operational efficiency through predictive maintenance and minimizing downtime (Almohana et al. 2022). These advancements position biodiesel as a key player in the renewable energy transition.

Biojet fuel, a sustainable aviation alternative, presents significant potential for AI applications. Despite being less explored, AI optimizes production pathways and improves efficiency in processes like hydroprocessing and Fischer–Tropsch synthesis (Voeten et al. 2024). ML refines reaction conditions, maximizing fuel yield and reducing unwanted byproducts, while advanced models analyze feedstocks' life cycles to assess greenhouse gas emissions and sustainability metrics (Jong et al. 2017; Wang & Yu 2024). As aviation focuses on decarbonization, AI-driven innovations in fuel synthesis and supply chain management are critical for reducing the carbon footprint of air travel and advancing sustainable practices.

The adoption of AI and ML in the biomass-based biofuels industry faces significant challenges, particularly regarding data availability and quality (DOE-US 2023). AI models require accurate and diverse datasets, yet information on biomass characteristics and conversion processes is often limited or inconsistent. The diversity of feedstocks also complicates model generalization, as each type of biomass has unique properties that require tailored approaches (Liao & Yao 2021). Additionally, integrating AI technologies with existing facilities demands costly retrofitting with IoT devices and advanced sensors, which can be prohibitive for smaller producers. High computational demands and a lack of specialized expertise further hinder the widespread implementation of these technologies.

Economic, regulatory, and environmental factors add to these challenges. Ensuring compliance with stringent environmental regulations while optimizing processes with AI can be complex, and the high initial costs of adoption may deter smaller players in the industry. Fluctuations in biomass availability and market prices introduce further uncertainty, limiting the reliability of AI predictions. Moreover, the energy consumption of AI systems raises concerns about their carbon footprint, potentially conflicting with the sustainability goals of biofuels. Addressing these barriers requires investments in infrastructure, workforce training, and regulatory support to fully unlock the potential of AI and ML in enhancing the efficiency and sustainability of biofuel production (Jeswani et al. 2020).

In summary, the biomass-based biofuels industry is undergoing a digital transformation driven by the integration of AI and ML. These technologies are revolutionizing the production of ethanol, biodiesel, biojet fuel, and biogas by optimizing processes, enhancing efficiency, and promoting sustainability. From fermentation monitoring and feedstock evaluation to supply chain management and advanced catalyst development, AI is enabling significant productivity gains while reducing environmental impact. However, challenges such as limited data availability, high implementation costs, and regulatory complexities hinder broader adoption. Addressing these barriers through investments in infrastructure, workforce training, and technological innovation will be critical to fully harnessing AI's potential in advancing the global transition to renewable energy.

12.2 AI AND ML APPLICATIONS BY BIOFUEL TYPE

12.2.1 Bioethanol

Bioethanol is a renewable biofuel produced by the fermentation of sugars present in different types of biomass. It can be classified into three categories based on its origin: first generation, obtained from food crops such as sugarcane, corn, and wheat, and is easy to produce but raises concerns about food safety; second generation, derived from lignocellulosic biomass, such as sugarcane bagasse and agricultural waste, and is more sustainable but requires more complex conversion processes; and third generation, produced from microalgae, a highly productive source that can be cultivated in adverse environments but still faces cost challenges for commercial viability (Lugani et al. 2020; Vasco-Correa et al. 2016; Alfonsín et al. 2019).

The bioethanol production process can be divided into four main stages (Owusu & Marfo 2023). The first one is pretreatment, which consists of reducing particle size and degrading the biomass to facilitate the release of fermentable sugars. The next step, called hydrolysis, is the conversion of cellulosic polymers into monosaccharides, which can be done using acids or enzymes. The third step is fermentation, which consists of transforming sugars into ethanol, generally using microorganisms such as *Saccharomyces cerevisiae* and *Zymomonas mobilis*. Finally, there is recovery and purification, where ethanol is separated from the fermented mixture by processes such as distillation, facing challenges due to its azeotropic nature. In the following, some examples of the use of AI and ML in the bioethanol production phases will be presented.

In the pretreatment stage, AI is used to optimize process parameters, reducing the need for empirical experimentation and increasing industrial efficiency. In Smuga-Kogut et al. (2021), ML algorithms (Random Forest [RF] and Artificial Neural Networks [ANN]) were applied to predict the amount of bioethanol obtained after pretreatment with ionic liquids. Another study (Luo et al. 2021) used an ANN model to predict the concentration of phenolic compounds derived from lignin and the amount of glucose released after pretreatment with dilute acid and enzymatic hydrolysis.

Enzymatic hydrolysis, responsible for converting cellulose and hemicellulose into fermentable sugars, also benefits from AI. Studies indicate that the combination of different enzymes can improve the efficiency of this process, depending on the characteristics of the biomass (Moreira et al. 2014). Among the AI techniques applied at this stage, the following stand out: ANN (Verrengia et al. 2020), Multivariate Regression (Haldar et al. 2023), Response Surface Methodology (Qi et al. 2009), and Genetic Algorithms (Das et al. 2014; Sivamani & Rajoo 2018), which allow predicting and optimizing the ideal conditions for the conversion of biomass into fermentable sugars.

During fermentation, microorganisms convert fermentable sugars into ethanol and carbon dioxide. The efficiency of this process depends on factors such as temperature, pH, substrate concentration, fermentation time, and the presence of inhibitors. Proper control of these variables is essential to maximize productivity and minimize the formation of undesirable byproducts, such as organic acids and phenolic compounds, which can inhibit yeast growth and reduce ethanol yield. AI techniques, such as ANN and particle swarm optimization (PSO), have been used to predict and automatically adjust these parameters (Fischer et al. 2017; Smuga-Kogut et al. 2021; Pereira et al. 2020).

In the last step, recovery and purification, ethanol is separated from the fermented mixture using methods such as distillation, adsorption, solvent extraction, and membrane technologies. Fractional distillation is the most widely used method to concentrate ethanol, while techniques such as pervaporation and azeotropic distillation are used to remove residual water and obtain anhydrous ethanol (Errico & Rong 2012). AI can be applied to optimize distillation and purification processes, ensuring greater energy efficiency and less waste of raw materials.

Furthermore, bibliometric analyses, such as the one carried out in Coşgun et al. (2023), demonstrate the growth of AI applications in bioethanol production, with research trends indicating greater integration between AI, deep learning and smart sensors, allowing for more precise and autonomous control of the production process. In summary, the application of AI at each stage of bioethanol production has shown promise for reducing operational costs, improving biofuel yield and minimizing environmental impacts.

12.2.2 Biogas

Biogas is a renewable fuel produced from anaerobic digestion (AD), a biological process in which microorganisms decompose organic matter in the absence of oxygen, resulting in the production of methane (CH_4), carbon dioxide (CO_2), and other trace gases (Rutland et al. 2023). This process occurs both in anoxic natural environments, such as sediments and moist soils, and in industrial biodigesters, where it is widely used to convert agricultural waste, sewage, and animal manure into energy (Gaida 2023). Biogas has several applications, including the generation of electricity, heat, and as a vehicle fuel, being a sustainable alternative to fossil fuels (Guo et al. 2021). In addition, the solid residue from the process, called digestate, can be reused as fertilizer due to its high nutrient content.

Biogas production occurs in a process composed of four fundamental steps (Cinar et al. 2021). Hydrolysis represents the first stage, in which complex macromolecules, such as proteins, carbohydrates, and lipids, are broken down into smaller molecules, such as amino acids, simple sugars, and fatty acids. Then, during acidogenesis, the products resulting from hydrolysis are transformed into volatile fatty acids, hydrogen and carbon dioxide by acidogenic microorganisms. In the third stage, known as acetogenesis, volatile fatty acids are converted into acetate, hydrogen, and carbon dioxide, which will serve as substrate for the last phase of the process. Finally, in methanogenesis, acetate and hydrogen are converted into methane and carbon dioxide by methanogenic microorganisms (Pessuto et al. 2016). To ensure maximum process efficiency, it is essential to control variables such as temperature, pH, hydraulic retention time, and composition of the substrate present in the biodigester (Zhang et al. 2021).

The application of AI has shown great promise in optimizing anaerobic digestion, helping to predict, adjust, and control biogas production (Cruz et al. 2022). ML models allow the interpretation of nonlinear relationships in the process, enabling more efficient and accurate operational adjustments (Gupta et al. 2023). The use of AI for predictive modeling of biogas production has been gaining prominence, with techniques such as ANN, RF, and Support Vector Machines (SVM) being used to estimate biogas production based on parameters such as temperature, pH, and organic load (Zhang et al. 2022). This enables process optimization, ensuring that biodigesters operate under ideal conditions, maximizing the conversion of organic matter into methane and reducing losses.

Acidogenesis is the second phase of anaerobic digestion, where hydrolysis products are converted into volatile fatty acids, hydrogen, and carbon dioxide by acidogenic microorganisms. AI has been used to model and optimize this step, mainly to predict the conversion of hydrolysis products into intermediate compounds. According to Zaher et al. (2009), ML-based models are applied to understand microbial activity and predict the optimal conditions for acidogenesis, taking into account factors such as availability of macronutrients (C, H, N, O, P, and S) and the thermodynamic reactions involved. During acetogenesis, volatile fatty acids are transformed into acetate, hydrogen, and CO_2, which will serve as precursors for methane production. ML models, such as ANNs and statistical regression, have been used to predict the conversion and optimize the efficiency of acetogenesis. A study cited by Rajendran et al. (2014) used a computational model based on Aspen Plus software to predict biogas production under various operating conditions, simulating the biochemical processes of anaerobic digestion, including acetogenesis. Methanogenesis is the final phase of the anaerobic digestion process, where methanogenic archaea convert acetate, hydrogen, and carbon dioxide into methane and CO_2. This step is crucial for the efficiency of biogas production. AI has been mainly applied to predict methane production and detect possible inhibitions in the process. A model described by Kana et al. (2012) used ANNs and genetic algorithms to model methanogenesis and predict methane production based on substrate composition and operating conditions. Furthermore, hybrid models combining neural networks and statistical optimization have been used to predict process inhibitions and automatically adjust operating conditions. According to Olabi et al. (2020), AI has been used to maximize methane production from different types of substrates, analyzing large volumes of operational and historical data to optimize biodigester performance. Another important benefit of AI in the biogas sector is in optimizing the feed of biodigesters. The choice of feedstock directly influences the quality and quantity of biogas generated, and ML algorithms analyze historical data to identify the ideal combinations of organic waste, ensuring greater efficiency in the conversion of carbon into methane (Shahsavar et al. 2021). This process reduces resource waste and improves production sustainability. Furthermore, AI has been widely applied in real-time monitoring and control of biodigesters. The integration of IoT sensors with ML systems allows for the continuous collection and analysis of variables such as temperature, pressure, and gas composition, automatically adjusting process conditions to avoid failures and maintain operational stability (Nguyen et al. 2022). This predictive control minimizes fluctuations in production and reduces operating costs.

12.2.3 Biodiesel

Biodiesel is a renewable biofuel produced from vegetable oils or animal fats through the transesterification process, in which triglycerides are converted into fatty acid methyl esters (FAMEs) in the presence of an alcohol and a catalyst. This fuel stands out as an alternative to fossil diesel due to its favorable environmental properties, such as reducing the emission of atmospheric pollutants. In addition, it is compatible with conventional diesel engines and can be used pure or mixed with petroleum diesel. Its production and quality are regulated by international standards, such as EN 14214 in Europe and ASTM D-6751 in the United States, which establish specifications for properties such as viscosity, oxidative stability, and cetane number (Singh et al. 2019).

The biodiesel production process involves several fundamental steps (Arif et al. 2025; Akande et al. 2025), which directly influence the efficiency and quality of the final product. The first step is the selection and preparation of the raw material, which includes the choice of the most suitable oils or fats, the removal of impurities and the chemical characterization to determine the viability of the process. Next, transesterification occurs, a chemical reaction catalyzed by acids or bases that converts the lipids into biodiesel and glycerol as a by-product. After this step, the fuel undergoes the separation and purification process, where glycerol is removed and impurities are eliminated through techniques such as water washing, adsorption, and distillation. To ensure the commercialization and adequate performance of biodiesel, it is essential to carry out rigorous quality control, in which parameters such as oxidation stability, water content and viscosity are evaluated. Finally, in the storage and distribution phase, measures such as the use of antioxidants are applied to prevent fuel degradation, ensuring its stability during transportation and use (Malode et al. 2021).

The application of AI in biodiesel production has proven to be a promising strategy for optimizing processes and improving fuel efficiency. Models such as ANNs and Adaptive Neuro-Fuzzy Inference System (ANFIS) are used to predict and optimize transesterification parameters, including temperature, reaction time, catalyst type, and concentration. The implementation of these techniques allows greater control over process variables, resulting in increased production efficiency and improved biodiesel quality (Wong & Wong 2017; Wong et al. 2013; Gupta et al. 2021; Amini et al. 2016; Moradi et al. 2013; Ghaderi et al. 2016). According to recent studies, ML has been effective in solving complex challenges related to the configuration, stability, and operation of reactors used in biofuel production (Zhang et al. 2017).

In addition to optimizing the conversion process, AI is also widely used in the selection of feedstocks. ML models can predict biodiesel yield and quality based on the fatty acid composition of vegetable oils and animal fats, allowing producers to choose more suitable and sustainable inputs. Techniques such as neural networks and probabilistic algorithms are applied to assess the oil content and the presence of impurities in the inputs, increasing the reliability of the production process (Ardabili et al. 2019). Thus, AI plays a fundamental role in identifying economically viable and ecologically sustainable biomass sources (Jahirul et al. 2013).

Another crucial aspect of the application of AI in biodiesel production is real-time monitoring and control. The integration of smart sensors with statistical models and deep learning algorithms allows the automatic adjustment of critical process variables, such as temperature and pH, improving energy efficiency and minimizing losses. In addition, CNNs have been used to improve the operation of photobioreactors in the production of biodiesel from microalgae, significantly reducing processing time compared to traditional simulations (Zhengxin Wang et al. 2022). The use of these technologies improves process predictability and reduces operating costs. The modeling of biodiesel physicochemical properties has also benefited from advances in ML. AI-based models can predict characteristics such as viscosity, density, and cetane number with high accuracy, allowing real-time adjustments to ensure the final fuel quality. Mathematical approaches based on neural networks and deep learning have proven to be efficient in reducing waste, avoiding time-consuming laboratory tests and unnecessary costs (Rio-Chanona et al. 2018; Salam et al. 2020; Kessler et al. 2017). In addition, hybrid models, which combine deep learning with statistical optimization methods, are being explored to improve the reliability of predictions and ensure tighter control over production parameters.

In summary, AI is revolutionizing the biodiesel industry by providing advanced solutions for process optimization, feedstock selection, operational control, and final product quality prediction. Although challenges such as data availability and model interpretability still exist, the continuous advancement of ML and deep learning techniques promises to further boost the efficiency and sustainability of biodiesel production in the coming years (Wang et al. 2024).

12.2.4 Biojet Fuel

Biojet fuel (Sustainable Aviation Fuel [SAF]) is a biofuel produced from renewable sources, such as biomass, agricultural waste, vegetable oils, and residual fats. Its main advantage is the significant reduction in greenhouse gas emissions, making it an essential alternative for the decarbonization of aviation. To be used commercially, biojet fuel must meet strict certification standards, ensuring its compatibility with traditional fossil fuels (Heyne et al. 2021). In addition to its contribution to sustainability, the adoption of biojet fuel is considered essential for meeting global climate goals, and technologies such as ML can be applied to improve its production and integration into the energy matrix (Donti & Kolter 2021).

The production of biojet fuel involves several technological stages, from the selection of the raw material to the certification of the final product. Raw materials can be classified into three generations: (1) first generation, which uses seeds, grains and sugars as a carbon source, but faces limitations due to competition with food production; (2) second generation, based on the use of urban, agricultural, and industrial waste, such as sewage sludge and harvest remains, reducing environmental and economic impacts; and (3) third generation, which uses microalgae and other high-lipid organisms, standing out for their high productivity and lower demand for natural resources (He et al. 2024; Undavalli et al. 2023). After the raw material is selected, biojet fuel undergoes different conversion processes, which may include hydroprocessing of fatty acids and esters (HEFA), Fischer–Tropsch synthesis (FT-SPK) and thermal pyrolysis followed by hydrodeoxygenation (HDCJ), each with different approaches to transform biomass into liquid hydrocarbons compatible with aviation engines (Mansy et al. 2025; Liu & Yu 2021). Furthermore, hybrid processes, such as the integration of biomass gasification with natural gas, have been studied to increase efficiency and reduce emissions (Xia et al. 2023).

The application of AI and ML has proven to be a promising strategy for optimizing each stage of biojet fuel production. In the initial pre-processing phase, ML models can be employed to assess the viability of raw materials and predict critical fuel properties prior to production. ML techniques assist in the automated screening of biomass, reducing the need for time-consuming laboratory tests and optimizing the selection of inputs with greater energy potential and lower environmental impact (Heyne et al. 2021; He et al. 2024).

During the conversion of biomass into biojet fuel, ML algorithms play a crucial role in optimizing catalytic processes. The choice of catalysts directly influences the yield and quality of the fuel produced. Predictive models are used to select the most efficient catalysts, reducing energy consumption and minimizing the formation of undesirable byproducts. In addition, techniques such as SVM and Gaussian Process Regression (GPR) are applied to predict biomass gasification efficiency, allowing the adjustment of variables such as temperature and pressure to maximize the conversion of inputs into biojet fuel (Osman et al. 2024; Xia et al. 2023; Mansy et al. 2025).

AI is also essential in optimizing energy efficiency and the production chain. Deep learning models can predict biomass conversion yield under different operating conditions, allowing automatic adjustments in real time to minimize waste and improve the use of available resources. This approach contributes to reducing operating costs and improves the integration of SAF production with environmental sustainability guidelines, ensuring a more efficient supply chain with a smaller carbon footprint (Donti & Kolter 2021; Guimarães et al. 2023).

Finally, AI can accelerate the SAF certification process, making the development of new biofuels more agile and accessible. ML models are used to simulate and predict fuel performance, reducing the need for extensive physical testing and the costs associated with regulatory certification. This enables new sustainable fuels to be rapidly evaluated and incorporated into the aviation sector, driving the transition to a cleaner and more efficient energy system (Heyne et al. 2021; He et al. 2024). Thus, the integration of AI and ML in the production of biojet fuel is essential for the viability of sustainable aviation fuels, ensuring greater predictability, efficiency and competitiveness in the sector (Becken et al. 2023).

12.3 CHALLENGES AND SOLUTIONS

12.3.1 Challenges

12.3.1.1 Bioethanol

The application of AI in bioethanol production faces complex challenges, ranging from data availability and quality to the efficient integration of these models in industrial environments. One of the main obstacles is obtaining high-quality data, since the variability of the biomass used directly impacts the efficiency of predictive models. Different types of feedstock have distinct chemical compositions that influence ethanol yield, making it essential to develop more robust algorithms capable of handling heterogeneous data and variable operating conditions (Liao & Yao 2021). In addition, the lack of unified standards for data collection and structuring makes it difficult to replicate and validate models, requiring investment in adequate computational infrastructure to store and process large volumes of information (DOE-US, 2023).

The complexity of the biochemical processes involved in bioethanol production, such as enzymatic hydrolysis and fermentation, presents another relevant challenge. These processes are dynamic and non-linear, making predictive modeling difficult and requiring large data sets for calibration and refinement. The integration of AI with traditional production processes also faces resistance, both due to the need to reconfigure industrial plants and the lack of professionals trained to operate and interpret predictive models. In addition, the gap between academic research and practical application is evident in this context, as many technological advances are developed in laboratory environments and are not easily scalable to large-scale industrial operations (Owusu & Marfo 2023).

Another critical challenge is supply chain optimization, as the logistics of biomass supply and storage directly influence the costs and efficiency of the production process. AI models can assist in the prediction and optimization of these processes, but their large-scale implementation still faces barriers, especially due to the lack of digital infrastructure in plants (Liao & Yao 2021). Furthermore, the automation of production processes through AI creates regulatory and ethical challenges, particularly with regard to the use of real-time data for autonomous decision-making (DOE-US 2023).

To overcome these challenges, a joint effort between industry, academia, and regulatory bodies is essential. This includes creating standards for data collection and structuring, improving AI models to deal with complex systems, and training professionals to operate these technologies effectively. Investments in technological infrastructure and the development of scalable solutions are essential for AI to be applied efficiently in bioethanol production, contributing to the sustainability and competitiveness of the bioenergy sector in the long term (Sinha & Lee 2024).

12.3.1.2 Biogas

In biogas production, the integration with AI faces significant challenges, ranging from the variability of anaerobic digestion systems to efficient integration with existing industrial technologies. One of the main obstacles is the difficulty in modeling nonlinear relationships between biological and operational variables, since biogas production depends on factors such as sludge flow rate, total solids content, and volatile fatty acids, making predictive models more complex (Gaida 2023). In addition, the instability of the microorganisms involved in the process adds a factor of uncertainty, since variations in temperature and raw material composition can directly affect process efficiency, making it difficult to create robust and adaptable models (Raja & Wazir 2017).

Another major challenge is the lack of standardized and real-time data for calibrating AI models. Many critical parameters, such as pH and alkalinity, are collected manually or by sensors that present delays in collection and inconsistencies in accuracy, compromising the reliability of predictions (Rutland et al. 2023). In addition, laboratory data are not always scalable to industrial operations, as the diversity of equipment and operating conditions prevents the generalization of AI models to different biogas plants. This issue is aggravated by the lack of standardization in sensors and monitoring equipment, which makes it difficult to obtain consistent data sets for training ML algorithms (Cinar et al. 2021). The integration of AI with existing automation and control systems also represents a considerable barrier. Many biogas plants still use conventional technologies that are not compatible with advanced AI-based solutions, requiring significant investments to modernize and adapt industrial infrastructure (Cinar et al. 2021). Furthermore, there is resistance from industrial operators and managers, who may not have the necessary training to implement and interpret predictive models, limiting the adoption of these technologies on a large scale (Gaida 2023).

Finally, supply chain optimization emerges as a critical challenge, since the logistics of biomass supply and storage directly influence the efficiency of biogas production. In addition, the automation of these processes through AI generates regulatory and ethical challenges, especially with regard to the use of real-time data for autonomous decision-making (DOE-US, 2023). Given these challenges, investment in data standardization, advanced sensors, and computational infrastructure is essential to enable the efficient implementation of AI in biogas production. In addition, professional training and greater integration between industry, academia, and regulatory bodies are essential to overcome technological and operational barriers, ensuring that AI can contribute to the optimization and sustainability of this energy sector.

12.3.1.3 Biodiesel

Regarding biodiesel production, the use of AI faces significant challenges, ranging from data quality and availability to industrial implementation and process optimization. Training deep learning models, such as ANNs, requires robust and standardized datasets. However, the diversity of raw materials, such as vegetable oils, animal fats, and industrial waste, results in variability in experimental data, making predictive modeling less reliable. In addition, the lack of standardization in the collection and structuring of these data makes it difficult to replicate and generalize models to different industrial scenarios (Xing et al. 2021; Akande et al. 2025).

Another challenge is the complexity of the transesterification process, which involves interdependent variables, such as temperature, oil-to-alcohol molar ratio, type of catalyst, and reaction time. Small variations in these parameters can significantly impact the conversion of lipids into biodiesel and its final properties. Modeling these chemical interactions remains a challenge for AI models, which need to be able to deal with operational uncertainties and fluctuations (Xing et al. 2021; Arif et al. 2025). In addition, optimizing the selection of raw materials is also complex, since the variability in their chemical composition directly influences the yield and quality of the biodiesel produced. Predictive models need to be refined to accurately estimate lipid conversion and the characteristics of the final fuel, such as viscosity and flash point (Arif et al. 2025).

The lack of interpretability of AI models is another obstacle. Many deep learning techniques operate as "black boxes", making it difficult to understand the relationships between process parameters and biodiesel quality. This lack of transparency generates resistance from industrial operators and experts, who need clearer explanations to validate the predictions made by the models (Akande et al. 2025). To overcome this limitation, new approaches, such as explainable AI (XAI), can be explored to provide greater clarity on how models work.

In addition to the technical challenges, the integration of AI with industrial monitoring and control systems still faces barriers. Many biodiesel plants use conventional technologies that were not designed to operate with AI. The implementation of advanced sensors to collect data in real time is limited due to the high cost and complexity of adapting these technologies to existing infrastructures. In addition, modernizing these systems requires significant investments and professional training to ensure that operators are able to interpret the results generated by the models (Xing et al. 2021; Cinar et al. 2021).

In view of these challenges, the adoption of AI in biodiesel production requires a multidisciplinary approach that combines data standardization, improvement of learning algorithms, and investments in technological infrastructure. The development of more accessible sensors, the application of explainable ML techniques, and the training of industry professionals are key measures to ensure the reliability and scalability of AI in the biodiesel sector. With these advances, AI could play a central role in optimizing production, reducing costs, and increasing the sustainability of the biofuels industry.

12.3.1.4 Biojet Fuel

In the production of biojet fuel, the challenges of using AI and ML involve issues related to data quality, integration with industrial processes, operational costs, and regulatory acceptance. The application of these technologies in the production of this biofuel is still in the development phase, and overcoming these difficulties is essential to ensure the economic and environmental viability of the sector.

One of the main challenges faced is the availability and quality of the data used to train AI models. The diversity of raw materials, which include agricultural waste, vegetable oils, and microalgae, results in significant variations in the chemical composition of the inputs. This heterogeneity makes it difficult to build robust and reliable predictive models to estimate the conversion and yield of biojet fuel (He et al. 2024). In addition, the lack of unified standards in data collection and structuring hinders the replicability of models in different industrial facilities (Heyne et al. 2021).

The complexity of the catalytic processes used in the conversion of biomass into biojet fuel also represents a challenge. Methods such as hydroprocessing of fatty acids and esters (HEFA) and Fischer–Tropsch synthesis require precise control of operational variables such as temperature, pressure, and catalyst composition (Liu & Yu 2021). ML models can be used to optimize these variables, but generalization of models to different operating conditions remains a critical problem (Osman et al. 2024; Xia et al. 2023).

Another relevant obstacle is the integration of AI into existing automation and control systems of biojet fuel refineries. Many industrial plants use conventional technologies that are not compatible with AI-based solutions, requiring significant investments in computational infrastructure and advanced sensors to enable real-time data collection and analysis (Donti & Kolter 2021). Industrial operators' resistance to adopting these technologies also limits large-scale implementation, since many professionals do not have adequate training to interpret the results provided by predictive algorithms.

In addition to technological barriers, the high costs of implementing AI in biojet fuel production are a concern for the sector. The need to modernize industrial plants and acquire specialized hardware increase initial investments, making adoption by small and medium-sized companies more difficult. In addition, the certification of new sustainable fuels poses additional challenges, as AI models need to be validated by regulatory bodies before they can be widely used (Heyne et al. 2021; He et al. 2024). Finally, regulatory acceptance of the use of AI in biojet fuel production is still in its early stages. Aviation fuel regulations require strict performance and safety standards, and the introduction of AI-based methods in the certification of new fuels may face resistance from regulatory agencies (Mansy et al. 2025). To overcome these challenges, it is essential to develop policies that encourage the digitalization of the sector, in addition to joint initiatives between industry, academia, and regulatory bodies to ensure the reliability and scalability of AI-based solutions.

12.3.1.5 Proposed Solutions

The application of AI and ML in bioethanol production faces challenges related to biomass variability, lack of data standardization, and complexity of biochemical processes. To address these issues, the development of hybrid models, which combine different ML techniques, such as ANNs and RF algorithms, has been proposed to improve prediction accuracy and reduce the impact of heterogeneous data (Smuga-Kogut et al. 2021). In addition, the creation of unified standards for data collection and storage can facilitate the replication and validation of models in different industrial contexts (DOE-US, 2023).

Optimizing biochemical processes, such as enzymatic hydrolysis and fermentation, is essential to increase production efficiency. ML algorithms can automatically adjust operational parameters, reducing the need for long calibration periods and maximizing biomass conversion to ethanol (Liao & Yao 2021). Strategies such as deep learning and Bayesian optimization have been tested to improve the modeling of these processes, ensuring more reliable predictions about yield and industrial efficiency (Coşgun et al. 2023). The need for large processing and storage capacity to handle large volumes of data can be addressed with cloud computing solutions and distributed architectures. The use of high-performance platforms allows AI models to be trained and applied in real time without compromising the efficiency of industrial operations (Owusu & Marfo 2023). In addition, the integration of adaptive control systems allows automatic adjustments to production parameters based on predictions generated by ML.

In biomass logistics and storage, AI can be used to predict demand and optimize the supply of raw materials. Forecasting models and optimization algorithms help reduce operating costs and ensure continuous and economical production (Liao & Yao 2021). The use of sensors connected to the IoT also helps to collect data in real time, improving the accuracy of predictive models. Resistance to the adoption of AI can be minimized through training professionals and partnerships between universities and companies. Specific training programs for industrial operators ensure that new technologies are implemented efficiently (Sinha & Lee 2024). In addition, the creation of clear regulatory guidelines for automation and the use of AI can guarantee the operational safety and sustainability of the sector (DOE-US, 2023).

In biogas production, the complexity of the relationships between biological and operational variables can be reduced by the use of hybrid models that combine deep neural networks and Bayesian optimization, improving adaptation to variations in operational conditions and biomass composition (Gaida 2023). Advanced data normalization and preprocessing strategies help to deal with the variability of environmental conditions and substrates used. To solve the lack of standardization in data collection and calibration, the use of advanced sensors and real-time monitoring systems is recommended. The use of IoT devices allows the continuous and accurate acquisition of variables such as pH, temperature and volatile fatty acid concentration, improving the reliability of AI models (Cinar et al. 2021). In addition, the adoption of unified standards for data structuring can ensure interoperability between different biogas plants. The modernization of industrial infrastructures can be challenging due to incompatibility with advanced AI solutions. To mitigate this problem, it is recommended to adopt modular systems that allow the progressive integration of digital technologies, reducing the need for high initial investments (Rutland et al. 2023). The use of adaptive control based on ML makes it possible to optimize production without requiring the complete replacement of equipment. The application of AI in demand forecasting and biomass inventory management can improve supply chain efficiency, reducing costs and avoiding waste (Li et al. 2024). Furthermore, the automation of decision-making processes in biogas plants raises regulatory and ethical challenges, which can be addressed by developing clear guidelines for the use of AI in industrial settings, ensuring compliance with safety and environmental sustainability standards (DOE-US 2023).

In biodiesel production, the variability of raw materials and the complexity of the transesterification process require advanced AI solutions. The use of deep neural networks combined with optimization algorithms allows dynamic adjustments in operational parameters, maximizing lipid conversion and biodiesel quality (Arif et al. 2025). Furthermore, the implementation of Explainable AI (XAI) can increase the transparency of models, facilitating the validation of predictions and the acceptance of these technologies in the productive sector (Akande et al. 2025). The modernization of biodiesel plants can be made possible by the gradual implementation of smart sensors and adaptive control systems, allowing the integration of AI with industrial automation without requiring high short-term investments (Xing et al. 2021; Cinar et al. 2021). ML models can also assist in choosing the most efficient and sustainable feedstock, reducing costs and environmental impacts by selecting alternatives such as waste oils and non-edible fats (Arif et al. 2025).

The production of biojet fuel faces additional challenges, such as the complexity of catalytic processes and the need for regulatory certification. The development of hybrid models that combine deep learning with statistical optimization methods has been explored to improve the reliability of predictions and ensure tighter control of production parameters (Osman et al. 2024). Furthermore, AI can be used to optimize the biojet fuel certification process, reducing the need for extensive physical testing and the costs associated with regulatory certification (Mansy et al. 2025). To ensure the acceptance of AI in biofuel production, strategies such as professional training, investments in digital infrastructure, and the development of unified standards for data collection and analysis are essential (Heyne et al. 2021). These actions will ensure that AI is applied efficiently and sustainably, promoting innovation and com petitiveness in the biofuels sector.

12.4 ECONOMIC AND ENVIRONMENTAL IMPACTS

12.4.1 Economic Impacts

The use of AI and ML in biofuel production has generated significant economic impacts, promoting the optimization of industrial processes and reducing operating costs. These technologies enable more precise control of production variables, improving production predictability and reducing input waste. However, their implementation also poses financial challenges, such as high initial costs and the continuous need for technological updates.

In the bioethanol sector, the use of AI has revolutionized the fermentation process, allowing for more accurate prediction of the final ethanol concentration and reducing the need for time-consuming and costly experimental tests. ML models, such as neural networks and RF algorithms, maximize the use of raw materials and minimize losses, contributing to the profitability of plants (Smuga-Kogut et al. 2021). In addition, automation promoted by AI reduces dependence on specialized labor, directly impacting labor and operating costs (Owusu & Marfo 2023).

In biogas production, ML models have been applied to predict gas yield based on operational variables, optimizing biomass conversion and reducing substrate waste. This improves energy efficiency and allows real-time adjustments to avoid process failures, ensuring greater operational stability and reducing corrective maintenance costs (Cinar et al. 2021). In addition, AI-based predictive analytics facilitates supply chain planning, avoiding excessive inventories and minimizing financial losses (Rutland et al. 2023).

The biodiesel sector also benefits from AI by optimizing the choice of raw materials and transesterification control. Advanced deep learning models allow the selection of more profitable and sustainable inputs, reducing dependence on high-cost vegetable oils and favoring alternatives such as residual oils and animal fats (Akande et al. 2025). Furthermore, AI-based predictive maintenance reduces equipment downtime, avoiding production interruptions and increasing operational efficiency (Arif et al. 2025).

In biojet fuel production, AI plays a key role in optimizing conversion routes and choosing catalysts, reducing the time and costs involved in developing new processes. The application of predictive models allows for better forecasting of the demand and supply of this biofuel, helping to stabilize prices and attract investment to the sector (Mansy et al. 2025). However, the need for robust computing infrastructure and specialized hardware, such as GPUs and TPUs, represents an economic challenge, especially for small and medium-sized companies (He et al. 2024).

Despite the costs associated with implementing these technologies, AI and ML offer substantial economic opportunities for the biofuels sector. The greater efficiency of production processes makes biofuels more competitive in relation to fossil fuels, encouraging the sustainable energy transition. Furthermore, the use of AI can enable new business models, such as the commercialization of carbon credits and the use of industrial byproducts to generate additional revenue (Gaida 2023). Thus, the adoption of AI and ML in the biofuels sector presents a balance between economic benefits and financial challenges. Although the initial investments may be high, the gains in efficiency, predictability, and competitiveness justify the implementation of these technologies. The economic viability of AI will depend on strategies that balance innovation and financial sustainability, ensuring that the benefits outweigh the costs in the long term.

12.4.2 Environmental Sustainability

The application of AI and ML in biofuel production has contributed to the optimization of industrial processes, increasing efficiency and reducing waste. However, these technologies also present significant environmental challenges, especially with regard to high energy consumption, intensive use of water resources and the generation of electronic waste.

One of the main environmental impacts of AI arises from the high energy demand required for the training and execution of advanced models, such as deep neural networks. These models require high-performance data centers, whose continuous operation consumes large amounts of electricity. In places where the energy matrix still depends on fossil fuels, this demand intensifies the carbon footprint associated with biofuel production (Liao & Yao 2021). In addition, the need to store and process large volumes of data leads to the growth of cloud computing, increasing the demand for globally distributed servers and, consequently, the environmental impact due to the transportation and maintenance of this infrastructure (Coşgun et al. 2023).

Another relevant factor is the consumption of water for cooling servers, which is essential to prevent overheating and ensure the efficient operation of AI systems. This consumption can be particularly problematic in regions with water scarcity, compromising the sustainability of projects that use AI to optimize biofuel production (Saraiva et al. 2024). The increasing need for reprocessing ML models also results in short life cycles for systems, which generates an increase in the disposal of electronic equipment and the demand for new materials to maintain the computing infrastructure (Liao & Yao 2021).

The rapid obsolescence of hardware represents another significant environmental challenge. The accelerated development of AI-optimized processors, such as GPUs and TPUs, increases the generation of electronic waste, which contains heavy metals and other toxic materials that can contaminate soil and water if not disposed of correctly. Furthermore, the extraction of minerals for the manufacture of these semiconductors and the global transportation of these components increase the environmental footprint of AI in the production of biofuels (Coşgun et al. 2023).

The integration of AI with the IoT for monitoring industrial processes also contributes to increased energy consumption, as it requires robust data transmission networks and intermediary devices, such as routers and antennas. This growth in interconnectivity intensifies the use of computing resources and increases the environmental impact of the sector (Câmara & Guimarães 2024). Furthermore, although AI is often promoted as a tool for optimizing industrial processes and reducing environmental impact, its effectiveness depends on the quality of the data used, the efficiency of the algorithms, and the available computing infrastructure. In some cases, the high energy consumption required to run complex models can outweigh the environmental gains obtained in optimizing production, generating a net negative effect (Góes-Favoni et al. 2018).

Despite these challenges, strategies can be adopted to minimize the environmental impacts of using AI in biofuel production. The use of renewable energy sources in data centers, the optimization of the life cycle of AI models, and the implementation of efficient electronic waste recycling programs are some of the solutions that can help reduce the environmental damage associated with the use of this technology. Therefore, the search for technological innovation must be balanced with sustainable practices, ensuring that the benefits of AI are maximized without compromising the preservation of natural resources.

12.5 FUTURE PERSPECTIVES

The integration of AI and ML into the biofuel production process has the potential to improve the conversion of biomass into energy, optimize the selection of raw materials and predict operational failures, making the production of bioethanol, biodiesel, biogas, and biojet fuel increasingly competitive and sustainable (Coşgun et al. 2023; Osman et al. 2024; Mansy et al. 2025).

In the coming years, AI-driven industrial automation is expected to transform the monitoring of production processes, enabling automated adjustments in real time. Smart sensors connected to machine learning systems will be able to continuously evaluate variables such as temperature, pH, and substrate concentration, ensuring maximum efficiency and reducing waste. In biogas production, IoT sensors are expected to advance to ensure the stability of anaerobic digestion and increase the predictability of energy production (Donti & Kolter 2021; Li 2024). In the case of bioethanol, predictive models could allow automatic adjustments in fermentation, minimizing the formation of undesirable by-products. In biodiesel and biojet fuel biorefineries, AI-based automation is expected to make predictive maintenance more accessible, reducing downtime and increasing energy efficiency (Heyne et al. 2021; Xia et al. 2023).

In addition, AI-based catalyst optimization is expected to revolutionize the efficiency of chemical processes, enabling the discovery of new, more sustainable and effective catalysts. In biodiesel production, machine learning models are expected to be widely used to develop more efficient heterogeneous catalysts, reducing the need for aggressive chemical processes and making the conversion of waste oils more economically viable (Arif et al. 2025; Osman et al. 2024). For biojet fuel, the future should bring significant advances in the application of AI to develop optimized catalysts for Fischer–Tropsch synthesis and hydroprocessing of esters and fatty acids (HEFA), improving the conversion of lignocellulosic biomass into aviation fuels (Mansy et al. 2025). In biogas production, AI is expected to be used to improve microbial biocatalysts, making the anaerobic digestion process more efficient and predictable (Li 2024).

Feedstock selection should become more efficient with the use of ML, allowing producers to choose more productive and sustainable inputs. In the future, algorithms are expected to advance to analyze large volumes of data on the chemical composition, availability and economic viability of different substrates. This approach could be essential to diversify the biomass matrix, reducing dependence on traditional sources and promoting sustainable production. In bioethanol and biodiesel, computational models are expected to predict with high accuracy the conversion yield of vegetable oils and fermentable sugars (Coşgun et al. 2023; Bosnjakovic 2020). For biogas, AI could help define optimal combinations of organic residues, maximizing energy generation (Li 2024). In biojet fuel, AI is expected to be widely used to assess the efficiency of new biomass sources, such as agricultural residues and algae, improving the availability of renewable feedstocks for aviation fuels (He et al. 2024).

Another expectation for the coming years is that AI will play an essential role in the logistics and sustainability of biofuel production. Advanced predictive models should be integrated into supply chain management, optimizing distribution and reducing waste in transportation (Heyne et al. 2021; Xia et al. 2023). Furthermore, AI is increasingly being used for life cycle analysis (LCA), enabling more detailed assessments of the environmental impacts of biofuels, from biomass harvesting to its final conversion (Liao & Yao 2021). This development could facilitate the implementation of more sustainable practices and assist in complying with increasingly stringent environmental regulations (Pasha et al. 2021).

With the expected technological advances, the application of AI and ML in biofuel production has the potential to profoundly transform this industry, promoting greater efficiency, cost reduction, and environmental gains. These technologies are likely to become an essential pillar for the competitiveness of biofuels in relation to fossil fuels. With continued investments in research and innovation, AI is expected to play a central role in the global energy transition, consolidating bioethanol, biodiesel, biogas and biojet fuel as strategic alternatives for a low-carbon future (Mansy et al. 2025; Osman et al. 2024; Coşgun et al. 2023).

REFERENCES

Akande, O., Okolie, J. A., Kimera, R., & Ogbaga, C. C. (2025). A Comprehensive review on deep learning applications in advancing biodiesel feedstock selection and production processes. *Green Energy and Intelligent Transportation*, 4(3), 100260. https://doi.org/10.1016/j.geits.2025.100260

Alfonsín, V., Maceiras, R., & Gutiérrez, C. (2019). Bioethanol production from industrial algae waste. *Waste Management*, 87, 791–797.

Almohana, A. I., Almojil, S. F., Kamal, M. A., Alali, A. F., Kamal, M., Alkhatib, S. E., Felemban, B. F., & Algarni, M. (2022). Theoretical investigation on optimization of biodiesel production using waste cooking oil: Machine learning modeling and experimental validation. *Energy Reports, 8*, 11938–11951. https://doi.org/10.1016/j.egyr.2022.08.265

Amini, Y., Gerdroodbary, M. B., Pishvaie, M. R., Moradi, R. & Monfared, S. M. (2016) Optimal control of batch cooling crystallizers by using genetic algorithm. *Case Studies in Thermal Engineering*, 8, 300–310. https://doi.org/10.1016/j.csite.2016.09.001

Ardabili, S., Mosavi, A., & Várkonyi-Kóczy, A. R. (2019). Systematic review of deep learning and machine learning models in biofuels research. In: Várkonyi-Kóczy, A. (eds.) *Engineering for Sustainable Future. INTER-ACADEMIA 2019. Lecture Notes in Networks and Systems*, v. 101. Springer, Cham. https://doi.org/10.1007/978-3-030-36841-8_2

Arif, M., Alalawy, A. I., Zheng, Y., Koutb, M., Kareri, T., Salama, E., & Li, X. (2025) Artificial intelligence and machine learning models application in biodiesel optimization process and fuel properties prediction. *Sustainable Energy Technologies and Assessments*, 73, 104097. https://doi.org/10.1016/j.seta.2024.104097

Becken, S., Mackey, B., & Lee, D. S. (2023). Implications of preferential access to land and clean energy for Sustainable Aviation Fuels. *Science of The Total Environment*, 886, 163883. https://doi.org/10.1016/j.scitotenv.2023.163883

Bošnjaković, M., & Sinaga, N. (2020). The perspective of large-scale production of algae biodiesel. Applied Sciences, 10(22), 8181. https://doi.org/10.3390/app10228181

Câmara, C. N., & Guimarães, C. C. (2024). Fermentação alcóolica: um estudo de caso sobre as ações adotadas para a otimização deste processo. Ciência & Tecnologia: FATEC-JB, 16(1), e16102. https://doi.org/10.52138/citec.v16i1.333

Cinar, S., Cinar, S. O., Wieczorek, N., Sohoo, I., & Kuchta, K. (2021). Integration of artificial intelligence into biogas plant operation. *Processes*, 9, 85. https://doi.org/10.3390/pr9010085.

Coşgun, A., Günay, M. E., & Yıldırım, R. (2023). A critical review of machine learning for lignocellulosic ethanol production via fermentation route. *Biofuel Research Journal*, *10*(2), 1859–1875. https://doi.org/10.18331/BRJ2023.10.2.5

Cruz, I. A., Chuenchart, W., Long, F., Surendra, K. C., Andrade, L. R., Bilal, M., Liu, H., Figueiredo, R. T., Khanal, S. K., & Ferreira, L. F. (2022). Application of machine learning in anaerobic digestion: Perspectives and challenges, *Bioresource Technology*, 345, 126433. https://doi.org/10.1016/j.biortech.2021.126433.

Das, S., Bhattacharya, A., Ganguly, A., Gu, S., & Chatterjee, P. K. (2014). Kinetics and genetic algorithm study of acid catalysed hydrolysis of water hyacinth biomass. *Journal of Renewable Sustainable Energy, 6, 063115.* https://doi.org/10.1063/1.4904723.

DOE-US (2023). Artificial Intelligence and Machine Learning for Bioenergy Research: Opportunities and Challenges, DOE/SC-0211. U.S. Department of Energy Office of Science and Office of Energy Efficiency and Renewable Energy. https://doi.org/10.2172/1968870.

Donti, P. L., & Kolter, J. Z. (2021). Machine learning for sustainable energy systems. *Annual Review of Environment and Resources*, 46, 719–747. https://doi.org/10.1146/annurev-environ-020220-061831

Errico, M., & Rong, B. (2012). Synthesis of new separation processes for bioethanol production by extractive distillation. *Separation and Purification Technology*, 96, 58–67. https://doi.org/10.1016/j.seppur.2012.05.022.

Fischer, J., Lopes, V. S., Cardoso, S. L., Filho, U. C., & Cardoso, V. L. (2017). Machine learning techniques applied to lignocellulosic ethanol in simultaneous hydrolysis and fermentation. *Brazilian Journal of Chemical Engineering*, 34, 53–63.

Gaida, R. (2023). Application of machine learning in biogas process. *Master's Thesis, University of South-Eastern Norway*, Norway. https://openarchive.usn.no/usn-xmlui/bitstream/handle/11250/3076266/no.usn%3Awiseflow%3A6838201%3A54569109.pdf?sequence=1&isAllowed=y

Ghaderi, H., Pishvaee, M. S., & Moini, A. (2016). Biomass supply chain network design: An optimization-oriented review and analysis. *Industrial Crops and Products*, 94, 972–1000. https://doi.org/10.1016/j.indcrop.2016.09.027

Góes-Favoni, S. P., Monteiro, A. C. C., Dorta, C., Crippa, M. G., & Shigematsu, E. (2018). Fermentação alcoólica na produção de etanol e os fatores determinantes do rendimento. *Revista Ibero-Americana de Ciências Ambientais*, 9(4), 285–296.

Guimarães, H. R., Bressanin, J. M., Motta, I. L., Chagas, M. F., Klein, B. C., Bonomi, A., Filho, R. M., & Watanabe, M. D. (2023). Decentralization of sustainable aviation fuel production in Brazil through biomass-to-liquids routes: A techno-economic and environmental evaluation. *Energy Conversion and Management*, 276, 116547. https://doi.org/10.1016/j.enconman.2022.116547

Guo, H., Wu, S., Tian, Y., Zhang, J., & Liu, H. (2021). Application of machine learning methods for the prediction of organic solid waste treatment and recycling processes: A review. *Bioresource Technology*, 319, 124114. https://doi.org/10.1016/j.biortech.2020.124114.

Gupta, K. K., Kalita, K., Ghadai, R. K., Ramachandran, M., & Gao, X.-Z. (2021). Machine learning-based predictive modelling of biodiesel production: A comparative perspective. *Energies*, 14(4), 1122. https://doi.org/10.3390/en14041122

Gupta, R., Zhang, L., Hou, J., Zhang, Z., Liu, H., You, S., Ok, Y. S., & Li, W. (2023). Review of explainable machine learning for anaerobic digestion. *Bioresource Technology*, 369, 128468. https://doi.org/10.1016/j.biortech.2022.128468.

Haldar, D., Shabbirahmed, A. M., & Mahanty, B. (2023). Multivariate regression and artificial neural network modelling of sugar yields from acid pretreatment and enzymatic hydrolysis of lignocellulosic biomass. *Bioresource Technology*, 370, 128519. https://doi.org/10.1016/j.biortech.2022.128519.

He, X., Wang, N., Zhou, Q., Huang, J., Ramakrishna, S., & Li, F. (2024). Smart aviation biofuel energy system coupling with machine learning technology. *Renewable and Sustainable Energy Reviews*, 189, 113914. https://doi.org/10.1016/j.rser.2023.113914

Heyne, J., Rauch, B., Clercq, P. L., & Colket, M. (2021). Sustainable aviation fuel prescreening tools and procedures. *Fuel*, 290, 120004. https://doi.org/10.1016/j.fuel.2020.120004

Ibarra-Junquera, V., Murguia, J. S., Escalante-Minakata, P., & Rosu, H. C. (2007). Application of multifractal wavelet analysis to spontaneous fermentation processes. *Physica A: Statistical Mechanics and its Applications, 387*(12), 2802-2808. https://doi.org/10.1016/j.physa.2008.01.083

Itto-Nakama, K., Watanabe, S., Kondo, N., Ohnuki, S., Kikuchi, R., Nakamura, T., Ogasawara, W., Kasahara, K., & Ohya, Y. (2022). AI-based forecasting of ethanol fermentation using yeast morphological data. *Bioscience, Biotechnology, and Biochemistry,* 86(1), 125–134. https://doi.org/10.1093/bbb/zbab188

Jahirul, M. I., Brown, R. J., Senadeera, W., O'Hara, I. M., & Ristovski, Z. D. (2013). The use of artificial neural networks for identifying sustainable biodiesel feedstocks. *Energies,* 6(8), 3764–3806. https://doi.org/10.3390/en6083764

Jeswani, H. K., Chilvers, A., & Azapagic, A. (2020). Environmental sustainability of biofuels: A review. *Proceedings of the Royal Society A*, 476 (2243). https://doi.org/10.1098/rspa.2020.0351

Jong, S., Antonissen, K., Hoefnagels, R., Lonza, L., Wang, M., Faaij, A., & Junginger, M. (2017). Life-cycle analysis of greenhouse gas emissions from renewable jet fuel production. *Biotechnology for Biofuels,* 10(64). https://doi.org/10.1186/s13068-017-0739-7

Kana, E. G., Oloke, J. K., Lateef, A., & Adesiyan, M. O. (2012). Modeling and optimization of biogas production on saw dust and other co-substrates using artificial neural network and genetic algorithm. *Renewable Energy*, 46, 276–281. https://doi.org/10.1016/j.renene.2012.03.027

Kessler, T., Sacia, E. R., Bell, A. T., & Mack, J. H. (2017) Artificial neural network based predictions of cetane number for furanic biofuel additives. *Fuel*, 206, 171–179. https://doi.org/10.1016/j.fuel.2017.06.015

Khatri, N., & Khatri, K. K. (2022). Artificial Intelligence for modeling and optimization of the biogas production. *Artificial Intelligence for Renewable Energy Systems*. https://doi.org/10.1002/9781119761686.ch4

Kwon, H., Shiu, J., Rivera, E. C., & Yamakaya, C. (2024). Soft sensor for ethanol fermentation monitoring through data-driven modeling and synthetic data generation. *Proceedings*, 104(1), 30. https://doi.org/10.3390/proceedings2024104030

Kyriklidis, C., Koutouvou, A., Moustakas, K., Karayannis, V., & Tsanaktsidis, C. (2025). Artificial Intelligence and nature-inspired techniques on optimal biodiesel production: A review—recent trends. *Energies*, *18*(4), 768. https://doi.org/10.3390/en18040768

Li, Y., Lu, M., Campos, L. C., & Hu, Y. (2024). Predicting biogas yield after microwave pretreatment using artificial neural network models: performance evaluation and method comparison. ACS ES&T Engineering, 4 (10), 2435–2448. https://doi.org/10.1021/acsestengg.4c00276

Liao, M., & Yao, Y. (2021). Applications of artificial intelligence-based modeling for bioenergy systems: A review. *GCB Bioenergy*, 13(5). https://doi.org/10.1111/gcbb.12816

Liu, W. -J., & Yu, H. -Q. (2021). Thermochemical conversion of lignocellulosic biomass into mass-producible fuels: Emerging technology progress and environmental sustainability evaluation. *ACS Environmental Au*, 2(2). https://doi.org/10.1021/acsenvironau.1c00025

Long, F., Wang, L., Cai, W., Lesnik, K., & Liu, H. (2021). Predicting the performance of anaerobic digestion using machine learning algorithms and genomic data. *Water Research*, 199, 117182. https://doi.org/10.1016/j.watres.2021.117182

Lopez, P. C., Feldman, H., Mauricio-Iglesias, M., Junicke, H., Huusom, J. K., & Gernaey, K. V. (2019). Benchmarking real-time monitoring strategies for ethanol production from lignocellulosic biomass. *Biomass and Bioenergy, 127*, 105296. https://doi.org/10.1016/j.biombioe.2019.105296

Lugani, Y., Rai, R., Prabhu, A. A., Maan, P., Hans, M., Kumar, V., Kumar, S., Chandel, A. K., & Sengar, R. S. (2020) Recent advances in bioethanol production from lignocelluloses: A comprehensive review with a focus on enzyme engineering and designer biocatalysts. *Biofuel Research Journal*, 28, 1267–1295.

Luo, H., Gao, L., Liu, Z., Shi, Y., Xie, F., Bilal, M., Yang, R., & Taherzadeh, M. (2021) Prediction of phenolic compounds and glucose content from dilute inorganic acid pretreatment of lignocellulosic biomass using artificial neural network modeling. *Bioresources and Bioprocessing*, 8, 134. https://doi.org/10.1186/s40643-021-00488-x.

Malode, S. J., Prabhu, K. K., Mascarenhas, R. J., Shetti, N. P., & Aminabhavi, T. M. (2021). Recent advances and viability in biofuel production. *Energy Conversion and Management: X*, 10, 100070. https://doi.org/10.1016/j.ecmx.2020.100070

Mansy, A. E., Daniel, S., Monguen, C. K., Wang, H., Osman, A. I., & Tian, Z.-Y. (2025). Catalytic production of aviation jet biofuels from biomass: A review. *Environmental Chemistry Letters*. 23, 419–461. https://doi.org/10.1007/s10311-024-01806-3

Moradi, G. R., Dehghani, S., Khosravian, F., & Arjmandzadeh, A. (2013). The optimized operational conditions for biodiesel production from soybean oil and application of artificial neural networks for estimation of the biodiesel yield. *Renewable Energy*, 50, 915–920. https://doi.org/10.1016/j.renene.2012.08.070

Moreira, R. F., Almeida, O., Machado, N. T., Rodrigues, E. M., Neto, A. M., Cordeiro, M. A., Chagas, B. R., Tenoro, E. L., Teixeira, M. L., & Silva, P. R. V. (2014). Produção de bioetanol a partir da hidrólise enzimática do bagaço de cana-de-açúcar. *XX Congresso Brasileiro de Engenharia Química*, 1(2).

Nguyen, V. T., Ta, Q. T., & Nguyen, P. K. (2022). Artificial intelligence-based modeling and optimisation of microbial electrolysis cell-assisted anaerobic digestion fed with alkaline-pretreated waste-activated sludge. *Biochemical Engineering Journal*, 187, 108670. https://doi.org/10.1016/j.bej.2022.108670

Olabi, A. G., Nassef, A. M., Rodriguez, C., Abdelkareem, M. A., & Rezk, H. (2020) Application of artificial intelligence to maximize methane production from waste paper. *International Journal Energy Research*, 44(12), 9598–9608. https://doi.org/10.1002/er.5446

Osman, A. I., Nasr, M., Farghali, M., Rashwan, A. K., Abdelkader, A., Al-Muhtaseb, A. H., Ihara, I., & Rooney, D. W. (2024). Optimizing biodiesel production from waste with computational chemistry, machine learning and policy insights: A review. *Environmental Chemistry Letters*, 22, 1005–1071. https://doi.org/10.1007/s10311-024-01700-y

Owusu, W. A., & Marfo, S. A. (2023). Artificial intelligence application in bioethanol production. *International Journal of Energy Research*. https://doi.org/10.1155/2023/7844835.

Pasha, M. K., Dai, L., Liu, D., Guo, M., & Du, W. (2021). An overview to process design, simulation and sustainability evaluation of biodiesel production. *Biotechnology for Biofuels*, 14(1), 129.

Pereira, R. D., Badino, A. C., & Cruz, A. J. (2020). Framework based on artificial intelligence to increase industrial bioethanol production. *Energy and Fuels*, 34(4), 4670–4677.

Pessuto, J., Scopel, B. S., Perondi, D., Godinho, M., & Dettmer, A. (2016). Enhancement of biogas and methane production by anaerobic digestion of swine manure with addition of microorganisms isolated from sewage sludge. *Process Safety and Environmental Protection*, 104(A), 233–239. https://doi.org/10.1016/j.psep.2016.08.020

Qi, B., Chen, X., Shen, F., Su, Y., & Wan, Y. (2009). Optimization of enzymatic hydrolysis of wheat straw pretreated by alkaline peroxide using Response Surface Methodology. *Industrial & Engineering Chemistry Research*, 48(15), 7346–7353. https://doi.org/10.1021/ie8016863.

Raja, I., & Wazir, S. (2017). Biogas production: The fundamental processes. *Universal Journal of Engineering Science*, 5, 29–37. https://doi.org/10.13189/ujes.2017.050202

Rajendran, K., Kankanala, H. R., Lundin, M., & Taherzadeh, M. J. (2014). A novel process simulation model (PSM) for anaerobic digestion using Aspen Plus. *Bioresource Technology*, 168, 7–13. https://doi.org/10.1016/j.biortech.2014.01.051

Rio-Chanona, E. A., Wagner, J. L., Ali, H., Fiorelli, F., Zhang, D., & Hellgardt, K. (2018). Deep learning-based surrogate modeling and optimization for microalgal biofuel production and photobioreactor design. *American Institute of Chemical Engineers*, 65(3), 915–923. https://doi.org/10.1002/aic.16473

Rutland, H., You, J., Liu, H., Bull, L., & Reynolds, D. A. (2023). A systematic review of machine-learning solutions in anaerobic digestion. *Bioengineering*, 10, 1410. https://doi.org/10.3390/bioengineering10121410.

Salam, S., Choudhary, T., Pugazhendhi, A., Verma, T. N. & Sharma, A. (2020). A review on recent progress in computational and empirical studies of compression ignition internal combustion engine. *Fuel*, 279, 118469. https://doi.org/10.1016/j.fuel.2020.118469

Saraiva, A. M., Osório, F. S., Colaço, A. F., Drucker, D. P., Mendiondo, E. M., Corrêa, F. E., Soares, F. M., Molin, J. P., Benso, M. R., Marques, P. A. A., Silva, R. F., Miranda, S. H. G., Costa, W. F., & Delbem, A. C. B. (2024). A inteligência artificial na pesquisa agrícola. *Revista USP*, 141, 91–106. https://doi.org/10.11606/issn.2316-9036.i141p91-106

Shahsavar, M. M., Akrami, M., Gheibi, M., Kavianpour, B., Fathollahi-Fard, A. M., & Behzadian, K. (2021). Constructing a smart framework for supplying the biogas energy in green buildings using an integration of response surface methodology, artificial intelligence and petri net modelling. *Energy Conversion and Management*, 248, 114794. https://doi.org/10.1016/j.enconman.2021.114794

Singh, D., Sharma, D., Soni, S. L., Sharma, S., & Kumari, D. (2019). Chemical compositions, properties, and standards for different generation biodiesels: A review. *Fuel*, 253, 60–71. https://doi.org/10.1016/j.fuel.2019.04.174

Sinha, S., & Lee, Y. M. (2024). Challenges with developing and deploying AI models and applications in industrial systems. *Discover Artificial Intelligence* 4, 55. https://doi.org/10.1007/s44163-024-00151-2

Sivamani, S., & Rajoo, B. (2018). Process design and optimization of bioethanol production from cassava bagasse using statistical design and genetic algorithm. *Preparative Biochemistry & Biotechnology*, 48(9), 834–841. https://doi.org/10.1080/10826068.2018.1514512.

Smuga-Kogut, M., Kogut, T., Markiewicz, R., & Słowik, A. (2021). Use of machine learning methods for predicting amount of bioethanol obtained from lignocellulosic biomass with the use of ionic liquids for pre-treatment. *Energies*, 14, 243. https://doi.org/10.3390/en14010243.

Undavalli, V., Olatunde, O. B., Boylu, R., Wei, C., Haeker, J., Hamilton, J., & Khandelwal, B. (2023). Recent advancements in sustainable aviation fuels. *Progress in Aerospace Sciences*, 136, 100876. https://doi.org/10.1016/j.paerosci.2022.100876

Vasco-Correa, J., Ge, X., & Li, Y. (2016). Biological pretreatment of lignocellulosic biomass. In *Biomass Fractionation Technologies for a Lignocellulosic Feedstock Based Biorefinery* (pp. 561–585). Elsevier.

Verrengia, M. M., Vargas, R., & Vecchietti, A. (2020). General modeling of the enzymatic hydrolysis process of lignocellulosic materials using approaches based on Artificial Neural Networks. In the *39th International Conference of the Chilean Computer Science Society* (pp. 1–6). https://doi.org/10.1109/SCCC51225.2020.9281171.

Voeten, R. L., Hendriks, F., & Bezemer, G. L. (2024). Fischer–Tropsch Synthesis for the Production of Sustainable Aviation Fuel: Formation of Tertiary Amines from Ammonia Contaminants. *ACS Omega*, 9(29), 31974-31985. https://doi.org/10.1021/acsomega.4c03734

Wang, M., & Yu, X. (2024). An overview of the sustainable aviation fuel: LCA, TEA, and the sustainability analysis. *Engineering Proceedings,* 80(1), 3. https://doi.org/10.3390/engproc2024080003

Wang, R., He, Z., Chen, H., Guo, S., Zhang, S., Wang, K., Wang, M., & Ho, S.-H. (2024). Enhancing biomass conversion to bioenergy with machine learning: Gains and problems. *Science of The Total Environment,* 927, 172310. https://doi.org/10.1016/j.scitotenv.2024.172310

Wang, Z., Peng, X., Xia, A., Shah, A. A., Huang, Y., Zhu, X., Zhu, X., & Liao, Q. (2022). The role of machine learning to boost the bioenergy and biofuels conversion. *Bioresource Technology*, 343, 126099. https://doi.org/10.1016/j.biortech.2021.126099

Wong, K. I., & Wong, P. K. (2017). Optimal calibration of variable biofuel blend dual-injection engines using sparse Bayesian extreme learning machine and metaheuristic optimization. *Energy Conversion and Management*, 148, 1170–1178. https://doi.org/10.1016/j.enconman.2017.06.061

Wong, K. I., Wong, P. K., Cheung, C. S., & Vong, C. M. (2013). Modeling and optimization of biodiesel engine performance using advanced machine learning methods. *Energy*, 55, 519–528. https://doi.org/10.1016/j.energy.2013.03.057

Xia, J., Yan, G., Abed, A. M., Nag, K., Galal, A. M., Deifalla, A., & Li, J. (2023). Machine learning approach to predict the biofuel production via biomass gasification and natural gas integrating to develop a low-carbon and environmental-friendly design: Thermodynamic-conceptual assessment. *Chemosphere*, 336, 138985. https://doi.org/10.1016/j.chemosphere.2023.138985

Xing, Y., Zheng, Z., Sun, Y., & Alikhani, M. A. (2021). A review on machine learning application in biodiesel production studies. International Journal of Chemical Engineering, (3), 1-12. https://doi.org/10.1155/2021/2154258

Zaher, U., Pandey, P., & Chen, S. (2009). A simple elemental continuity based model application to study the anaerobic microbial activity for the treatment of dairy manure. *Applied Mathematical Modelling*, 33(9), 3553–3564. https://doi.org/10.1016/j.apm.2008.11.019

Zhang, F., Wang, J., Liu, S., Zhang, S. & Sutherland, J. W. (2017). Integrating GIS with optimization method for a biofuel feedstock supply chain. *Biomass and Bioenergy*, 98, 194–205. https://doi.org/10.1016/j.biombioe.2017.01.004

Zhang, W., Wang, X., Xing, W., Li, R., & Yang, T. (2021). Responses of anaerobic digestion of food waste to coupling effects of inoculum origins, organic loads, and pH control under high load: Process performance and microbial characteristics. *Journal of Environmental Management*, 279, 111772. https://doi.org/10.1016/j.jenvman.2020.111772.

Zhang, Y., Li, L., Ren, Z., Yu, Y., Li, Y., Pan, J., Lu, Y., Feng, L., Zhang, W., & Han, Y. (2022). Plant-scale biogas production prediction based on multiple hybrid machine learning technique. *Bioresource Technology*, 363, 127899. https://doi.org/10.1016/j.biortech.2022.127899

13 AI-Driven Optimization Strategies for Enhanced Biobutanol Production

Aditi Dey, Soumya Pandit, and Elvis Fosso-Kankeu

13.1 INTRODUCTION

13.1.1 Biobutanol: An Overview

Biobutanol is a four-carbon alcohol, specifically known as n-butanol or butyl alcohol (Segovia-Hernández et al. 2022). It has a chemical formula $C_4H_{10}O$ (chemical structure is shown in Figure 13.1). The fuel is obtained through fermentation of biomass from feedstock, which could be agricultural residues and sugars from organic materials. So, biobutanol itself is unique because it is one of the potential substitutes for fossil fuels and bioethanol as a renewable fuel (Amiri and Karimi 2019). It is mainly produced by fermentation of carbohydrates by specific microorganisms, such as *Clostridium acetobutylicum*, by a process called Acetone-Butanol-Ethanol (ABE) fermentation. This process produces biobutanol together with acetone and ethanol by-products. Biobutanol has several advantageous characteristics. It has a greater energy density, for instance, than ethanol and thus behaves more like gasoline in internal combustion engines (Ndaba, Chiyanzu, and Marx 2015). In addition, its vapor pressure is lower, and it can also be mixed with gasoline in larger amounts without any major changes to current engines. But there are challenges as well. For example, at elevated levels, its production balance is lower and toxic to the fermenting microorganisms.

13.1.2 Role of Artificial Intelligence

Most important is the fact that it makes fermentation processes more efficient and sustainable in the production of modern biobutanol. Utilizing artificial neural networks (ANNs) through machine learning algorithms, researchers were able to optimize parameters such as temperature, pH, and even the composition of the substrate with which they achieve very high yields of butanol. These AI-based models are metamodels, which analyze the complex interactions within the fermentation system without making any assumptions and developing prior knowledge about the underlying processes involved. It can predict outcomes with high accuracy based on historical data. The possibility of new fermentation techniques by AI can also minimize

DOI: 10.1201/9781003571223-13

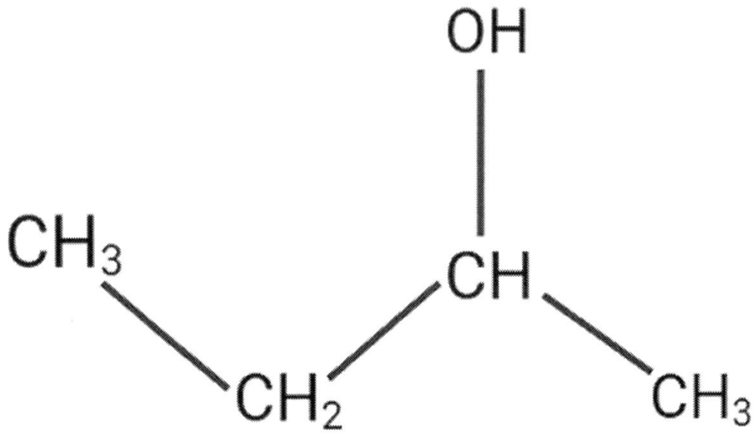

FIGURE 13.1 Structure of butanol.

carbon loss during production and hence reduce greenhouse gas emissions, while improving the carbon conversion efficiency. By integrating AI in the production of biobutanol, industries would obtain greater yields, lower costs, and a reduced footprint on the environment, thereby making this biofuel a more competitive alternative to traditional fossil fuels and leading to a greener energy future.

13.1.3 Relevance as a Renewable Energy Source

Due to the several benefits of biobutanol over conventional fossil fuels and other biofuels such as ethanol, biobutanol is getting more attention as an important renewable fuel (Jin et al. 2011). Such renewable fuel can be produced from various biomass feedstocks, reducing reliance on limited fossil fuel reserves and, consequently, enhancing energy security and sustainability. It has a higher energy density compared to ethanol, which provides more power for the same volume of fuel, and its energy density is only 10–20% lower than that of gasoline, which makes it a better choice for internal combustion engines (Timung, Singh, and Annu 2021). Biobutanol can be blended with gasoline without requiring major changes to existing fuel distribution systems or vehicle engines, making it easier to fit into existing infrastructure and encouraging its use as a mainstream fuel (Panahi et al. 2023). Moreover, the use of biobutanol reduces greenhouse gas emissions because plants used for its production absorb carbon dioxide during growth, which offsets emissions when the fuel is burned. Its less corrosive character than ethanol makes it less hazardous to groundwater contamination in the event of spills and allows it to be transported via existing pipelines without any concerns for material compatibility (Segovia-Hernandez, Behera, and Sanchez-Ramirez 2022). The production of biobutanol can also generate employment and boost local economies by using locally sourced feedstocks, while holding promise for use as a feedstock for chemicals and solvents besides fuel. Summarily put, the relevance of biobutanol to the pursuit of sustainable energy alternatives is important due to environmental reasons, compatibility with the current system, and as a potential driver of economic benefit (Rathour et al. 2018).

13.2 THE DEVELOPMENT OF BUTANOL PRODUCTION TECHNIQUES

There has been renewed interest in biobutanol production lately, due to concerns of sustainability. Modern developments in this regard include the use of lignocellulosic biomass, which requires pre-treatment and hydrolysis to generate fermentable sugars (Oloyede et al. 2024). Innovations in the development, such as SSFR (Simultaneous Saccharification and Fermentation), cell recycling, and continuous fermentation systems, have improved yields and reduced toxicity in production. Other methods are also being investigated to increase butanol production efficiency and lessen reliance on food-based feedstocks, such as gas fermentation from syngas and metabolic engineering techniques (Rathour et al. 2018).

13.2.1 Transition from Fossil Fuels to Sustainable Bio-Based Solutions

The shift from fossil fuels to biobased approaches, especially regarding the production of biobutanol, will be a considerable step toward producing sustainable energy (Ab Rasid et al. 2021). Biobutanol is increasingly being highlighted as a significant alternative to the traditional fossil-based fuels due to its desirable features and environmentally friendly benefits (Jin et al. 2011). Major recent advances in biobutanol production targeted the improvement in fermentation processes as well as techniques for separation. For example, scientists have synthesized novel MOFs that can effectively extract biobutanol from fermentation broths, which is one of the major challenges in its commercial viability (Kushwaha et al. 2022). This innovation may open the way for large-scale production, thereby reducing dependence on fossil fuels and contributing to energy security. Furthermore, biobutanol is more energy-dense than ethanol, making it directly compatible with conventional gasoline engines, without the need for engine modification (Jin et al. 2011). It has lower emissions than ethanol; hence, it is quite attractive in cutting down greenhouse gas output. Another support for this transition is in the exploration of diverse feedstocks, including second- and third-generation biomass, thus avoiding competition with food resources and enhancing sustainability (Guo et al. 2022). As research continues to improve production methods and reduce costs, biobutanol stands poised to play a crucial role in the shift toward renewable energy sources, helping to combat climate change while meeting global energy demands.

13.2.2 Current Developments in the Study of Biobutanol

The current trends in biobutanol research include wide-ranging production technology developments, growth in the marketplace, and growing utilization areas. Innovations in fermentation technology are also at the forefront of the process, improving processes to enhance yield and product purity (Ab Rasid et al. 2021). Of course, these involve the strain engineering efforts that would lead to the design of more productive microbial strains that can tolerate high concentrations of butanol to boost overall productivity. There is increasing interest in utilizing diverse biomass feedstocks, which include agricultural residues and lignocellulosic materials (Amiri and Karimi 2019). It reduces competition for food resources but also promotes sustainability in production methods. The global biobutanol market is expected to expand monumentally, which is projected at USD 2.3 billion by 2033. This reflects a compound annual growth rate (CAGR) of about 4.56% from 2025 to 2033. Increasing demand for renewable energy sources, combined with the beneficial government policies supporting bio-based products, will drive this growth. Besides that, biobutanol is increasingly recognized as a versatile molecule beyond fuel applications. Today, it is an intermediate chemical product, for instance, in manufacturing butyl acrylates to be used as paints, coatings, and adhesives. With increasing consciousness about greenhouse gases, a smaller carbon footprint in the case of biobutanol compared to fossil fuels also makes it very appealing. Many governments implement pro-support policies toward the consumption of biofuels such as biobutanol, creating inducements that encourage the continuation of R&D efforts into this end (Jin et al. 2011). In general, high-demand industries driving the adoption rate in the Asia-Pacific region include paints and coatings, the textile sectors, and especially the automotive industry. The current trend in biobutanol research reflects a dynamic interplay of technological advancement, expansion of the market, diverse application, and supporting policies to encourage sustainable energy solutions.

13.2.3 Methods of Production

13.2.3.1 An Overview of the Acetone, Butanol, and Ethanol (ABE) Process

The Acetone-Butanol-Ethanol (ABE) fermentation method, first patented by Chaim Weizmann, is highly important in generating solvents from the conversion of starch, glucose, and other carbohydrates (Koller 2023). This type of anaerobic fermentation utilizes unique strains of bacteria belonging to Clostridium, where mostly *Clostridium acetobutylicum* is applied since it converts the sugars into an acetone and butanol product with a mean ratio of about 3:6:1 ethanol. The ABE fermentation has two unique phases: the initial acidogenesis, which is where sugar is converted into acidic metabolites such as acetate and butyrate, followed by the solventogenesis phase, which is where these acids get re-assimilated into the desired solvents (Guan, 2016). This two-phase nature of ABE fermentation is essential for optimizing product yields, but it presents problems, too, like product inhibition due to a toxic build-up of high levels of butanol. Among all fermentation methods, ABE fermentation stands out from others for its special product profile and metabolic pathways. Although the product of yeast fermentation is primarily ethanol, with high demand in alcoholic beverages and biofuel, ABE fermentation has a wide variety of solvents of economic importance. Furthermore, compared to ethanol fermentation, which is mainly aerobic in nature, ABE uses strict anaerobes, thus requiring particular environmental conditions for growth (Rathour et al. 2018). The capability of *Clostridium* species to live in an environment with low pH levels during the acidogenic phase makes the conversion of substrates efficient, though a switch from acidogenesis to solvent production requires careful management of fermentation conditions to avert inhibitory effects of products that may have accumulated in the process (Kushwaha et al. 2022). Overall, ABE fermentation represents a promising avenue for sustainable biobased chemical production with ongoing research focused on enhancing its efficiency and economic viability.

13.2.3.2 Fermentation of Syngas

Syngas, more formally known as synthesis gas, is a mixture that primarily contains hydrogen (H_2), carbon monoxide (CO), and carbon dioxide (CO_2) produced from the gasification of biomass and other fossil fuels (Awang et al. 1988). The conversion of syngas to biobutanol can be carried out through a bio-process called syngas fermentation, in which specific microorganisms, known to be acetogenic bacteria, break down the gaseous substrates into the formation of butanol and other alcohols (Ab Rasid et al. 2021). This fermentation process usually operates under mild conditions, which is a benefit compared to thermochemical methods such as Fischer-Tropsch synthesis, which require high temperatures and pressures. The advantages of syngas fermentation include its ability to use a wide variety of feedstocks, including waste materials and industrial off-gases, thereby making it sustainable and environmentally friendly (Procentese 2015a). The process can also achieve high conversion efficiencies and is less dependent on the chemical composition of the biomass feedstock. However, such challenges still face the system. These include low alcohol productivity and mass transfer limitations of the gas-liquid interface, which impede overall efficiency. Syngas also comprises impurities that inhibit microbial activity, thus meaning that effective cleaning processes have to be carried out before fermentation occurs (Guo et al., 2022). Overcoming such challenges is crucial for improving economic viability and scale-up for syngas-fed biobutanol production.

13.2.3.3 Catalysis by Chemicals

Among the various processes in the conversion of biomass into chemicals and fuels of value, processes like Fischer-Tropsch synthesis (FTS) for thermochemical production of biobutanol are at the forefront. The FTS process is a catalytic conversion of syngas produced through the gasification of biomass, synthesizing liquid hydrocarbons: a mixture of carbon monoxide and hydrogen (Prakash K. Sarangi and Nanda 2018). The great advantage of this method lies in the range of hydrocarbons produced during carbon monoxide hydrogenation. Some of them are butanol. The interest of Fischer-Tropsch synthesis, therefore, relates to producing clean fuels based on renewable sources as a remedy to the adverse impacts of using fossil fuels. Feedstock flexibility is another prominent advantage offered by FTS: it uses diverse carbonaceous material sources that fit into biomass-to-liquid processes. Products formed from FTS also offer the advantages of very low levels of sulfur and nitrogen species that normally find their way into petroleum-derived fuel products. Nonetheless, major setbacks remain in costs, being the expensive technology; it has challenges associated with poor catalyst efficiencies toward high selectivity in maximizing butanol formation without a plethora of byproduct compounds (Outram 2018). Despite these challenges, research in catalyst development and process optimization continues to improve the feasibility of thermochemical routes for biobutanol production, which makes them a promising alternative in the quest for sustainable energy solutions.

13.3 INGREDIENTS USED FOR BIOBUTANOL PRODUCTION

13.3.1 The Lignocellulosic Material

Lignocellulosic biomass has many advantages as a feedstock for biobutanol production due to its abundance and sustainability. It is primarily composed of cellulose, hemicellulose, and lignin, which are three components that make up this biomass. Cellulose is a major polysaccharide, while hemicellulose is a binding agent and lignin provides rigidity. Lignocellulosic biomass is recalcitrant to degradation due to its complex structure (Guo et al. 2022). Therefore, it requires efficient pretreatment processes for its effective conversion into fermentable sugars. Different pretreatment methods can be applied, including physical (e.g., steam explosion), chemical (e.g., acid or alkaline hydrolysis), and biological treatments. Physical pretreatments enhance surface area, while chemical methods break down hemicellulose and alter the structure of lignin (Hanh 2009). Biological pretreatment utilizes microorganisms for the degradation of biomass under mild conditions. These pretreatment strategies are quite important in hydrolysis, which helps maximize the yield during fermentation.

13.3.2 Feedstocks Based on Starch

Starch feedstocks, among them crops of corn and sugar beets, are the preferred feedstocks that have been considered for biobutanol manufacture due to their high carbohydrate nature and availability in abundance (Yadav et al. 2022). As a primary crop in the United States, corn starch is highly easily convertible into fermentable sugars, and so is with sugar beets as a readily fermentable direct source of sucrose. These starches would then be hydrolyzed into sugars using enzymatic hydrolysis, which breaks down the more complicated carbohydrates to simpler sugars, primarily glucose (Prakash K. Sarangi and Nanda 2018). The process generally employs enzymes such as amylases and glucoamylases, catalyzing the hydrolysis of starch granules to dextrins and eventually into glucose(Jin et al. 2011). Subsequently, the produced sugars would be fermented by microorganisms like Clostridium acetobutylicum to produce biobutanol. Enzymatic hydrolysis has been the preferred choice for specificity and efficiency, with reduced by-products and improved overall yield in the production of biobutanol from starch-based feedstocks(Guan 2016).

13.4 MACHINE LEARNING MODELS

Machine learning models are increasingly applied in the production of biobutanol to improve the fermentation process. Several are artificial neural networks, support vector machines, regression models, among others, having strengths in dealing with larger datasets (Kurani et al. 2023). Some of the applications, for instance, include general regression neural networks (GRNN) along with long short-term memory networks (LSTM), which would predict the result of fermentation; they have proved to capture input parameters like temperatures, pH levels, and substrates' nonlinear relationships (Nalepa and Kawulok 2019). Hybrid approaches, including both mechanistic modeling and data-driven techniques, are found to generate more robust predictions and optimizations. These applications of machine learning have enabled the fermentation conditions to be monitored and controlled in real-time, improving the yields and resource savings. This model is even more than a simple production process simplification—it promotes a means of manufacturing biobutanol that is a bit more sustainable (Cervantes et al. 2020).

13.4.1 Data Analytics

Historical data analysis and predictive analytics can both be used to get a process edge through improved process enhancement. By using past fermentation data, researchers are able to note trends or direct relationships between various parameters such as substrate composition, fermentation conditions, and microbial performance, among others, which may help optimize these factors for better efficiency and productivity (Khaleghi et al. 2021). This would be added by predictive analytics that use machine learning models to predict fermentation results by the values of input variables. There can be various simulation scenarios in which optimal conditions would lead to maximum yields of biobutanol while reducing costs and resource utilization (Zhu et al. 2020). Data analytics integration thus facilitates the use of underutilized substrates besides streamlining the production processes for sustainable practices, as it reduces impacts on the environment associated with conventional biofuel production methods.

13.4.2 Acidogenic and Solventogenic Phases

The acidogenic phase and the solventogenic phase are two separate stages into which the fermentation process for producing biobutanol is divided (Jin et al. 2011). In the acidogenic phase, a microorganism like *Clostridium acetobutylicum* metabolizes sugars derived from feedstocks like starch or lignocellulosic biomass to produce organic acids, mainly in the form of acetic and butyric acids, along with hydrogen and carbon dioxide, characterized by rapid bacterial growth and the accumulation of these acids, which serve as precursors for solvent production (Xue et al. 2019). The accumulation of butyric acid plays a crucial role in signaling the transition to the next phase, as it helps maintain a favorable pH environment necessary for the solventogenic phase. When the bacteria shift from an acid production mode into a solvent production mode as fermentation continues to enter the solventogenic phase, butanol, acetone, and ethanol are produced (Serrano, Fajardo, and Castro 2024). Often, the initiation of this phase is linked with the fall in pH and a rise in the concentration of organic acids that could be toxic at certain concentrations to the cells. The phase is characterized by the use of reserve energy for the synthesis of solvents, as it continues to be hampered by solvent toxicity, specifically butanol. After this, fermentation runs its course either when nutrients are depleted or when the cells die or sporulate due to toxic action from accumulating products. Optimizing conditions throughout the phases will enhance the yield and productivity of biobutanol (Panahi et al. 2023).

13.4.3 Fermentation Efficiency—Influencing Variables

The efficiency of fermentation in the production of biobutanol is significantly influenced by critical factors such as pH, temperature, and nutrient availability, all of which are involved in the metabolic activity of the microorganisms used. pH is an important parameter that affects the growth and metabolic pathways of *Clostridium acetobutylicum* and other fermentative bacteria. In the acidogenic phase, a slightly acidic environment (pH 5 to 6) favors the growth of bacteria and organic acid production (Timung, Singh, and Annu 2021). However, with advancement in the fermentation phase toward the solventogenic phase, lower pH inhibits solvent production and brings stress to cells as a result of acid accumulation. Therefore, there is an imperative need for optimal pH in both phases of fermentation for achieving higher yields of butanol. Temperature also greatly influences fermentation efficiency. Most *Clostridium* species prefer mesophilic temperatures within the range of 30°C and 37°C (Kumar and Gayen 2011). Temperature shifts away from these levels are undesirable since they cause poor growth rates and metabolism rates for the microbial. High temperature accelerates kinetic reactions but makes it vulnerable to thermal inhibition as well as degradation of important fermentative enzymes; conversely, low temperature has the potential to slow down metabolism, hence yielding lower product quantities (Yadav et al. 2022). The other critical factor influencing the efficiency of fermentation is the availability of nutrients. Essential nutrients such as nitrogen sources, for example, ammonium salts or amino acids, vitamins, and trace minerals are necessary for the optimal growth and metabolism of microorganisms. Poor levels of nutrients can lead to slow fermentation rates and lower yields of biobutanol (Guo et al. 2022). The balance of carbon sources also influences the metabolic pathways; for example, high sugar concentration promotes rapid acid production but may lead to product inhibition if not controlled (Awang et al. 1988). In summary, optimizing pH, temperature, and nutrient availability is essential for enhancing fermentation efficiency and achieving higher yields of biobutanol in industrial applications.

13.4.4 Product Inhibition and Mitigation Technique

One of the main problems of biobutanol production is product inhibition (Amiri and Karimi 2019). The major source of product inhibition in this context is the buildup of butanol and other fermentation metabolites that interfere with microbial activity and thereby lower overall yields. The level of butanol builds up in fermentation to concentrations that are inhibitory to the growth and metabolic activity of *Clostridium acetobutylicum* and other solventogenic bacteria as fermentation progresses (Ibrahim et al. 2017). This toxicity not only limits the maximum achievable butanol concentration but also leads to a decline in cell viability, finally stopping the fermentation process. Various strategies have been developed to combat product inhibition. One effective strategy involves optimizing the fermentation conditions to enhance microbial resistance against solvent toxicity (Ashokkumar et al. 2022). This may include optimizing the pH and temperature for fermentation. Also, continuous fermentation processes combined with cell recycling can be used to keep high biomass concentrations in the reactor, which could improve productivity with minimal inhibitory effects of products accumulated. Additionally, techniques like gas stripping and pervaporation may be used in real-time butanol removal from the fermentation broth, thereby decreasing its concentration and toxicity. More on metabolic engineering, this time by improving the microbial strains to become more tolerant of butanol, thus increasing metabolic efficiency (Baral and Shah 2014). In total, all these strategies help to improve biobutanol production efficiency and overcome the hurdles of product inhibition.

13.4.5 AI-Driven Optimization Technique

Techniques such as Response Surface Methodology (RSM) and Design of Experiments (DOE) have played very crucial roles in optimizing the biobutanol production processes by computer-driven optimization. RSM is a statistical technique aimed at exploring multiple variables and their impact on various desired outcomes by defining optimal fermentation conditions. RSM, by response surface modeling, reveals how the alteration of parameters such as temperature, pH, and substrate concentration has an impact on the biobutanol yields (Veza et al. 2023). In contrast, DOE provides an exploratory approach in conducting experiments in an ordered way by systematically varying input factors to assess their impact on production outcomes. In doing this, fewer experiments are needed than in the factorial method, and the results become more reliable, because interactions among the variables are properly assessed (Asrul et al. 2022). Together, these AI-driven techniques contribute to the establishment of strong fermentation processes that have maximum yield, minimal resource utilization, and least environmental degradation; hence, biobutanol production will become more sustainable.

13.4.6 AI's Role in Improving Cell Recycling

Optimizing the various processes involved in recycling and separating cells related to the biobutanol production process is being carried out with considerable intensity by the utilization of AI. Machine learning algorithms are being used by researchers for the development of predictive models in order to optimize conditions related to cell retention and product recovery during fermentation. For example, AI may determine the most favorable parameters under which high cell density can be achieved with a minimal amount of cell washout that is key to maximizing productivity (Veza et al. 2023). Even ultrafiltration membranes can be enhanced with artificial intelligence monitoring systems that predict fouling on membranes to optimize cleaning cycles to enhance the operational lifetime while lowering maintenance costs. Additionally, it will be possible to design integrated systems where fermentation will be combined with product separation in real-time by integrating the process and making continuous operation possible and increasing yield. Such integration is not only productive but also eco-friendly because, in comparison to the traditional separation methods, the waste and utilization of energy are reduced. Overall, AI-driven optimization of cell recycling and separation processes contributes considerably to making biobutanol production more efficient and sustainable (Asrul et al. 2022).

13.5 BARRIERS TO THE PRODUCTION OF BIOBUTANOL

Many challenges in biobutanol production can prevent its large-scale adoption and economic viability. Chief among these is the low yield of butanol during the fermentation process, especially when produced using traditional methods such as ABE fermentation (Procentese 2015b). High concentrations of butanol can be toxic to the fermenting microorganisms, thus inhibiting growth and metabolic activity and thereby lowering total productivity. Therefore, tolerant microbial strains must be developed, and optimized fermentation conditions are found to improve yields (Prakash K. Sarangi and Nanda 2018). The economic viability of the process is still a challenge because the cost of production of biobutanol is relatively higher compared to fossil fuels and other biomass-derived biofuels, such as ethanol. The competition for starch-based feedstocks, such as crops like corn and sugar beets, brings ethical concerns concerning food availability and prices (Rathour et al. 2018). Fuel produced from these foodstuffs could exacerbate the world's food deficit and also push up the price of commodities even further. Furthermore, technology issues are faced when trying to increase production levels to industrial standards. Separation and purification processes of biobutanol have not yet reached efficiency levels, as existing methods can be expensive and energy-intensive. All these must be addressed in order to improve the viability of biobutanol as a sustainable substitute for conventional fuels.

13.6 SUCCESSFUL APPLICATION OF AI IN BIOBUTANOL

Artificial intelligence has been successfully applied in a wide range of biobutanol production areas, thus making the efficiency and sustainability much better. An example is the optimization of the fermentation process with machine learning analysis of historical data for extracting optimal working conditions. For instance, AI has predicted the performance of different Clostridium strains using under-exploited substrates such as agricultural waste and microalgae because it comes with increased yields and environmental friendliness (Khaleghi et al. 2021). AI-driven system has also employed real-time monitoring and control over fermentation parameters for adaptive responses to changing conditions for reduced wastage of resources. Combining AI with conventional bioprocessing methods such as Response Surface Methodology (RSM) and Design of Experiments (DOE) is another effective use of the technology that allows for a more methodological investigation of the interactions of different fermentation factors (Zhu et. al., 2020). These developments demonstrate how AI might revolutionize biobutanol production optimization, which is currently seen as a more practical and competitive alternative to fossil fuels and which is crucial for tackling urgent environmental problems.

13.7 CURRENT DEVELOPMENTS

The recent developments in biobutanol production have been marked by substantial innovations in metabolic engineering, process integration, and novel fermentation technologies (Morone and Pandey 2014). Approaches in metabolic engineering have involved genetic modification of microorganisms, especially *Clostridium* species, to improve their tolerance against butanol toxicity and optimize the utilization of substrate for increased biobutanol yield (Ab Rasid et al. 2021). These changes allow the strains to produce butanol effectively from a variety of feedstocks, such as lignocellulosic biomass and waste streams. Process integration approaches, for example, SSF and CBP, are being used to facilitate production. The SSF process has the advantage of saccharifying complex carbohydrates and fermenting at the same time, which can reduce the overall time and cost of the product needed for the production of biobutanol. In addition, there are modern reactor designs, such as continuous fermentation systems that utilize immobilized cells, which enhance productivity due to high cell biomass density and increased recovery of solvents (Villacreses-Freire, Ketzer, and Rösch 2022). Collectively, these innovations overcome traditional barriers in the production of biobutanol and make it economically viable and sustainable. With ongoing research, these developments promise to establish biobutanol as a significant component in the renewable energy spectrum.

13.8 SUSTAINABILITY AND THE EFFECT ON THE ENVIRONMENT

The environmental effect and sustainability of biobutanol production have grown in importance as the focus on renewable energy sources has expanded (Segovia-Hernández et al. 2022). Life cycle assessments (LCA) have indicated that biobutanol, especially that produced from biomass like corn or lignocellulosic materials, can result in a drastic reduction in greenhouse gas emissions compared to fossil fuels. For example, according to GREET model-based studies, biobutanol based on corn provides considerable energy savings, with life cycle emissions significantly less when used as fuel for transportation (Amiri and Karimi 2019). The major contribution is through carbon dioxide absorption in plants' life cycles. This has also been critical for energy security since biobutanol offers a renewable alternative that can be integrated into already prevailing fuel infrastructures with minimal modifications. Again, with most countries trying to meet the climate goals that minimize reliance on fossil fuels, biobutanol is seen as an option that not only reduces any environmental impact but also contributes toward sustainable energy systems that enhance both economic and ecological resilience (Wang et al. 2017).

13.9 COST–BENEFIT EVALUATION OF AI APPLICATIONS IN THE MANUFACTURE OF BIOBUTANOL

Cost–benefit analysis of AI implementations in biobutanol production (demand currently in the market is shown in Table 13.1) links AI applications in biobutanol production to better efficiency and economic feasibility. Optimizing AI-driven techniques, including machine learning models and predictive analytics, can drastically mitigate the cost for producers for fermentation processes as well as resource utilization management (Khaleghi et al. 2021). For instance, AI can optimize substrate selection and fermentation conditions, thus achieving higher yields with lower raw material costs, which will account for more than 70% of the production cost. Besides, AI can also be utilized for monitoring and controlling production parameters in such a way that the consumption of energy and waste formation is minimized, hence saving costs. According to the studies, even though the cost of AI technology is pretty high at the initiation stage, it can have greater long-term advantages in terms of efficiency, environment, and product quality. In addition, with these factors, biobutanol can be made to compete closely with fossil fuels. AI may be used to maximize resource availability and minimize ecological footprints, resulting in the production of biobutanol that is both affordable and sustainable.

13.10 FUTURE ASPECTS

Future prospects in biobutanol production appear bright with the help of artificial intelligence (AI) and advanced biotechnological techniques, where the industry is looking for some productive alternatives to replace fossil fuels. The use of underutilized substrates such as agricultural waste and microalgae can present significant

TABLE 13.1
Market Demand for Biobutanol

Market Demand for Biobutanol Production	Key Statistics
Current market size (2023)	USD 1.81 billion
Estimated market size (2024)	USD 1.93 billion
Projected market size (2030)	USD 3.09 billion
CAGR (2024–2030)	7.92%
Market drivers	Increasing demand for cleaner fuels
Key applications	Fuel additives, solvents, bio-plastics
Major end-use industries	Transportation, consumer goods
Regional insights	Asia-Pacific dominates the market
Challenges	High production costs
Opportunities	Enhanced production efficiency through genetic engineering

opportunities for increased production efficiency and environmental sustainability (Sikiru et al. 2024). Machine learning models, predictive analytics, and other optimization techniques that make use of artificial intelligence are predicted to be instrumental in optimizing biobutanol production fermentation, yield, and costs. In addition, improvements in genetic engineering and strain development of *Clostridium* species will likely increase their ability to use a variety of substrates effectively, mitigating some of the current issues related to butanol toxicity and slow growth rates. As research continues to explore innovative pathways for biobutanol synthesis, such as syngas fermentation and catalytic conversion, the integration of AI technologies will further streamline operations, making biobutanol a more economically viable and environmentally friendly biofuel option (Wagh et al. 2024). Global sustainability goals will be greatly aided by the convergence of biotechnology and artificial intelligence in the manufacture of biobutanol, especially in the shift to renewable energy sources.

13.11 CONCLUSION

Optimization of biobutanol through the development of a machine learning model allows big gains in terms of yield increase, cost cutting, and lowering adverse environmental effects. For such industrial-scale usage, this model can determine optimized operating conditions via historical data obtained on fermentation time, substrate concentration, and various processing parameters, predict performance outcomes, and provide real-time assistance (Fan et al. 2021). Scalability as well as robustness are enormous issues, in particular, including robust techniques concerning data preprocessing and feature engineering alongside validations. Advanced methods, such as deep learning and reinforcement learning, can further refine prediction while optimizing the control in processes through use of hybrid models. A well-designed machine learning framework will, in the end, revolutionize biobutanol production through efficiency, waste reduction, and support for sustainable biofuel development. Long-term success in industrial applications will be achieved through continuous model updates, integration with IoT and automation, and collaboration between data scientists and biochemical engineers. Besides optimizing biobutanol production efficiency, machine learning promotes sustainability through the minimum waste of resources and emissions. Optimized conditions for fermentation and processing result in reduced energy input and the elimination of chemical additives, thus further making biobutanol more feasible than fossil fuels (Abdelouahed et al. 2024). On the other hand, combining machine learning with the Internet of Things and automation brings about smart monitoring, self-correcting in real time, in order to gain maximum efficiency at biorefineries. Continuous collaboration with data scientists, biochemical engineers, and industry stakeholders will be key to refining the models and thus achieving long-term effects on the biofuel industry. AI-based process optimization is ultimately poised to be the harbinger of mass-scale adoption of biobutanol as a source of renewable energy in facilitating a global transition to cleaner sources of fuel (Xu and Saleh 2021).

REFERENCES

Ab Rasid, Nurul Suhada, Amnani Shamjuddin, Athran Zuhail Abdul Rahman, and Nor Aishah Saidina Amin. 2021. "Recent Advances in Green Pre-Treatment Methods of Lignocellulosic Biomass for Enhanced Biofuel Production." *Journal of Cleaner Production* 321:129038.

Abdelouahed, Sabri My, Rahmouni Abla, Ennaji Asmae, and Aarab Abdellah. 2024. "Harnessing Feature Engineering to Improve Machine Learning: A Review of Different Data Processing Techniques." In *2024 International Conference on Intelligent Systems and Computer Vision (ISCV)*, 1–6. IEEE. https://ieeexplore.ieee.org/abstract/document/10620105/.

Amiri, Hamid, and Keikhosro Karimi. 2019. "Biobutanol Production." In *Advanced Bioprocessing for Alternative Fuels, Biobased Chemicals, and Bioproducts*, 109–33. Elsevier. https://www.sciencedirect.com/science/article/pii/B9780128179413000061.

Ashokkumar, Veeramuthu, Radhakrishnan Venkatkarthick, Shanmugam Jayashree, Santi Chuetor, Selvakumar Dharmaraj, Gopalakrishnan Kumar, Wei-Hsin Chen, and Chawalit Ngamcharussrivichai. 2022. "Recent Advances in Lignocellulosic Biomass for Biofuels and Value-Added Bioproducts: A Critical Review." *Bioresource Technology* 344:126195.

Asrul, Mohamad Afiq Mohd, Mohd Farid Atan, Hafizah Abdul Halim Yun, and Josephine Chang Hui Lai. 2022. "A Review of Advanced Optimization Strategies for Fermentative Biohydrogen Production Processes." *International Journal of Hydrogen Energy* 47(38):16785–804.

Awang, Gregor M., G. A. Jones, W. M. Ingledew, and A. M. B. Kropinski. 1988. "The Acetone-Butanol-Ethanol Fermentation." *CRC Critical Reviews in Microbiology* 15(supp 1):S33–67. https://doi.org/10.3109/10408418809104464.

Baral, Nawa Raj, and Ajay Shah. 2014. "Microbial Inhibitors: Formation and Effects on Acetone-Butanol-Ethanol Fermentation of Lignocellulosic Biomass." *Applied Microbiology and Biotechnology* 98(22):9151–72. https://doi.org/10.1007/s00253-014-6106-8.

Cervantes, Jair, Farid Garcia-Lamont, Lisbeth Rodríguez-Mazahua, and Asdrubal Lopez. 2020. "A Comprehensive Survey on Support Vector Machine Classification: Applications, Challenges and Trends." *Neurocomputing* 408:189–215.

Fan, Cheng, Meiling Chen, Xinghua Wang, Jiayuan Wang, and Bufu Huang. 2021. "A Review on Data Preprocessing Techniques toward Efficient and Reliable Knowledge Discovery from Building Operational Data." *Frontiers in Energy Research* 9:652801.

Guan, Wenjian. 2016. "Bioconversion of Lignocellulose into Acetone-Butanol-Ethanol (ABE): Pretreatment, Enzymatic Hydrolysis and Fermentation." PhD Thesis, Auburn University. https://search.proquest.com/openview/373cfa55a63cdcad880ff8d2f323a332/1?pq-origsite=gscholar&cbl=18750&diss=y.

Guo, Yuan, Yi Liu, Mingdong Guan, Hongchi Tang, Zilong Wang, Lihua Lin, and Hao Pang. 2022. "Production of Butanol from Lignocellulosic Biomass: Recent Advances, Challenges, and Prospects." *RSC Advances* 12(29):18848–63.

Hanh, Tran Thi My. 2009. "Optimization of Acetone-Butanol-Ethanol Production from Cassava Starch by a Mixed Culture of Clostridium Sp. and Bacillus Sp. in Batch and Fed-Batch Fermentation." PhD Thesis, Prince of Songkla University. https://core.ac.uk/download/pdf/14980634.pdf.

Ibrahim, Mohamad Faizal, Norhayati Ramli, Ezyana Kamal Bahrin, and Suraini Abd-Aziz. 2017. "Cellulosic Biobutanol by Clostridia: Challenges and Improvements." *Renewable and Sustainable Energy Reviews* 79:1241–54.

Jin, Chao, Mingfa Yao, Haifeng Liu, F. Lee Chia-fon, and Jing Ji. 2011. "Progress in the Production and Application of N-Butanol as a Biofuel." *Renewable and Sustainable Energy Reviews* 15(8):4080–4106.

Khaleghi, Mohammad Karim, Iman Shahidi Pour Savizi, Nathan E. Lewis, and Seyed Abbas Shojaosadati. 2021. "Synergisms of Machine Learning and Constraint-based Modeling of Metabolism for Analysis and Optimization of Fermentation Parameters." *Biotechnology Journal* 16(11):2100212. https://doi.org/10.1002/biot.202100212.

Koller, Martin. 2023. "Biotechnological Approaches to Generate Biogenic Solvents and Energy Carriers from Renewable Resources." *The EuroBiotech Journal* 7(2):96–120. https://doi.org/10.2478/ebtj-2023-0007.

Kumar, Manish, and Kalyan Gayen. 2011. "Developments in Biobutanol Production: New Insights." *Applied Energy* 88(6):1999–2012.

Kurani, Akshit, Pavan Doshi, Aarya Vakharia, and Manan Shah. 2023. "A Comprehensive Comparative Study of Artificial Neural Network (ANN) and Support Vector Machines (SVM) on Stock Forecasting." *Annals of Data Science* 10(1):183–208. https://doi.org/10.1007/s40745-021-00344-x.

Kushwaha, Anamika, Shivani Goswami, Afreen Sultana, Nirmal Kumar Katiyar, Mantasha Athar, Lucky Dubey, Lalit Goswami, Chaudhery Mustansar Hussain, and Mohammad Abdul Kareem. 2022. "Waste Biomass to Biobutanol: Recent Trends and Advancements." *Waste-to-Energy Approaches towards Zero Waste*, 393–423.

Morone, Amruta, and R. A. Pandey. 2014. "Lignocellulosic Biobutanol Production: Gridlocks and Potential Remedies." *Renewable and Sustainable Energy Reviews* 37:21–35.

Nalepa, Jakub, and Michal Kawulok. 2019. "Selecting Training Sets for Support Vector Machines: A Review." *Artificial Intelligence Review* 52(2):857–900. https://doi.org/10.1007/s10462-017-9611-1.

Ndaba, Busiswa, Idan Chiyanzu, and Sanette Marx. 2015. "N-Butanol Derived from Biochemical and Chemical Routes: A Review." *Biotechnology Reports* 8:1–9.

Oloyede, Christopher Tunji, Oyetola Ogunkunle, Simeon Olatayo Jekayinfa, and Opeyeolu Timothy Laseinde. 2024. "Butanol Production from Lignocellulosic Biomass Wastes." In *Waste Valorization for Bioenergy and Bioproducts*, 227–50. Elsevier. https://www.sciencedirect.com/science/article/pii/B9780443191718000225.

Outram, Victoria. 2018. "*In Situ* Product Recovery of Butanol from the Acetone Butanol Ethanol Fermentation." PhD Thesis, Newcastle University. http://theses.ncl.ac.uk/jspui/handle/10443/4152.

Panahi, Hamed Kazemi Shariat, Mona Dehhaghi, Gilles J. Guillemin, Christopher Chukwudi Okonkwo, James E. Kinder, and Thaddeus Chukwuemeka Ezeji. 2023. "Bio-Butanol Production: Scope, Significance, and Applications." In *Advances and Developments in Biobutanol Production*, 1–45. Elsevier. https://www.sciencedirect.com/science/article/pii/B9780323911788000084.

Procentese, Alessandra. 2015a. "Processes for Biobutanol Production from Renewable Resources." Universita di Napoli Federico II. https://www.academia.edu/download/93666551/TesiDottorato_AlessandraProcentese_240315.pdf.

Rathour, Ranju Kumari, Vishal Ahuja, Ravi Kant Bhatia, and Arvind Kumar Bhatt. 2018. "Biobutanol: New Era of Biofuels." *International Journal of Energy Research* 42(15):4532–45. https://doi.org/10.1002/er.4180.

Sarangi, Prakash K., and Sonil Nanda. 2018. "Recent Developments and Challenges of Acetone-Butanol-Ethanol Fermentation." In *Recent Advancements in Biofuels and Bioenergy Utilization*, edited by Prakash Kumar Sarangi, Sonil Nanda, and Pravakar Mohanty, 111–23. Singapore: Springer Singapore. https://doi.org/10.1007/978-981-13-1307-3_5.

Segovia-Hernandez, Juan Gabriel, Shuvashish Behera, and Eduardo Sanchez-Ramirez. 2022. *Advances and Developments in Biobutanol Production*. Woodhead Publishing.

https://books.google.com/books?hl=en&lr=&id=STVxEAAAQBAJ&oi=fnd&pg=PP1&dq=Biobutanol+is+a+four-carbon+alcohol+specifically+known+as+n-butanol+or+butyl+alcohol&ots=Hl4MfeA9Zw&sig=7j6411_6uu-1_J5mPBhEU4noYrc.

Segovia-Hernández, Juan Gabriel, Eduardo Sanchez-Ramirez, Heriberto Alcocer-Garcia, Ana Gabriela Romero-Garcia, and Juan José Quiroz-Ramirez. 2022. "Biobutanol." In *Sustainable Production of Biofuels Using Intensified Processes*, edited by Juan Gabriel Segovia-Hernández, Eduardo Sanchez-Ramirez, Heriberto Alcocer-Garcia, Ana Gabriela Romero-Garcia, and Juan José Quiroz-Ramirez, 51–89. Green Energy and Technology. Cham: Springer International Publishing. https://doi.org/10.1007/978-3-031-13216-2_4.

Serrano, Victor Alejandro, Carlos Alberto Guerrero Fajardo, and Karol Tatiana Castro. 2024. "A Review on Biobutanol: Eco-Friendly Fuel of the Future, History, Current Advances, and Trends." https://www.preprints.org/manuscript/202409.2094/download/final_file.

Sikiru, Surajudeen, Kunmi Joshua Abioye, Habeeb Bolaji Adedayo, Sikiru Yesirat Adebukola, Hassan Soleimani, and M. Anar. 2024. "Technology Projection in Biofuel Production Using Agricultural Waste Materials as a Source of Energy Sustainability: A Comprehensive Review." *Renewable and Sustainable Energy Reviews* 200:114535.

Timung, Seim, Harsimranpreet Singh, and Anshika Annu. 2021. "Bio-Butanol as Biofuels: The Present and Future Scope." In *Liquid Biofuels*, edited by Krushna Prasad Shadangi, 1st ed., 467–85. Wiley. https://doi.org/10.1002/9781119793038.ch13.

Veza, Ibham, Martin Spraggon, IM Rizwanul Fattah, and Muhammad Idris. 2023. "Response Surface Methodology (RSM) for Optimizing Engine Performance and Emissions Fueled with Biofuel: Review of RSM for Sustainability Energy Transition." *Results in Engineering* 18:101213.

Villacreses-Freire, Daniela, Franziska Ketzer, and Christine Rösch. 2022. "Advanced Metabolic Engineering Approaches and Renewable Energy to Improve Environmental Benefits of Algal Biofuels: LCA of Large-Scale Biobutanol Production with Cyanobacteria Synechocystis PCC6803." *BioEnergy Research* 15(3):1515–30. https://doi.org/10.1007/s12155-021-10323-y.

Wagh, Mrunal S., S. Sowjanya, Pinku Chandra Nath, Arnab Chakraborty, Rajshree Amrit, Bishwambhar Mishra, Awdhesh Kumar Mishra, and Yugal Kishore Mohanta. 2024. "Valorisation of Agro-Industrial Wastes: Circular Bioeconomy and Biorefinery Process–a Sustainable Symphony." *Process Safety and Environmental Protection*. https://www.sciencedirect.com/science/article/pii/S0957582024000648.

Wang, Yue, Shih-Hsin Ho, Hong-Wei Yen, Dillirani Nagarajan, Nan-Qi Ren, Shuangfei Li, Zhangli Hu, Duu-Jong Lee, Akihiko Kondo, and Jo-Shu Chang. 2017. "Current Advances on Fermentative Biobutanol Production Using Third Generation Feedstock." *Biotechnology Advances* 35(8):1049–59.

Xu, Zhaoyi, and Joseph Homer Saleh. 2021. "Machine Learning for Reliability Engineering and Safety Applications: Review of Current Status and Future Opportunities." *Reliability Engineering & System Safety* 211:107530.

Xue, Chuang, Youduo Wu, Y. Gu, W. Jiang, H. Dong, Y. Zhang, C. Zhao, and Y. Li. 2019. "3.07-Biofuels and Bioenergy: Acetone and Butanol☆." *Comprehensive Biotechnology*, 79–100.

Yadav, Anjali, Praveen Purohit, Abhishek Vashishtha, P. D. Charan, and Gautam Kumar Meghwanshi. 2022. "Microbial Assisted Production of Alcohols, Acetone and Glycerol." In *Bioprospecting of Microbial Diversity*, 47–92. Elsevier. https://www.sciencedirect.com/science/article/pii/B9780323909587000261.

Zhu, Xianglin, Khalil Ur Rehman, Bo Wang, and Muhammad Shahzad. 2020. "Modern Soft-Sensing Modeling Methods for Fermentation Processes." *Sensors* 20(6):1771.

14 Harnessing the Potential of Microbial Electrochemical Systems with AI and ML

Garima Singh, Soumya Pandit, and Elvis Fosso-Kankeu

14.1 INTRODUCTION

14.1.1 Overview of Microbial Electrochemical Systems (MES)

Microbial Electrochemical Systems (MES) are a promising expeditiously growing technology that unites microbial systems, electrochemistry, and material science to produce energy and sensing devices (Hassan et al. 2021). MES are bioelectrochemical devices that utilize the metabolic activities of microorganisms to convert chemical energy into electrical energy or to synthesize chemicals through electrochemical reactions (Wang and Ren 2013). The primary mechanism involves the oxidation of organic substrates at the anode by microorganisms, leading to the release of electrons that can be harnessed as an electrical current (Hernandez and Osma 2020). The adaptability of MES enables various configurations, including Microbial Fuel Cells (MFCs) for energy generation and Microbial Electrolysis Cells (MECs) for chemical production (Hassan et al. 2021). The interaction of microorganisms and the electrode surface plays a central role in achieving high efficiency due to its essential contribution to an efficient electron-transfer process and the overall system function (Hernandez and Osma 2020).

The most important aspect of MES includes its ability to face various environmental challenges at the same time. They offer a sustainable alternative to traditional approaches of energy generation and waste treatment through the integration of energy recovery with pollutant degradation. It increases resource recovery and decreases dependence on fossil fuels, thus becoming an attractive technology for a circular economy (Schröder et al. 2015).

DOI: 10.1201/9781003571223-14

14.1.2 Applications

14.1.2.1 Production of Energy

MES, in particular MFCs, is used to generate electricity from organic waste materials. In these systems, the microbes at the anode oxidize organic compounds that produce electrons to flow through an external circuit into the cathode, generating electrical power. In this process, energy is generated while wastewater is treated, thereby making it cost-effective for the management of organic wastes while generating renewable energy (Wang and Ren 2013; Hernandez and Osma 2020).

14.1.2.2 Wastewater Treatment

One major application of MES is in wastewater treatment (Al-Sahari et al. 2021). Organic pollutant oxidation capability of MES with simultaneous electricity generation makes it highly relevant to municipal and industrial wastewater treatments (Pant et al. 2012). Complex organic compounds that the microbial communities in MES break down can lead to cleaner effluent and less environmental impact (Hassan et al. 2021).

14.1.2.3 Bioremediation

MES also plays an important role in bioremediation, where they are used to detoxify contaminated environments by utilizing specific strains of microbes that can degrade hazardous substances or heavy metals. This therefore enables the breakdown of these pollutants in situ. Therefore, this method is applicable to various industrial waste contaminated areas and agricultural runoff sites where traditional remediation methods are either less effective or cost too much (Hernandez and Osma 2020; Hassan et al. 2021). MES integrated with bioremediation strategies significantly increases the rate of pollutant removal and accelerates the recovery of valuable resources from waste streams.

14.2 ARTIFICIAL INTELLIGENCE

Artificial Intelligence (AI) is the ability of a machine to imitate intelligent human behavior; it includes all the widespread technologies that allow systems to perform tasks typically requiring human intelligence. Reasoning, learning, problem-solving, and decision-making are some of the activities associated with human intelligence, making them similar to and potentially suitable for AI. Furthermore, AI can be categorized into different forms. Narrow AI is used for specific tasks, while general AI seeks to build an artificial brain, trying to replicate human cognitive abilities across a wide range of activities (Xu et al. 2021; Staff 2023).

14.3 MACHINE LEARNING

Machine Learning (ML) is one of the parts of AI and is based on the creation of algorithms that help computers to learn from and predict data. Systems can enhance their performance through experience as more data is introduced into them, though they are not programmed for the same. Instead, this system identifies patterns or makes a choice based on the statistical analysis without following predetermined rules (Emmert-Streib et al. 2020; Staff 2023).

14.4 ROLE OF AI AND MACHINE LEARNING IN MICROBIAL ELECTROCHEMICAL SYSTEMS

14.4.1 Importance of AI in Enhancing MES Performance

Data-driven insights help enhance MES performance, with these technologies allowing the improvement of the system design and its operation using AI and machine learning (Figure 14.1). They aid in predictive modeling, forecasting the dynamics of the microbial community, substrate utilization, and reactor performance based on the analysis of historical data for optimized operational conditions (Li et al. 2023a). Machine learning is used for determining parameters with respect to maximizing energy restoration, pollutant degradation, and reduced costs (Gadkari 2022). AI allows real-time monitoring with adaptive control strategies that dynamically react to changing conditions to ensure sustainable operation of the MES (Du et al. 2022).

The significance of AI and ML lies in the ability to address the complexity and variability inherent in biological systems (Alber et al. 2019). Different algorithms such as ANN, GP, and SVR are utilized by researchers for the improvement of prediction accuracy concerning microbial behavior and system outputs so that the processes involved in MES can be better controlled and optimized (Ibrahim et al. 2024). These AI-based approaches help to optimize the electrode and reactor designs based on the performance data, with a high recovery efficiency in terms of energy as

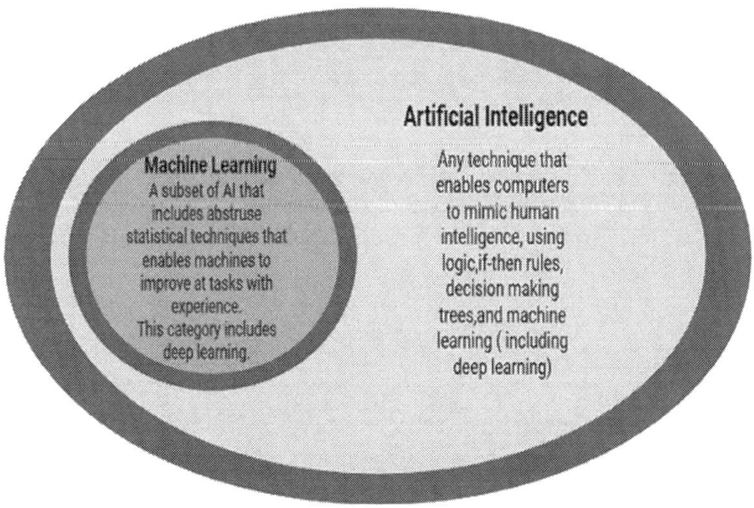

FIGURE 14.1 Artificial intelligence and machine learning.

well as pollutant degradation. In addition, AI deepens the knowledge regarding the interaction among microbes within the electrochemical environment by examining extensive datasets and then extracting relevant patterns, thereby enabling the design of more efficient and adaptive MES systems. By these advancements, AI contributes to the general betterment of MES technology to become more reliable, scalable, and effective in a number of applications (Li et al. 2023a).

14.4.2 AI-Driven Applications Used in Microbial Systems

The incorporation of AI into microbial systems has undergone extensive changes through the years, from basic and simple classification activities to more complex modeling. Initially, AI was only used for basic classification purposes in microbiology based on conventional statistical techniques (Qu et al. 2019). However, the development of high-throughput sequencing technologies has led to the generation of considerable amounts of microbial data, which in turn calls for the use of more sophisticated machine learning techniques in the analysis and interpretation of that data (Qu et al. 2019). This development enabled scientists to identify complex microbial dynamics and interactions that were not easy to detect. In recent years, AI has been incorporated into biochemical systems. Here, models have been developed to predict the performance of a system based on the structure of the microbial community and metabolic pathways. These AI-based methods have improved the productivity and optimization of bioelectrochemical operations, thereby enhancing energy restoration and pollutant breakdown in microbial electrochemical systems (Gadkari 2022).

14.5 FUNDAMENTALS OF MICROBIAL ELECTROCHEMICAL SYSTEMS

14.5.1 Mechanism of Microbial Electrochemical Systems (MES)

Microbial Electrochemical Systems (MES) is a new concept that brings together microbiology and electrochemistry in a way that helps in transforming chemical energy into electrical energy or valuables (Roy and Pandit 2019). The mechanisms of MES comprise diverse types and KPIs. The core mechanism of MES is the ability of microorganisms to convert chemical energy into electrical energy via electron transfer activity. In these systems, microorganisms oxidize organic substrates with the release of electrons that subsequently get transferred to an electrode that produces a quantifiable electric current (Wang and Ren 2013). This process may be divided into various pathways that are based on how the transfer of electrons is occurring. A pathway is the Extracellular Electron Transfer (EET) through which microorganisms transfer electrons to electrodes or even via mediators to interact with non-permeable molecules. This mechanism is critical to the function of MFCs and MECs, where electron transfer efficiency is necessary for energy conversion (Schröder et al. 2015; Hernandez and Osma 2020). Moreover, MES is characterized by different anodic and cathodic activities that determine its functions. In MFCs, microorganisms oxidize substrates at the anode, transferring electrons to the electrode and generating electricity. Instead, MFCs function by supplying electrons to the microorganisms in the cathode, allowing the microbes to reduce the substrates and produce important chemicals like hydrogen. These phenomena thus illustrate the complex relationship between microbial communities and electrochemical reactions as a basis for the expansion and optimization of MES technology (Hernandez and Osma 2020).

14.5.2 Types of MES

Microbial Fuel Cells (MFCs): MFCs produce electricity by the process of oxidation of organic matter by the microorganisms (Xiao and He 2014). Generally, they have two compartments divided by a membrane; the anodic compartment harbors microbes that oxidize organic substrate (Hoang et al. 2022).

Microbial Electric Cells (MECs): They employ an external power source in driving electrochemical reactions to help generate hydrogen, among other desirable chemicals, through organic substrates. They essentially operate like an MFC, except that their aim is at chemical synthesis and not energy recovery (Schröder et al. 2015).

Microbial Electrosynthesis Cells (MESs): These are microbial metabolism systems using electrical energy for the synthesis of complex compounds from the substrates (Saratale et al. 2017).

14.5.3 Key Performance Indicators

Several important metrics are commonly used to evaluate MES performance.

Current density: The effectiveness of electron transfer is shown by the current density, which quantifies the amount of electric current generated per unit area of the electrode.

Resistance: The sum of the resistance that the current encounters while flowing through the system; lesser resistance usually leads to greater efficiency.

Capacitance: It is the amount of charge stored by the system, and, thus, it varies with the electrode area and material.

Exchange Current Density: An essential parameter for comprehending reaction kinetics in MS; this value shows the rate at which electrochemical reactions take place at equilibrium (Hernandez and Osma 2020).

14.5.4 Influencing Factors on MES Performance Metrics

The performance metrics of MES systems can be influenced by numerous parameters, including microbial attachment to the electrodes, environmental conditions such as pH and temperature, and system design (Sharma et al. 2024). For the optimization of MES systems for applications in energy production, wastewater treatment, and bioremediation, these influences are critically important.

Microbial attachment to the electrodes: The interaction between the microorganisms and the electrode surface is vital in the performance of MES (Saratale et al. 2017). Attachment of microorganisms to the electrodes enhances the transfer of electrons and directly influences the current density, which reflects in the performance of the entire system. The configuration and composition of electrodes influence biofilm formation. For example, porous electrodes will increase the available surface area for microbial colonization and enhance electrochemical interaction that might improve the metrics of performance like current density and capacitance (Vishwanathan 2021).

Again, growth parameters of microbial communities can be optimized to further contribute to the development of electrochemically active biofilms. This involves nutrient availability and synthetic biology approaches for improving the performance of microbes (Vishwanathan 2021).

Environmental conditions: One environmental parameter that largely influences microbial activity and subsequently MES performance is pH. The same is with the temperature, because its influence extends to the efficiency of MES by a large percentage. For example, pH alters the rate at which protons transfer across the proton exchange membrane in MFCs and MECs, thereby influencing their performance (Rahimnejad et al. 2014). Suitable pH levels lead to improved rates of COD removal and power production. It has been established that deviations from optimum pH ranges may result in low microbial activity as well as reduced efficiency (Li and Chen 2018; Hassan et al. 2021).

Temperature also plays an important role, in that increases in temperature can enhance the rate of reaction but may also stress microbial communities if they exceed optimal ranges. Thus, the selection of appropriate environmental conditions is necessary for improving performance in MES (Hassan et al. 2021).

System designs: The reactor configuration as well as the materials of the electrode have a serious impact on their efficiency (Liu et al. 2023). Such materials are subjected to various rates of biocompatibility, stability, and electrical conductivity that alter biofilm development and electron rates (Vishwanathan 2021). Carbon-based materials, for example, can be found applied to the anode because they demonstrate good conductivity with good biocompatibility (Aryal et al. 2017). The external resistance in the system also impacts performance because it determines the flow of electrons from the anode to the cathode. Higher external resistance can limit the output of current and reduce the power density; therefore, optimizing the level of resistance improves system efficiency (Li et al. 2020).

14.5.5 AI and Machine Learning Algorithms in MES

Microbial electrochemical systems integrate more and more artificial intelligence techniques and machine learning to enhance their performance and optimize operations. There are multiple algorithms commonly used in these contexts, some of which come with specific properties and applications.

14.5.5.1 Commonly Used Algorithms

Artificial Neural Network (ANN): ANNs are computational models that mimic the neural networks found in the human brain, in which they identify patterns and can predict future states from entry data (Abiodun et al. 2018). In MES, ANNs are used for predicting microbial community patterns, reactor operation, and energy generation efficiency (Xiao and He 2014). Their capability to learn from historical data allows them to optimize MES operations for varying conditions (Can and Heavey 2016).

Genetic Programming (GP): GP is an evolutionary computation that is applied for the evolution of programs or models with a purpose (Tang and Wu 2011). In MES, it is applied to optimize the system attributes and configurations through various simultaneous operational scenarios. It leads to identifying optimal strategies to improve reactor operation and microbial activity (Liu et al. 2023).

Fuzzy Logic (FL): FL mirrors human thinking, dealing with uncertainties and inaccuracies in data (McNeill et al. 1994). In MES, FL can be used to describe complicated relationships among variables that are difficult to measure, such as microorganism interactions and environmental aspects. This approach supports decision-making tasks in situations in which simple logic is insufficient (Li et al. 2023a).

Support Vector Regression (SVR): SVR is a supervised machine learning used to analyze data for regression tasks (Kostopoulos et al. 2018). It works best in the prediction of continuous outcomes given by entry variables. In MES, it can be applied to predict performance metrics like current density and substrate breakdown rates to gain insight into system efficiency (Li et al. 2023a).

Adaptive Neural-Fuzzy Inference System (ANFIS): It has the learning of neural networks together with the reasoning capabilities of fuzzy logic systems (Karaboga and Kaya 2019). This allows for a much more accurate representation of complex systems like MES when both numerical and qualitative information are integrated. ANFIS can improve predictive precision about microbial action and system response (Du et al. 2022). Table 14.1 summarizes a comparison of all the algorithms based on their strengths and weaknesses.

14.5.6 Applications of AI and ML in MES

Incorporating AI in Microbial Electrochemical Systems has notably improved their efficiency using predictive modeling and optimization approaches. These aspects will lead to enhanced energy generation, better use of substrates, and even efficient management of microbial communities. More research developments in these directions will lead to even more sustainable energy production technologies and wastewater treatment.

14.5.6.1 Predictive Modeling

Predicting microbial community patterns and substrate usage: The models and prediction of microbial community dynamics in MES are made with the help of AI algorithms, specifically ANNs and ANFIS. These models study historical data to predict how microbial populations will respond to changes in substrate concentration and environmental aspects. For example, the research by Dr. Lakhveer Singh illustrates the application of AI for predicting the appearance of microbial communities,

TABLE 14.1
Comparison between Algorithms on the Basis of Strengths and Weaknesses

Algorithms	Strengths	Weaknesses	References
Artificial Neural Network (ANN)	Effective for modeling complex non-linear relationships	Requires large datasets; can overfit	Coy-Aceves and Corona-Vasquez (2022), Li et al. (2023b)
Genetic Programming (GP)	Optimizes system parameters through evolutionary algorithms	Computationally intensive; slow to evolve optimal solutions	Coy-Aceves and Corona-Vasquez (2022), Li et al. (2023b)
Fuzzy Logic (FL)	Handles uncertainty and inaccuracy well; intuitive decision-making	Relies on expert- defined rules, which can introduce bias	Coy-Aceves and Corona-Vasquez (2022), Li et al. (2023b)
Support Vector Regression (SVR)	Accurate for regression tasks; models continuous outcomes effectively	Struggles with very large datasets; kernel selection can be challenging	Coy-Aceves and Corona-Vasquez (2022), Li et al. (2023b)
Adaptive Neural-Fuzzy Inference System (ANFIS)	Combines neural networks and fuzzy logic for accurate modeling	Complex turning of parameters required; may need expertise in both fields	Coy-Aceves and Corona-Vasquez (2022), Li et al. (2023b)

particular to specific substrates, in bioelectrochemical systems and allows for improved reactor performance and efficiency control (Mahapatra et al. 2022).

Improving the performance of the reactor using data-driven insights: The integration of AI-driven insights will help in optimizing operational parameters such as pH, temperature, and substrate loading for improved reactor performance. Real-time adjustments based on predictive models are feasible in data-driven approaches; therefore, energy produced and substrate breakdown rates can be improved. Data analysis of the complex datasets provides insights to the interaction between microbial functionality and reactor conditions that can lead to better decision-making (Farahani et al. 2024).

14.5.6.2 Optimization Strategies

Parameter optimization for enhanced efficiency: Successful optimization of MES by using AI techniques like Particle Swarm Optimization (PSO) in identifying optimal operating parameters. It has been well established that optimizing the level of sulfonation and the quantity of cathode catalysts can elevate the power output and COD removal efficiency in a microbial fuel cell. Optimization not only results in energy recovery but also promotes operations of MES (Farahani et al. 2024).

14.6 CASE STUDIES: SUCCESSFUL APPLICATIONS

AI-improved MFCs: Research evidence has indicated that AI tends to significantly enhance the performance of MFCs. A study achieved a power density of 61 mW/m², coupled with 96.21% COD removal efficiency, through multi-objective programming using AI techniques. Such outcomes reflect the impact of AI in optimizing operational efficiencies (Farahani et al. 2024).

Electrochemical sensors: Machine learning in electrochemical sensors has allowed the simultaneous qualification of several toxicants. This example shows how AI can extend the scope of MES applications from what they have been used for traditionally. The integration has led to a more complete environmental quality monitoring, which is a testament to the flexibility of AI in MES applications (Du et al. 2022).

14.7 CHALLENGES AND LIMITATIONS

14.7.1 Data Quality and Availability

The quality of the data used for training and validation decides the effectiveness of AI and ML algorithms, so the quality of data is an important factor in MES applications. Since noise, inconsistencies, or errors in measurement always arise from deviation in experimental attributes, microbial behavior, and measurement techniques used for measurement, there could be a problem with the data in MES, and it might bring low reliability to predictions, thus poor performance in the system. Availability of data also represents a major obstacle since robust data records that capture the intricacy of microbial interaction and reactor operation at various time intervals are typically missing. Many publication centers in specific facets of MES, where no vast datasets are produced for analysis. This severely hampers the possibility of developing sturdy models that establish well in other settings. Without high-quality and easily available data, the full potential of AI and ML in optimizing MES cannot be realized effectively (Hernandez and Osma 2020; Coy-Aceves and Corona-Vasquez 2022).

14.7.2 Issues with Data Collection and Processing in MES Environments

The process of MES involves complex biological functions, determined by a range of factors that include microbial community dynamics, availability of substrates, and environmental parameters such as pH and temperature. The challenge for structured data is to express this complexity as these interacting components behave dynamically, often in unforeseeable manners, since online monitoring is actually required for reliable data that allows good reflection on system behavior, many existing MES settings lack in-built sensors and technologies that record and continuously read data (Hassan et al. 2021). Lacking these usually means that information lags behind; as a result, thus, optimal training of these models becomes somewhat less effective through AI-driven optimization (Hernandez and Osma 2020). Data integration still poses a huge challenge since the research of MES involves multiple sources of data. These include chemical testing and microbial community outlining, which are gathered in different forms and through diverse methodologies, which makes it very challenging to combine them into one dataset for overall analysis and development of models. All these challenges can only be overcome with better data gathering, sensor technology, and the unification of data formats to ensure more reliable and integrative AI-driven MES optimization (Hassan et al. 2021).

Overfitting is one of the common problems of machine learning models applied to MES. The model gained an understanding of the data too well, capturing noise rather than the true basic patterns. This makes the model very poor in generalization, so it is unproductive when introduced to new data and has limited predictive power in real-world applications (Baca et al. 2016; Du et al. 2022). Variability in biological systems further compounds generalization, as models are familiarized with particular datasets and may not function well under different circumstances or microbial communities. Differences in environmental factors, microbial dynamics, and system configurations may also lead to a mismatch between predicted and actual performance (Hassan et al. 2021; Du et al. 2022). In addition, another challenge involves the demand for computational resources required by high-order algorithms used in AI, since advanced model training and validation can consume substantial amounts of computing resources. Limited availability of such resources could prevent researchers from harnessing these models in full capability and hence to their full potential when robust AI-driven MES optimization solutions are needed (Wang and Ren 2013; Baca et al. 2016).

14.8 FUTURE DIRECTIONS AND CONCLUSION

The future of Microbial Electrochemical Systems (MES) looks very promising due to emerging trends in Artificial Intelligence (AI) and Machine Learning (ML) and ongoing research opportunities. It is considered a new route to boost the efficiency, scalability, and applicability of MES for various applications in environmental and energy-related areas. One of the most important new trends in MES research is using advanced machine learning techniques, namely deep learning, which is becoming increasingly dominant. They can handle extensive data and discern complex patterns underlying microbial interactions as well as electrochemical processes, leading to better models of reactor performance and microbial action. For example, deep learning may be utilized for finding appropriate operational parameters, given historical records from different setups of the MES systems. Besides deep learning, hybrid models that combine the traditional mechanistic approach with AI techniques are used to improve accuracy in predictions associated with microbial dynamics and system performance. By capitalizing on both methods, scientists can better grasp the underlying mechanisms in MES. Another significant trend is the possible real-time monitoring and adaptive control systems that are now made possible with the integration of IoT (Internet of Things) technologies. This allows critical parameters such as pH, temperature, and current density to be continuously monitored for adaptive control systems to dynamically alter operating conditions based on real-time data. These advancements will improve the robustness to variations in environmental conditions and microbial activity, making operations in MES systems more stable and efficient. Besides, the innovation of sophisticated sensors, such as biosensors and electrochemical sensors, gives immediate feedback on system performance, enabling proactive management and optimization.

Apart from these technological developments, there are more areas for research: understanding the role of environmental factors on microbial activity. Nutrient availability, temperature fluctuations, and substrate composition are some of the factors that must be explored in order to optimize reactor design and operational strategies. Specific environmental conditions can reveal the resilience and functionality of communities and improve system performance. Standard methodologies for MES studies are another important area of research. Standardized protocols in data collection, analysis, and reporting will ensure cross-study comparisons, provide a basis for building on established knowledge, and promote collaboration among researchers. In addition, the standard methodologies developed will lead to the development of models that will be applicable everywhere and can predict system behavior in various conditions. Research in novel materials and new technologies, in addition to new electrode materials such as nanomaterials and bio-derived electrodes, will enable enhanced efficiencies in electron transfers. Innovations may enhance power output, efficiency, and stability in MES. Likewise, integration may also contribute to improved power output, efficiency, and stability. Integration of MES into other technologies, such as anaerobic digestion and constructed wetlands, will open additional resource recovery and waste treatment opportunities that are highly effective and sustainable.

Commercialization and scale-up: This shall be the ultimate step toward implementing MES technologies globally. Research and development efforts on these aspects focus on economic feasibility, design improvement on large scales, and also the business models to support full-scale commercialization. Case studies where MES has been implemented at a much larger scale would be of great use in understanding best practices and pitfalls in real applications. Overcoming these challenges and capitalizing on emerging technological advancements will provide a brighter future for MES to be a transformative solution in sustainable energy production, waste treatment, and environmental management.

REFERENCES

Abiodun OI, Jantan A, Omolara AE, et al. (2018) State-of-the-art in artificial neural network applications: a survey. *Heliyon* 4:

Alber M, Buganza Tepole A, Cannon WR, et al. (2019) Integrating machine learning and multiscale modeling—perspectives, challenges, and opportunities in the biological, biomedical, and behavioral sciences. *NPJ Digital Medicine* 2:115

Al-Sahari M, Al-Gheethi A, Mohamed RMSR, et al. (2021) Green approach and strategies for wastewater treatment using bioelectrochemical systems: a critical review of fundamental concepts, applications, mechanism, and future trends. *Chemosphere* 285:131373

Aryal N, Ammam F, Patil SA, Pant D (2017) An overview of cathode materials for microbial electrosynthesis of chemicals from carbon dioxide. *Green Chemistry* 19:5748–5760

Baca M, Singh S, Gebinoga M, et al. (2016) Microbial electrochemical systems with future perspectives using advanced nanomaterials and microfluidics. *Advanced Energy Materials* 6:1600690. https://doi.org/10.1002/aenm.201600690

Can B, Heavey C (2016) A demonstration of machine learning for explicit functions for cycle time prediction using MES data. In: *2016 Winter Simulation Conference (WSC)*. IEEE, pp. 2500–2511

Coy-Aceves LE, Corona-Vasquez B (2022) Bibliometric analysis of artificial intelligence algorithms used for microbial fuel cell research. *Water Practice and Technology* 17:2071–2082. https://doi.org/10.2166/wpt.2022.116

Du L, Yan Y, Li T, et al. (2022) Machine learning enables quantification of multiple toxicants with microbial electrochemical sensors. *ACS ES&T Engineering* 2:92–100. https://doi.org/10.1021/acsestengg.1c00287

Emmert-Streib F, Yli-Harja O, Dehmer M (2020) Artificial intelligence: A clarification of misconceptions, myths and desired status. *Frontiers in Artificial Intelligence* 3:524339

Farahani H, Ghasemi M, Sedighi M, Raut N (2024) Employing artificial intelligence for enhanced microbial fuel cell performance through wolf vitamin solution optimization. *Sustainability* 16:6468

Gadkari S (2022) Chapter 23 - Application of artificial intelligence methods for the optimization and control of bioelectrochemical systems. In: Jadhav DA, Pandit S, Gajalakshmi S, Shah MP (eds) *Scaling Up of Microbial Electrochemical Systems*. Elsevier, pp 437–455

Hassan RYA, Febbraio F, Andreescu S (2021) Microbial electrochemical systems: principles, construction and biosensing applications. *Sensors (Basel)* 21:1279. https://doi.org/10.3390/s21041279

Hernandez CA, Osma JF (2020) Microbial electrochemical systems: deriving future trends from historical perspectives and characterization strategies. *Frontiers in Environmental Science* 8:44

Hoang AT, Nižetić S, Ng KH, et al. (2022) Microbial fuel cells for bioelectricity production from waste as sustainable prospect of future energy sector. *Chemosphere* 287:132285

Ibrahim M, Haider A, Lim JW, et al. (2024) Artificial neural network modeling for the prediction, estimation, and treatment of diverse wastewaters: a comprehensive review and future perspective. *Chemosphere* 142860

Karaboga D, Kaya E (2019) Adaptive network based fuzzy inference system (ANFIS) training approaches: a comprehensive survey. *Artificial Intelligence Review* 52:2263–2293. https://doi.org/10.1007/s10462-017-9610-2

Kostopoulos G, Karlos S, Kotsiantis S, Ragos O (2018) Semi-supervised regression: a recent review. *Journal of Intelligent & Fuzzy Systems* 35:1483–1500

Li C, Guo D, Dang Y, et al. (2023a) Application of artificial intelligence-based methods in bioelectrochemical systems: recent progress and future perspectives. *Journal of Environmental Management* 344:118502. https://doi.org/10.1016/j.jenvman.2023.118502

Li C, Guo D, Dang Y, et al. (2023b) Application of artificial intelligence-based methods in bioelectrochemical systems: recent progress and future perspectives. *Journal of Environmental Management* 344:118502. https://doi.org/10.1016/j.jenvman.2023.118502

Li C, He W, Liang D, et al. (2020) Power density of microbial electrochemical system responds to mass transfer characters of non–ion–selective microbial separator. *Bioresource Technology* 311:123478

Li S, Chen G (2018) Factors affecting the effectiveness of bioelectrochemical system applications: data synthesis and meta-analysis. *Batteries* 4:34

Liu Z, Xue X, Cai W, et al. (2023) Recent progress on microbial electrosynthesis reactor designs and strategies to enhance the reactor performance. *Biochemical Engineering Journal* 190:108745

Mahapatra DM, Mishra P, Thakur S, Singh L (2022) Leveraging artificial intelligence in bioelectrochemical systems. *Trends in Biotechnology* 40:535–538

McNeill D, McNeill D, Freiberger P (1994) *Fuzzy Logic: The Revolutionary Computer Technology that Is Changing Our World.* Simon and Schuster

Pant D, Singh A, Van Bogaert G, et al. (2012) Bioelectrochemical systems (BES) for sustainable energy production and product recovery from organic wastes and industrial wastewaters. *RSC Advances* 2:1248–1263

Qu K, Guo F, Liu X, et al. (2019) Application of Machine Learning in Microbiology. *Frontiers in Microbiology* 10:. https://doi.org/10.3389/fmicb.2019.00827

Rahimnejad M, Bakeri G, Najafpour G, et al. (2014) A review on the effect of proton exchange membranes in microbial fuel cells. *Biofuel Research Journal* 1:7–15

Roy S, Pandit S (2019) Microbial electrochemical system: principles and application. In: *Microbial Electrochemical Technology*. Elsevier, pp 19–48

Saratale RG, Saratale GD, Pugazhendhi A, et al. (2017) Microbiome involved in microbial electrochemical systems (MESs): a review. *Chemosphere* 177:176–188

Schröder U, Harnisch F, Angenent LT (2015) Microbial electrochemistry and technology: terminology and classification. *Energy and Environmental Science* 8:513–519. https://doi.org/10.1039/C4EE03359K

Sharma R, Solanki P, Chaudhary M, et al. (2024) Unveiling the potential of microalgae for bioplastic production from wastewater – current trends, innovations, and future prospects. *Biotechnology for Sustainable Materials* 1:10. https://doi.org/10.1186/s44316-024-00010-1

Staff C (2023) What is artificial intelligence? Definition, uses, and types

Tang WH, Wu QH (2011) Evolutionary Computation. In: *Condition Monitoring and Assessment of Power Transformers Using Computational Intelligence.* Springer London, London, pp 15–36

Vishwanathan AS (2021) Microbial fuel cells: a comprehensive review for beginners. *3 Biotech* 11:248. https://doi.org/10.1007/s13205-021-02802-y

Wang H, Ren ZJ (2013) A comprehensive review of microbial electrochemical systems as a platform technology. *Biotechnology Advances* 31:1796–1807. https://doi.org/10.1016/j.biotechadv.2013.10.001

Xiao L, He Z (2014) Applications and perspectives of phototrophic microorganisms for electricity generation from organic compounds in microbial fuel cells. *Renewable and Sustainable Energy Reviews* 37:550–559

Xu Y, Liu X, Cao X, et al. (2021) Artificial intelligence: a powerful paradigm for scientific research. *Innovation (Camb)* 2:100179. https://doi.org/10.1016/j.xinn.2021.100179

Index

Made in the USA
Las Vegas, NV
19 September 2026

2525435a-e658-4f35-be85-259fe6f59016R02